161 071959 6

T
P

D1556059

Remote Sensing for Geologists

Remote Sensing for Geologists
A Guide to Image Interpretation
Second Edition

Gary L. Prost
*Gulf Canada Resources Ltd.
Calgary, Alberta*

Gordon and Breach Science Publishers

Australia • Canada • France • Germany • India • Japan • Luxembourg • Malaysia
The Netherlands • Russia • Singapore • Switzerland

USA	Publishing Office:	TAYLOR & FRANCIS 29 West 35th Street New York, NY 10001 Tel: (212) 216-7800 Fax: (212) 564-7854
	Distribution Center:	TAYLOR & FRANCIS 7625 Empire Drive Florence, KY 41042 Tel: 1-800-624-7064 Fax: 1-800-248-4724
UK		TAYLOR & FRANCIS 27 Church Road Hove E. Sussex, BN3 2FA Tel: +44 (0) 1273 207411 Fax: +44 (0) 1273 205612

REMOTE SENSING FOR GEOLOGISTS: A Guide to Image Interpretation, Second Edition

Copyright © 2001 Taylor & Francis. All rights reserved. Printed in the United States of America. Except as permitted under the United States Copyright Act of 1976, no part of this publication may be reproduced or distributed in any form or by any means, or stored in a database or retrieval system, without prior written permission of the publisher.

1 2 3 4 5 6 7 8 9 0

Printed by Sheridan Books, Ann Arbor, MI, 2001.

A CIP catalog record for this book is available from the British Library.
The paper in this publication meets the requirements of the ANSI Standard Z39.48-1984 (Permanence of Paper).

Library of Congress Cataloging-in-Publication Data
CIP information available from publishers.

ISBN 90-5702-629-5

To Nancy

Contents

Preface — xiii
Acknowledgments — xvii
List of Tables — xix

SECTION 1   INITIATING PROJECTS — 1

1 Project Flow and Obtaining Data — 3

Chapter Overview — 3
Choosing Imagery — 3
Acquiring Imagery — 10
References — 11
Additional Reading — 11
Questions to Consider — 11

2 Photointerpretation Tools and Techniques — 20

Chapter Overview — 20
Tools — 20
Image Characteristics ✓ — 23
Interpretation Mechanics ✓ — 23
References — 25
Additional Reading — 26
Questions to Consider — 26

3 Remote Sensing Systems — 27

Chapter Overview — 27
Airphotos — 27
Airborne Multispectral/Hyperspectral Scanners — 33
Airborne Thermal Scanners — 35
Airborne Radar — 40
Satellite Photography — 44
Satellite Digital Imagery — 47
Side-Scan Sonar — 69
Laser Altimeters — 70
References — 70
Additional Reading — 71
Questions to Consider — 72

SECTION 2   EXPLORATION REMOTE SENSING   75

    Introduction   76
    Frontier Area Analysis   76
    Lease Evaluation   77

4   Recognizing Rock Types   78

    Chapter Overview   78
    Igneous Rocks   78
    Metamorphic Rocks   81
    Sedimentary Rocks   81
    Alteration   85
    References   91
    Additional Reading   92
    Questions to Consider   92

5   Recognizing Structure   93

    Chapter Overview   93
    Undeformed Terrain   93
    Recognizing Dip   93
    Anticlines, Domes, and Horsts   98
    Circular Features   127
    Fractures   128
    Interpreting the Timing of Structural Development   166
    The Field Check   169
    References   170
    Additional Reading   172
    Questions to Consider   173

6   Spectral Stratigraphy   174

    Chapter Overview   174
    Rock Spectra   174
    Hyperspectral Imagery   179
    Lithologic Mapping   184
    Applications-Hydrocarbons   191
    Applications-Minerals   193
    References   197
    Additional Reading   198
    Questions to Consider   199

7   Exploration Case Histories   200

    Chapter Overview   200
    Frontier Petroleum Exploration: Structural Traps at Trap Springs,
        Nevada   200
    Mature Basin Petroleum Exploration: Paradox Basin Stratigraphic
        Traps, Utah   203

Mature Basin Petroleum Exploration: Denver Basin Fractured Reservoirs, Colorado-Wyoming	206
Mineral Exploration: Alteration Associated with Gold at Goldfield, Nevada	210
Structural Mapping as a Guide to Porphyry Copper Deposits, Northeast China	214
Exploration for Kimberlites and Diatremes, Utah, Colorado, and Wyoming	216
Mineral Exploration in Mongolia	221
References	224
Additional Reading	224
Questions to Consider	225

SECTION 3 EXPLOITATION AND ENGINEERING REMOTE SENSING — 227

Introduction	228
Exploitation Applications	228
Hydrologic Applications	229
Engineering Applications	229

8 Exploitation Remote Sensing — 230

Chapter Overview	230
Choosing Infill And Stepout Well Locations	230
Bravo Dome Case History	231
Fractured Reservoirs	233
Coalbed Methane	244
Piceance Basin Coalbed Methane Case History	244
Secondary Recovery and Waterfloods	246
Cottonwood Creek Field Case History	246
Extending Known Mineral Deposits	249
Huancavelica Mining District Case History	251
Mine Safety Issues	255
References	256
Additional Reading	257
Questions to Consider	257

9 Hydrology — 258

Chapter Overview	258
Locating Sources of Water	258
Monitoring Surface Water	263
Flood Control	264
Erosion	267
Geothermal Resources	269
References	270
Additional Reading	271
Questions to Consider	271

10	**Logistics and Engineering**	**272**
	Chapter Overview	272
	Using Appropriate Base Maps	272
	Geographic Information Systems	273
	Case History: Mine Management Using a Geographic Information System	274
	Case History: Pipeline Routing Using Remote Sensing and a GIS	275
	Reconnaissance	276
	Site Selection	279
	Transportation	284
	Power Plants, Pipelines, Tunnels, and Dams	286
	Kakrapar Atomic Power Project Site Selection Case History	287
	References	287
	Additional Reading	288
	Questions to Consider	288

SECTION 4   ENVIRONMENTAL REMOTE SENSING   291

	Introduction	292
11	**Environmental Baselines and Monitoring**	**293**
	Chapter Overview	293
	Environmental Audits	293
	Habitat Mapping	294
	Surface Disturbance and Change Detection Mapping	296
	Coal Mine Fires	300
	Jharia, India Coal Fire Case History	300
	Mine Reclamation	301
	Couer d'Alene Mine Reclamation Case History	301
	Water Pollution	306
	Air Pollution	316
	References	318
	Additional Reading	320
	Questions to Consider	320
12	**Environmental Hazards, Legal Aspects, and Public Relations**	**321**
	Chapter Overview	321
	Landslides	321
	Earthquakes	323
	Floods	324
	Volcanos	324
	Subsidence	328
	Subsidence over Abandoned Coal Mines Case History	328
	Marine Hazards	331
	Legal Issues	333
	Public Relations	336
	References	337

Additional Reading	338
Questions to Consider	338

Closing Thoughts — **339**

Responses to Questions to Consider — **341**

Index — 355

Preface

Remote sensing technology can be traced at least to the thirteenth-century invention of eyeglasses by Roger Bacon. This technology literally got off the ground in 1858 when Gaspard Tournachon took photographs of Paris, France, from a hot air balloon in order to produce topographic maps [1]. Modern remote sensing can be traced to extensive airphoto surveys begun in the 1930s and 1940s, and with the advent of satellites and multispectral scanners the science of remote sensing has become increasingly useful in geologic exploration, engineering and logistical planning, and environmental monitoring. The objective of this book is to show the interested reader how to interpret or extract vital information from remote sensing imagery.

Remote sensing can be defined as technologies and techniques used to obtain information about distant objects using reflected or emitted electromagnetic radiation (Fig. 1), acoustic energy, potential fields (gravity, magnetics), or geochemical measurements. We will concern ourselves only with interpreting electromagnetic, and in some cases acoustic (sonar), images. We will for the most part not discuss data that is measured at points and then contoured (geochemistry, gravity) or along profiles (lidar, aeromagnetics, seismic). **Image interpretation** can be defined as the process of extracting useful information from remote sensing images, whether they are digital or analog, hardcopy (paper, film) or on a computer screen. This information may be compiled as a map or report, or annotated directly on the imagery itself.

The process of interpretation draws heavily on field experience, that is, it is important for the interpreter to have observed features on the ground in order to understand what is observed on images. In geologic remote sensing this draws heavily on structural geology and geomorphology; in environmental remote sensing it is important to know about plant communities and what can affect them, wildlife behavior, weather patterns, and surface geologic and hydrologic processes; in engineering and logistics it helps to know about slope stability, soil types, etc. Whereas photogrammetry is the precise measurement or surveying of ground features from photos to make, for example, topographic maps, in image interpretation we are dealing with the study of landforms, surface cover, and cultural features to learn about both the surface and subsurface geology, environmental sensitivity, or the suitability of an area for building.

To some extent the interpretation of images also requires a knowledge of the instruments used to acquire the image and the computer processing techniques used to generate the image. This book deals with these topics only as they directly apply to interpreting imagery. For a more complete understanding of instrumentation and processing refer to the *Manual of Remote Sensing* [2] as well as books by Condit and Chavez, Drury, Gupta, Jensen, Sabins, and Siegal and Gillespie, among others [3–8]. The purpose of this book is to serve as a manual of image interpretation that shows, by examples, what to look for on imagery when engaged in mineral or hydrocarbon exploration, mine and oil field development, engineering projects, and environmental monitoring.

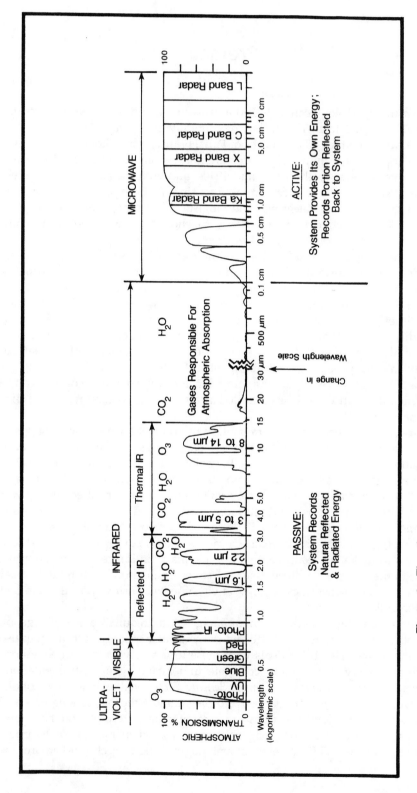

Figure 1 The electromagnetic spectrum as referred to in this text. From Sabins [7].

This book is organized into four sections: section one on initiating a project (chapters 1–3), section two on exploration techniques (chapters 4–7), section three on exploitation and engineering remote sensing (chapters 8–10), and section four on environmental concerns (chapters 11 and 12). The reader should obtain a basic understanding of what is involved in image interpretation and, with some practice, should be able to recognize features of interest and use the imagery in projects for the petroleum, mining, or groundwater industries.

Illustrations have been credited to the company or agency that originally acquired or processed them. There are many service companies that provide remote sensing surveys and build and operate remote sensing instruments. It is not the intention of this book to promote any particular company or vendor, nor would it be possible to list them all, both because of the large number and the transitory nature of the list. Primarily government agencies and principal distributors of data are listed in this text. For information regarding companies that provide services, the reader should consult local professional photogrammetric societies and journals.

REFERENCES

1. R.G. Reeves, A. Anson, and D. Landen, *Manual of Remote Sensing*. (American Society of Photogrammetry, Falls Church, 1975) 1st ed., Chap. 2, p. 27.
2. R.A. Ryerson, *Manual of Remote Sensing*. (American Society of Photogrammetry, Falls Church, 1999) 3rd ed.
3. C.D. Condit and P.S. Chavez, Jr., *U.S. Geol. Survey Bull.* 1462 (1979): 16 p.
4. S.A. Drury, *Image Interpretation in Geology*. (Allen & Unwin, London, 1987) Chap. 5, p. 118–148.
5. R.P. Gupta, *Remote Sensing Geology*. (Springer-Verlag, Berlin, 1991) Chap. 12, p. 183–221.
6. J.R. Jensen, *Introductory Digital Image Processing*. (Prentice-Hall, Englewood Cliffs, 1986) 379 p.
7. F.F. Sabins, Jr. *Remote Sensing Principles and Interpretation* (W.H. Freeman and Co., New York, 1987) 2nd ed., Chap. 7, p. 235–277.
8. B.S. Siegal and A.R. Gillespie, Remote Sensing Geology (John Wiley and Sons, New York, 1980): 702p.

Acknowledgments

Thanks to Amoco Production Company for allowing publication of this manuscript, to Conchita Sobarzo for typing (and retyping) the original text, and to Jean Munshi, Morton Lovestad, Stephen Hansen, and Tiffany Cortez for drafting the diagrams in this work. I am also grateful to the many individuals and companies that did the image processing and allowed me to use their work to illustrate the concepts presented here. I am indebted to Larry Lattman, Keenan Lee, and Eric Nelson for my background in geomorphology, remote sensing, and structural geology, respectively. Finally, many of the ideas put forward here are the result of discussions and many hours of head scratching with coworkers over the years, including Steve Nicolais, Bill DiPaolo, Ralph Baker, Dave Cole, Glen Steen, Don Erickson, Dave Koger, Sandra Perry, and John Berry. Thanks to Ron Marrs and Rebecca Dodge for suggestions on improving the second edition, and to Fred Kruse, Phoebe Hauff, Bob Agar, Doug Peters, Sandra Feldman, and Jim Ellis for contributing case histories to the second edition.

List of Tables

Table 1.1   Comparison of some common satellite imaging systems.

Table 1.2   Partial list of government and commercial sources of remote sensing data.

Table 1.3   Partial list of remote sensing web sites.

Table 2.1   List of some basic equipment needed to begin photointerpretation.

Table 2.2   Commonly used photographic map symbols.

Table 3.1   Film characteristics.

Table 3.2   US government-operated aircraft scanner systems.

Section 1
Initiating Projects

1

Project Flow and Obtaining Data

CHAPTER OVERVIEW

There are many factors to consider when ordering imagery. These factors include the purpose of the imagery, the size of the area requiring analysis, the amount of detail that is required, and the cost of the imagery. The answers to these questions will determine, for example, whether the imagery is color or black and white, if it is on film or digital, the available scales, whether it is from an aircraft or satellite, and whether it is in the visible range or thermal or microwave (radar) portion of the spectrum. There are several providers of imagery and image databases that are available once it is known what type of imagery is required. The principle sources and web sites are given.

CHOOSING IMAGERY

All remote sensing projects begin with a problem. For example, we may want to find out where the oil is, where to drill, the best way to get to a drill site, where to build a pipeline for the least cost with minimum environmental disruption, or how to lay out a seismic program most efficiently. We may wish to locate a source of water for drilling, for a coal slurry pipeline, or for keeping dust down in a mining operation. We are looking for any evidence of mineralization in a new mineral province, or where to extend a known deposit. In order to determine the best imagery to evaluate, we must know what it is we are looking for. Is the area large or small? Does our problem require us to see fine details (up to 10 meter resolution), moderate detail (20–30 m resolution), or regional features (80 m to 1 km resolution or more)? What scale do we wish to work with? Do we need to detect color changes (e.g., lithology, alteration) or vegetation stress? Is the area always under clouds? Is it in the polar regions during the dark season? Are we looking for changes in moisture conditions? Is the area under water? Do we want a certain date, or specific time of year? Do we need multitemporal (repetitive) coverage, or historical coverage? And finally, how much time do we have, and what kind of budget do we have to work with? Should we go to a vendor, the government, or process the data ourselves? The answers to these questions will determine the products that are acquired and the types of analyses that are possible.

The first step in any project (Fig. 1.1) is to clearly state the problem that needs to be solved, and determine the best approach to solving it. Although it sounds obvious, too

4   REMOTE SENSING FOR GEOLOGISTS

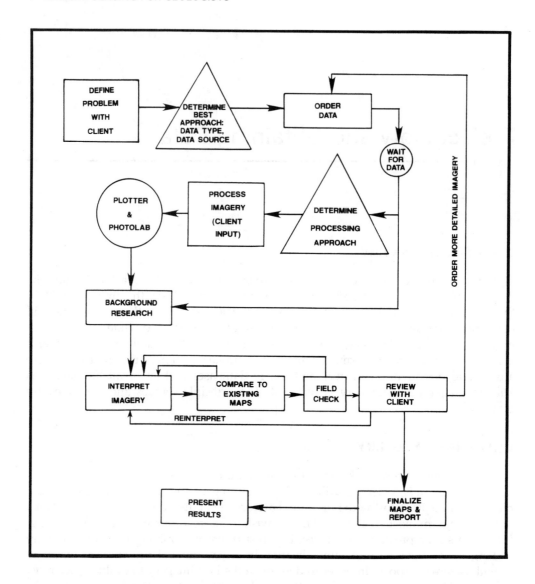

Figure 1.1   Project flowchart for a typical remote sensing study.

often not enough thought goes into this step and imagery is obtained that does not answer the question we have, or costs more than is necessary. A large number of data types are available to suit various needs (Table 1.1). If we are studying thrust belts we may wish to order 1:1,000,000 black/white (B/W) single band Landsat multispectral scanner (MSS) images with 80 m resolution and large area coverage (185 × 185 km) to make a mosaic of part of a continent. If we are doing a basin analysis, then a color MSS image at 1:250,000 or 1:500,000 should be acceptable. If we wish to map details of lithologic and facies changes, or vegetation patterns and wildlife habitats, we can use color SPOT multispectral (XS) imagery with 20 m resolution and 60 × 60 km (vertical view) coverage at a scale of 1:100,000 or 1:50,000. Color Landsat thematic mapper (TM)

Table 1.1 Comparison of Some Common Satellite Imaging Systems

Satellite Systems	Orbital Altitude	Area Coverage	Repeat Cycle	Ground Resolution	Spectral Bands
Digital Satellite Products					
Landsat 1,2,3*	918 km		18 day		
RBV		99 × 99 km		40 m	green, red, near-IR
MSS		185 × 185 km		79 m	0.5–0.6 (green), 0.6–0.7 (red), 0.7–0.8 microns
Landsat 4,5	705 km		16 days		
MSS		185 × 185 km		79 m	0.5–0.6, 0.6–0.7, 0.7–0.8, and 0.8–1.1 microns
TM				28.5 m	0.45–0.52, 0.51–0.60, 0.63–0.69, 0.76–0.90, 1.55–1.75, 2.08–2.35 microns
				120 m	Thermal: 10.4–12.5 microns
Landsat 7	705 km	185 × 170 km	16 days	15 m	0.52–0.90 (panchromatic)
ETM				30 m	0.45–0.52, 0.52–0.60, 0.63–0.69, 0.76–0.90, 1.55–1.75, 2.08–2.35 microns.
				120 m	10.4–12.5 microns
SPOT 1, 2, 3, 4	832 km	60 × 60 km	26 days		
P				10 m	0.51–0.73
XS				20 m	0.50–0.59, 0.61–0.68, and 0.79–0.89 microns.
HCMM*	620 km	716 × 716 km	16 days	500 m (reflected)	0.55–1.10,
				600 m (thermal)	10.5–12.5 microns
SeaSat*	790 km	100 km swath	152 days	25 m	23.5 cm (L-band radar)
SIR-A*	250 km	50 km		38 m	23.5 cm (L-band radar)
SIR-B*	225 km	40 km		25 m	23.5 cm (L-band radar)
SIR-C	225 km	15–90 km swath	1 day	15–40 m	3 cm (X-band), 6 cm (C-band), and 23 cm (L-band radar)
GOES	35,000 km	full hemisphere	stationary	1 km (reflected)	0.55–0.70,
				8 km (thermal)	10.5–12.6 microns
AVHRR	850 km	2700 km swath	1 day	1.1 km	0.55–0.68, 0.73–1.10, 3.55–3.93, 10.5–11.5, and 11.5–12.5 microns
CZCS	955 km	1800 km swath	6 days	800 m	0.43–0.45, 0.51–0.53, 0.54–0.56, 0.66–0.68, 0.70–0.80, and 10.5–12.5 microns

Table 1.1 (continued)

Satellite Systems	Orbital Altitude	Area Coverage	Repeat Cycle	Ground Resolution	Spectral Bands
ERS-1, 2	777 km	100 km swath	3,35, and 176 days	30 × 30 m	5.6 cm (C-band)
SAR					
JERS-1 (Fuyo-1)	568 km		44 days		
SAR		75 km		18 × 18 m	23.5 cm (L-band)
OPS		85 km		18.3 × 24.2 m	0.52–0.60, 0.63–0.69, 0.76–0.86 (stereo), 1.60–1.71, 2.01–2.12, 2.13–2.25, and 2.27–2.40 microns
RESURS-01		600 km swath	4 days	70 m	0.5–0.6, 0.6–0.7, 0.7–0.80.8–1.1 microns
				600 m	10.4–12.6 microns
ALMAZ	300 km	200–350 km swath	1–3 days	13 × 30 to 27 × 30 m	10 cm (S-band)
SAR					
RADARSAT 1	792 km	50–500 km	24 days	8 × 8 to 100 × 100 m	5.6 cm (C-band)
IRS 1A & 1B	904 km		22 days		
LISS-I		148 km		72 × 72 m	0.45–0.52, 0.52–0.59, 0.62–0.68, 0.77–0.86 microns
LISS-II		146 km		36 × 36 m	0.45–0.52, 0.52–0.59, 0.62–0.68, 0.77–0.86 microns
IRS 1C & 1D	900 km		24 days		
LISS-3		141 km		25 m	0.52–0.59, 0.62–0.68, 0.77–0.86 microns
				70 m	1.55–1.70
PAN		70 km		5 m	0.5–0.75 microns
WiFS		774 km		180 m	0.62–0.68, 0.77–0.86 microns
MOS-1 MESSR	909 km	100 to 185 km	17 day	50 × 50 m	0.51–0.59, 0.61–0.69, 0.72–0.80, and 0.80–1.10 microns
MOMS-1	225 km	140 km		20 × 20 m	0.57–0.62 and 0.82–0.92 microns
ORBVIEW-1	740 km	1300 km swath	2 days	10 km	1 band centered at 0.777 microns
ORBVIEW-2	705 km	2800 km swath	1 day	1 km	0.402–0.422, 0.433–0.453, 0.48–0.50, 0.50–0.52, 0.545–0.565, 0.66–0.68, 0.745–0.785, and 0.845–0.885 microns

Table 1.1 (continued)

Satellite Systems	Orbital Altitude	Area Coverage	Repeat Cycle	Ground Resolution	Spectral Bands
Ikonos Pan Multispectral	680 km	11 km swath	11 days	1 m 4 m	0.45–0.90 microns blue, green, red, near-infrared
ASTER Visible Near Infrared Thermal	705 km	60 km swath	4–16 days	15 m 15 m 30 m 90 m	3 bands in range 0.52–0.86 microns 0.77–0.86, for stereo 6 bands in range 1.6 to 2.43 microns 5 bands in range 8.125 to 11.65 microns
EO-1 Hyperion	705 km	7.5 × 100 km swath	16 days	30 m	220 channels in range 0.4–2.5 microns
Photographic Satellite Products					
Skylab* S190A	380 km	140 × 140 km	5 day	60 m or less	0.4–0.7 (true color), 0.50–0.88 (color infrared), 0.49–0.6 (green), 0.6–0.7 (red), 0.70–0.83 (near infrared 1), and 0.7–0.9 microns (near infrared 2) films
S190B		95 × 95 km		10 m or less	0.4–0.7 (true color) and 0.50–0.88 microns (color infrared) films.
KFA 1000		68 × 68 to 85 × 85 km		4–7 m	0.56—0.67 and 0.67–0.81 microns (2 emulsion color "spectrozonal" film)
MK-4		120 × 120 to 270 × 270 km		6 m 8 m	0.4–0.7, 0.46–0.505, 0.515–0.565, 0.580–0.800, 0.635–0.690, and 0.870–0.900 microns (B/W and color spectrozonal products)
KATE-200		180 × 180 km		15–30 m	0.5–0.6, 0.6–0.7, 0.7–0.85 microns (B/W and color spectrozonal products)
KVR 1000		34 × 57 km		2–3 m	0.51–0.76 microns
TK 350		175 × 257 km		10 m	0.51–0.76 microns
DD-5		13 × 13 km		2 m	0.45–0.59 (panchromatic; originally photos, now being marketed as digital data)

*No longer in operation.

Note: Hand held photos are available from various programs starting with Gemini and continuing with the Space Shuttle. These have variable sizes and resolutions as well as film types.

imagery, with 30 m resolution, and covering 185 × 185 km at a scale of 1:100,000 would also work. For mapping alteration associated with mineral deposits the ideal choice is high resolution (1–10 m) airborne hyperspectral imagery. If that is not available, or if the area is large and remote, we may wish to use Landsat TM for a reconnaissance look. If we wish to know where there are trails that can be used for access to remote areas, or where a well was drilled several years ago in a poorly mapped part of the world, we may chose to use a SPOT high resolution (10 m) panchromatic (P) image, a Soyuzkarta KFA-1000 photograph with 5 m resolution, or an Ikonos image with 1 m resolution. These can be enlarged to 1:25,000 and still appear clear and sharp.

The following factors must be considered when ordering data to be able to determine the type of data that best suits the needs of the project (see also Dekker [1]).

1. Cost

    If cost is no object the best procedure, and that which provides the most flexibility in end products, is to purchase a digital tape or CD and process the image on one's own system. The drawback is that one requires an image processing system, the knowledge to operate it and keep it updated, and the time to do the processing. For those with some latitude in the amount they can spend, there are a wide variety of products available from vendors with a wide range of costs. If the amount that can be spent is limited, the least expensive option is to purchase imagery off-the-shelf from a government agency or primary distributor.

    Another factor to keep in mind is that the smaller the area covered, the higher the cost per unit area. Airphotos or airborne imagery will almost always cost more per unit area than satellite images.

2. Timing

    If the imagery is needed immediately, one must purchase off-the-shelf data from the government or a vendor. Large image archives exist, and data can often be obtained quickly. The cost will increase if a "rush" job is requested. Purchasing custom images from vendors or processing digital data in your own shop can take up to several weeks.

3. Coverage

    Large area coverage can be obtained using weather satellites such as GOES (covers a full hemisphere) or the Advanced Very High Resolution Radiometer (AVHRR), which covers a 2700 km swath on the earth's surface. Moderate sized areas can be covered using some hand-held Shuttle (and other mission) photos (variable area coverage), as well as Landsat images (MSS and TM), which cover 185 × 185 km. Next in area coverage are MK-4 photos (120 to 270 km), KATE-200 photos (180 × 180 km), KFA-1000 photos (68 to 85 km), and the SPOT systems (XS and P) that cover 60 × 60 km. Satellite images generally cover larger areas than airborne photos or images, and their synoptic view is one of their greatest advantages. For regional studies, one usually requires large area coverage; for field or local studies airborne surveys or small area satellite images will save cost and/or provide more detail.

4. Resolution

    Generally speaking, the larger the area covered, the coarser the resolution, whereas smaller area images can resolve finer detail. Satellites available to civilian users always have less resolution than airborne photos and images, which are flown closer to the ground. Photographs always have better resolution than digital images, since the grain of the film is finer than the array of picture cells or "pixels" produced by the spacing of detectors.

The scale of the image will to some extent be a function of the resolution, in that one cannot enlarge, say, an image with 80 m resolution to a scale of 1:100,000 without the image becoming "pixelated," that is, breaking up into the individual resolution elements that appear as an array of colored squares.

5. Stereo Coverage

   If stereo coverage is needed for mapping geologic structures, for example, one must rely on specialized satellites or airphoto surveys. Airborne scanners can now provide stereo using "frame-grabber" technology. This technology utilizes video cameras to acquire images at intervals along the flight path such that one obtains the overlap required for parallax. Satellites such as Landsat provide stereo only where adjacent orbits overlap. Thus, there may be only 10% overlap near the equator, but up to 60% or more overlap in polar regions where orbits converge. For example, SPOT offers stereo by using two sensors, one forward-looking and the other near-vertical. EROS 1A is able to rotate the sensor to obtain overlap for stereo.

6. Color vs. Black/White

   All multispectral data and most films can be processed to provide color images. B/W airphotos, SPOT P, and most radar and thermal data consist of a single band or channel and lack color. Color often reveals subtle details regarding lithologies or vegetation, etc., that are not available from panchromatic images. Any B/W data can, however, be artificially colored by digitally assigning colors to various densities (gray levels) during image processing. This can enhance features not otherwise obvious to the human eye.

7. Haze Suppression

   As humidity increases, moisture condenses on atmospheric particulates and the amount of light that is scattered increases. Junge [2] showed that as humidity increased from 70–95%, horizontal visibility decreased by a factor of up to six times. Longer wavelengths penetrate haze more than short wavelengths. Thus, color infrared or B/W infrared images show less haze than true color or visible panchromatic images. Haze can also be diminished by flying closer to the ground and by using haze-reducing filters [3].

8. Cloud Cover

   Only radar imagery can penetrate clouds because of the difference between sensor wavelength and water vapor particle size. This is especially useful in tropical climates or areas known to have cloudy periods. Other ways to minimize clouds include timing overflights to occur around mid-morning, before clouds begin building, or after passage of a cold front.

9. Night Surveys

   Thermal and radar surveys can be flown effectively at night because neither system relies on reflected sunlight and the radar provides its own energy source. Predawn thermal imagery reveals, among other things, lithologic contrasts related to differing densities or tones (light vs. dark), shallow groundwater (generally warmer than background), and can detect oil spills on water under most conditions. Radar illuminates the ground with microwaves and can be flown at night to map oil spills, for example, or during polar night to map the movement of ice floes that could threaten an offshore rig.

10. Seasonal/Repetitive Coverage
    Certain seasons can be better for surveys than others. For example, an airphoto survey in an area covered by temperate forest would see more of the ground in spring before deciduous plant leaf-out, or in the fall after leaves have dropped. A high sun elevation angle (summer) provides images with the best color saturation, which can be useful when mapping lithologies in low contrast areas. On the other hand low sun angle images (morning or winter), especially with a light snow cover, enhance geologic structures in low relief terrain.
    If repetitive coverage is needed to monitor natural (e.g., ice floe) or man-made (e.g., drilling, roads) changes, it is often most cost-effective to use satellites because of their regular repeat cycles. Scheduling repeated aircraft surveys can be expensive.
11. Relief
    Low relief terrain may require low sun angle or grazing radar imagery to enhance subtle topographic and structural features. Whereas most airphoto surveys are flown with a 15.25 cm (6 in) lens, airphotos obtained with a 7.6 cm (3 in) focal length lens will increase vertical exaggeration and thus amplify subtle features. High relief terrain poses the problem of large shadowed areas obscuring important details. These areas should be flown during midday or using radar with a steep depression angle to minimize shadows.
12. Vegetation Mapping
    Color infrared images are very sensitive to changes in vegetation vigor, since the peak reflection in vegetation is in the near-infrared. Combinations of infrared and visible wavelengths have been used to map changes in vegetation related to underlying rock types and even hydrocarbon seepage [4]. Longer radar bands will penetrate a few layers of leaves (longer wavelengths penetrate farther), but radar in general is not particularly well suited for mapping vegetation. It will, however, faithfully map topography in vegetation-covered areas where the canopy follows topographic contours.
13. Water-covered Areas
    In areas that have clear water, a uniform bottom material, and little or no suspended sediment one may be able to map submerged features by measuring the amount of light reflected off the bottom. Shorter wavelengths penetrate water farther than longer wavelengths. The euphotic, or "light-penetrating" zone is known to extend to 30 m in clear water [5], and penetration of light to 20 m has been reported [6]. Infrared light is absorbed by water and cannot provide information on bottom features. Landsat TM, with its blue band, is excellent for mapping shallow water features such as shoals, reefs, or geologic structures. Likewise, true color and special water penetration films such as Aerocolor SO-224 have excellent water penetration capabilities [3]. In the last decade side-scan sonar has become readily available for shallow and deep water mapping, and produces images of the sea bottom using acoustic energy much like radar uses microwave energy to produce an image.
14. Digital or Film?
    Aside from issues such as resolution and cost (see above), digital imagery requires processing but provides tremendous flexibility in end product. In addition to applying geometric corrections, one can also adjust contrast, brightness, hue, saturation, and can filter to enhance sharpness, remove speckle, etc.

Table 1.2 Partial List of Government and Commerical Sources of Remote Sensing Data

	Agency and Address	Data Type
1.	AEROFILM Gates Studios, Station Road, Boreham Wood, Herts MD6 1EJ England Tel: 011-44-1-81-207-0666 FAX: 011-44-1-81-207-5433 email: library@aerofilms.com	airphotos of UK and some former colonies
2.	A.G.A.R.S.S. PTY. LTD. 32 Wheelwright Road Lesmurdie, WA 6076 Australia Tel: 61-8-9291-7929 FAX: 61-8-9291-8566 email: bob@agarss.com.au	Hyperspectral imagery
3.	AGENZIA SPAZIALE ITALIANA (ASI) Viale Regina Margherita, 202 00198 Roma, Italy Tel: 39-6-8567-1 FAX: 39-6-8567-267 http://www.asi.it/00HTL/eng/asiroma/address.html and 400 Virginia Ave. NW, Suite 320 Washington, DC 20024 Tel: 202-863-1298 FAX: 202-554-2435 email: eletico@msn.com	MIVIS, SIR-C, X–SAR
4.	AUSTRALIAN CENTRE FOR REMOTE SENSING Australian Surveying & Land Information Group Dept. Industry, Science, & Resources Scrivner Bldg., Dunlop Court, Fern Hill Park, Bruce ACT 2617 Box 2 Belconnen ACT 2616 Tel: 800-800-173, 61-2-6201-4201 FAX: 61-2-6201-4366 http://www.auslig.gov.au/acres/index.html	Satellite imagery
5.	AUSTRALIAN SURVEYING & LAND INFORMATION GROUP (ASLIG) Order airphotos from United Photo and Graphic Services Unit 4, P.O. Box 407, 2 Apollo Court Blackburn VIC 3130 Tel: 1-800-800-173 or 02-6201-1332 email: louiseelliott@auslig.gov.au http://www.auslig.gov.au Also Tel: 61-3-9877-3922 FAX: 61-3-9894-2971 email: upgs@rie.net.au http://melbourne.citysearch.com.au/E/V/MELB/0043/28/18/	Australian airphotos, maps, satellite images
6.	BUREAU OF LAND MANAGEMENT Airphoto Library, Division of Surveying and Mapping, Bldg. 46 P.O. Box 25047, Federal Center, Denver, Colorado 80225-0047. Tel: 303-236-7991 http://www.blm.gov	U.S. airphotos
7.	CANADA CENTRE FOR REMOTE SENSING (CCRS) 588 Booth St, Ottawa, Ontario Canada K1A 0Y7 Tel: 613-947-1218 or 613-947-1272 FAX: 613-947-1382 Email: info@ccrs.nrcan.gc.ca http://www.ccrs.nrcan.gc.ca	LANDSAT, RADARSAT, airphotos, SFSI

Table 1.2 (*continued*)

	Agency and Address	Data Type
8.	CANADIAN SPACE AGENCY 6767 Route de l'Aéroport Saint-Hubert, Québec Tel: 1-450-926-4800 FAX: 1-450-926-4352 http://www.space.gc.ca	RADARSAT
9.	CENTRE NATIONAL D'ETUDES SPATIALES (CNES) 2 Place Maurice Quentin 75 039 Paris CEDEX 01, France Tel: 33 (0) 1-44-76-7500 FAX: 33 (0)1-44-76-7676 http://www.cnes.fr/	SPOT
10.	DEUTCHES ZENTRUM FUR LUFT-UND RAUMFAHRT (DLR) Presse-und Öffentlichkeitsarbeit Linder Höhe 51147 Köln-Porz Postanschrist 51170, Köln info@dlr.de pressestelle@dlr.de http://www.dlr.de/pressestelle/ http://www.dlr.de/DLR-Homepage	SIR-C, X-SAR
11.	EARTH REMOTE SENSING DATA ANALYSIS CENTER (ERSDAC) Forefront Tower 14F 3-12-1, Kachidoki, Chuo-ku, Tokyo 104, Japan Tel: 81-3-3533-9380 FAX: 81-3-3533-9383 email: ersdesk@ersdac.or.jp http://www.iijnet.or.jp/ersdac/	JERS, ASTER
12.	EARTHWATCH 1900 Pike Rd. Longmont, CO 80501-6700 Tel: 303-882-3800 FAX: 303-882-3848 http://www.digitalglobe.com	Commercial satellite products
13.	EDAC Bandelier West, rm. 111, Albuquerque, NM 87131-6031 Tel: 505-277-3622 FAX: 505-277-3614 Email: laura@spock.unm.edu http://edac.unm.edu	Gemini, Apollo, Skylab, Shuttle photos
14.	EROS Data Center Mundt Federal Bldg. Sioux Falls, South Dakota 57198 Tel: 605-594-6151 1-800-252-4547 x2065 FAX: 605-594-6589 email: wilsonj@edcmail.cr.usgs.gov http://edcwww.cr.usgs.gov/ http://edcwww.cr.usgs.gov/webglis	National High Alt. Program (NHAP), U.S. Geol. Survey & NASA photos, National radar prog., Gemini, Apollo, Skylab & Shuttle photos.
15.	EUROPEAN SPACE AGENCY (ESA) 8-10 Rue Mario Nikis, 75738 Paris Cedex 15, France Tel: 33-1-5369-7654 FAX: 33-1-5369-7560 http://www.esa.int/hq/	ERS
16.	EURIMAGE Tel: 39-06-94180-751 FAX: 39-06-94161-09 Email: didomenico@eurimage.com http://www.eurimage.com	Landsat TM, ERS-1 & 2 (including interfero- metric sets), JERS, Resurs, KSA, KVR, JERS-1 & 2, IRS, Quickbird

Table 1.2 (*continued*)

	Agency and Address	Data Type
24.	NASA–AMES RESEARCH CENTER Aircraft Data Facility MS 240-6, Moffett Field, CA Tel: 650-604-5000 FAX: 650-604-4003 http://www.arc.nasa.gov	Airphoto/ imagery acquisition
25.	NASA GODDARD SPACE CENTER (GSFC) Greenbelt, MD 20771 Tel: 301-286-2000, 301-286-8103 http://pao.gsfc.nasa.gov	Global change, data processing, hydrosphere processes, atmospherics
25.	NASA JOHNSON SPACE CENTER (JSC) Human Space Flight Images, Office of Public Affairs Mail Stop AP4, 2101 Nasa Road 1, Houston, Texas 77058. Tel: 281-483-2911 FAX: 281-483-2000, 281-483-2911 email: Kamlesh.P.Lulla@jsc.nasa.gov http://eol.jsc.nasa.gov	Gemini, Apollo, Skylab & Shuttle photos
27.	NATIONAL OCEANIC AND ATMOSPHERIC ADMINISTRATION (NOAA) National Environmental Satellite Data and Information Service, National Climatic Center, Satellite Data Services Division, Federal Bldg., 151 Patton Ave. Ashville, NC 28801-5001 Tel: 828-271-4800 FAX: 828-271-4876 Email: satorder@ncdc.noaa.gov http://www.ncdc.noaa.gov	SEASAT, Coastal Zone Color Scanner
28.	NATIONAL REMOTE SENSING CENTRE LTD. (NRSC) Delta House, Southwood Crescent Farnborough, Hampshire GU14 0NL, England Tel: 44 (0) 1252-362000 FAX: 44 (0) 1252-375016 email: kthornton@nrsc.co.uk http://www.nrsc.co.uk	Airphotos, satellite imagery, remote sensing products
29.	NATIONAL SPACE DEVELOPMENT AGENCY OF JAPAN (NASDA) World Trade Center Bldg. 2-4-1 Hamamatsu-Cho Minato-ku, Tokyo 105-8060 Tel: 81-3-3438-6111, 81-3-3438-6000 FAX: 81-3-5402-6512 email: proffice@nasda.go.jp http://www.tksc.nasda.go.jp http://www.eoc.nasda.go.jp/homepage.html	ADEOS, JERS
30.	ORBITAL IMAGING 21700 Atlantic Blvd. Dulles, VA 20166 Tel: 703-406-5800 Fax: 703-404-8061 http://www.orbimage.com	ORBVIEW-1, -2, RADARSAT
31.	RADARSAT INTERNATIONAL 13800 Commerce Parkway MacDonald Dettwiler Bldg. Richmond, BC V6V 2J3 Tel: 604-244-0400 FAX: 604-244-0404 http://www.rsi.ca	RADARSAT, LANDSAT, SPOT, ERS, IRS JERS, High- res. optical

Table 1.2 (*continued*)

	Agency and Address	Data Type
17.	Furgo Airborne Surveys Pty. Ltd. Perth, Australia Tel: 61-892-736-400 FAX: 61-892-736-466 email: info@fugroairborne.com.au	Argus hyperspectral profiling spectrometer, radiometrics, aeromagnetics
18.	GEOPHYSICAL ENVIRONMENTAL RESEARCH (GER) One Bennett Common Millbrook, New York 12545 Tel: 914-677-6100 FAX: 914-677-6106 email: info@ger.com http://www.ger.com	Hyperspectral scanner imagerv
19.	ITRES RESEARCH LTD. Suite 155, 2635 37 Ave. NE Calgary, AB T1Y 5Z6 Tel: 403-250-9944 x 240 FAX: 403-250-9916 email: stevem@itres.com http://www.itres.com or 9785 Maroon Circle, Suite 150 Englewood, CO 80112 Tel: 303-792-0884 FAX: 303-792-9114 email: klittle@itres.com	CASI multispectral scanner
20.	IMAGE SAT INTERNATIONAL N.V. Textile House, 17th Floor 2 Kaufman Tel Aviv, Israel 68012 Tel: 9723-516-3429 Fax: 9723-516-3430 email: information@imagesatintl.com http://www.imagesatintl.com	EROS A1
21.	INSTITUT GÉOGRAPHIQUE NATIONAL (IGN) Photothèque Nationale 2/4, Avenue Pasteur 94165 Saint-Mandé CEDEX Paris, France Tel: 1-43-98-8000 FAX: 1-43-98-8401 email: phototheque@ign.fr http://www.ign.fr	photos over France and former territories
22.	InterMap 9785 Maroon Circle #150 Englewood, CO 80112 Tel: 303-708-0955 FAX: 303-708-0952 email: info@intermaptechnologies.com http://www.intermaptechnologies.com	airborne radar
23.	JET PROPULSION LAB Earth-Space Science Division, Terrestrial Science Unit 4800 Oak Grove Drive, Pasadena, California 91103 Tel: 818-354-0294 FAX: 818-354-0966 or 818-354-9476 http://www.makalu.jpl.nasa.gov	SIR-A, SIR-B, SIR-C Radar, Thermal, AVIRIS, Aster

Table 1.2 (*continued*)

	Agency and Address	Data Type
32.	REMOTE SENSING TECHNOLOGY CENTER OF JAPAN (RESTEC) Data Distribution Dept. 2F, Roppongi First Bldg., 1-9-9, Roppongi, Minato-Ku Tokyo 106-0032, Japan Tel: 81-3-5561-9777 FAX: 81-3-5574-8515 email: mayumi@restec.or.jp http://www.restec.or.jp or for data search: http://eus.eoc.nasda.go.jp	JERS, ADEOS, RADARSAT
33.	RUSSIAN SPACE AGENCY 129857 Russia, Moscow, 42 Shchepkin St. email: www@rka.ru http://www.rka.ru/english/eindex.htm	
34.	SPACE IMAGING/EOSAT 12076 Grant St. Thornton, CO 80241 Tel: 303-254-2000, 1-800-232-9037 FAX: 303-254-2215 email: customerservice@spaceimaging.com info@spaceimaging.com http://www.spaceimaging.com	Commercial products derived from IKONOS, JERS, ERS, IRS, RADARSAT, LANDSAT, air photos & airborne imagery
35.	SPOT IMAGE 1897 Preston White Drive, Reston, Virginia 22091-4368 Tel: 703-715-3100 FAX: 703-648-1813 Or 5, rue des Satellites, BP 4359 F31030, Toulouse CEDEX 4, France Tel: 33 (0) 562-194040 FAX: 33 (0) 562-194011 http://spot.com/spot/home/contact/welcome.htm	SPOT, ERS, RADARSAT, ALMAZ
36.	U.S. DEPT. OF AGRICULTURE Airphoto Field Office, 2222 West, 2300 South, P.O. Box 30010, Salt Lake City, Utah 84119-2020 Tel: 801-975-3503 FAX: 801-975-3532 email: sales@apfo.usda.gov	airphotos from U.S. Forest Service, Soil Conservation Service, NHAP
37.	U.S. GEOLOGICAL SURVEY Earth Science Information Center 12201 Sunrise Valley Dr. M.S. 507, Reston, Virginia 20192 Tel: 703-648-5920 1-888-ask-usgs FAX: 703-648-5548 email: esicmail@usgs.gov http://ask.usgs.gov	LANDSAT

The above listings are current to August 1999.
For commercial airphoto surveys in the United States, contact the
  American Society for Photogrammetry and Remote Sensing
  5410 Grosvenor Lane, Suite 210, Bethesda, Maryland 20814-0208,
  Tel: 301-493-0290, FAX: 301-493-0208,
  email: asprs@asprs.org
  http://www.asprs.org
or check under "Airphoto Surveys" or "Photographers-Aerial" in your local phone book. State departments of natural resources, mines and minerals, or fish and game may also have airphoto archives. Several universities have centers for remote sensing, such as the Geophysical Institute at the University of Alaska, Fairbanks (see Table 1.3).

16 REMOTE SENSING FOR GEOLOGISTS

Table 1.3 Partial List of Remote Sensing Web Sites

U.S. Government Sites

Bureau of Land Management	http://www.blm.gov/
Earth Data Handling System	http://edhs1.gfsc.nasa.gov/
Environmental Sciences Division	http://www.esd.ornl.gov/
EROS Data Center	http://edcwww.cr.usgs.gov/
EROS/USGS Product Search	http://edcwww.cr.usgs.gov/webglis/
Federal Geographic Data Products	http://info.er.usgs.gov/fgdc-catalog/title.html
Geophysical Data Center	http://www.ngdc.noaa.gov/ngdc.html
GRASS	http://www.cecer.army.mil/grass/GRASS.main.html
Jet Propulsion Lab	http://makalu.jpl.nasa.gov
NASA	http://hypatia.gsfc.nasa.gov/NASA_homepage.html
NASA Ames Flight Center	http://www.arc.nasa.gov
NASA Goddard Space Center	http://pao.gsfc.nasa.gov http://ltpwww.gsfc.nasa.gov/
NASA Johnson Space Center	http://eol.jsc.nasa.gov
NASA/JPL radar	http://southport.jpl.nasa.gov/
National Operational Hydrologic Remote Sensing Center	http://www.nohrsc.nws.gov/
NCAR	http://www.ucar.edu/
NOAA	http://www.ncdc.noaa.gov
Sandia Labs	http://www.sandia.gov/GIS/gis.html
Stennis Space Center	http://www.srsc.org/
USGS Spatial Data Library	http://waisqvarsa.er.usgs.gov/wais/home.html
US Geological Survey	http://ask.usgs.gov
US Geological Survey airphotos	http://terraserver.microsoft.com

International Government Sites

Agenzia Spaziale Italiana	http://www.asi.it
Australian Center Remote Sensing	http://www.auslig.gov.au/acres/index.html
Australian Surveying & Land Info. Group	http://melbourne.citysearch.com.au
Canada Centre Remote Sensing (CCRS)	http://www.ccrs.nrcan.gc.ca
Canadian Space Agency	http://www.space.gc.ca
Centre National d'Etudes Spatiales (CNES)	http://cnes.fr/
Consiglio Nazionale delle Richerche	http://ntserver.iia.mlib.cnr.it/index.htm
ERSDAC Japan	http://www.iijnet.or.jp/ersdac/
German Remote Sensing Data Center	http://cils.dlr.de/ http://www.dlr.de/DLR-Homepage
Indian Remote Sensing	http://www.fas.org/spp/guide/india/earth/irs.htm
Indian Nat. Remote Sensing Agency	http://www.hrsa.gov.in
Institut Geographique National (IGN)	http://www.ign.fr
National Remote Sensing Centre U.K.	http://www.nrsc.co.uk
National Space Development Agency Japan	http://www.tksc.nasda.go.jp http://www.eoc.nasda.go.jp
Remote Sensing Technology Center Japan	http://hdsn.eoc.nasda.go.jp/guide/satellite/satdata/jers_e.html
Russian Space Agency	http://www.rka.ru/

U.S. Academic Sites

Boston University	http://crs-www.bu.edu/
Cal Tech Jet Propulsion Lab	http://makalu.jpl.nasa.gov/
Colorado School of Mines	http://gn.mines.colorado.edu/
ERIM	http://www.erim.org/
George Mason University	http://geog.gmu.edu/
MIT	http://www-erl.mit.edu/
Ohio State	http://www.cis.ohio-state.edu/
Penn State	http://www.gis.psu.edu/

Table 1.3 (*continued*)

U.S. Academic Sites

Rutgers University	http://deathstar.rutgers.edu/welcome.html
SUNY	http://zia.geog.buffalo.edu/
Texas A & M	http://ageninfo.tamu.edu/geoscience.html
Univ. Arkansas	http://www.east.uark.edu/
Univ. California Berkeley	http://www.regis.berkeley.edu/index.html
Univ. Colorado	http://kurgan.colorado.edu/the_ville.html
Univ. Delaware	http://triton.cms.udel.edu/crs/crs.html
Univ. Maryland	http://kabir.umd.edu/Welcome.html
Univ. Minnesota Image View	http://www.gis.umn.edu/fornet/ids/imageview
Univ. Nebraska Lincoln	http://ianrwww.unl.edu/ianr/calmit/calmit.html
Univ. New Mexico EDAC	http://edac.unm.edu
Univ. North Carolina	http://itre.uncecs.edu/
Univ. South Carolina	http://lorax.geog.scarolina.edu/
Univ. Texas Austin	http://www.lib.utexas.edu/Libs/PCL/Map_collection/Map_collection.html
Univ. Texas El Paso	http://www.geo.utep.edu
Univ. Wyoming	http://faculty.gg.uwyo.edu/marrs/RWMarrs.htm

International Academic Sites

Australian National University	http://life.anu.edu.au/
CRISP Center Rem. Sens. Univ. Singapore	http://www.crisp.nus.edu.sg/
ITC, Netherlands	http://www.itc.nl
McGill University	http://stoner.eps.mcgill.ca/
Phototeque Nationale, Paris	http://www.ign.fr
Univ. Edinburgh GIS	http://www.geo.ed.ac.uk/home/gishome.html
Univ. Greenwich	http://www.gre.ac.uk/directory/earthsci/gis1.html
Univ. New South Wales Center Rem. Sens.	http://saturn.gas.unsw.oz.au/rsgis.html
Univ. Southampton	http://www.geodata.soton.ac.uk/html
Univ. Stuttgart	http://www.ifp.uni-stuttgart.de/
Univ. Toronto Center Landscape Research	http://www.clr.toronto.edu/clr.html
Univ. Vienna	http://info.arclab.tuwien.ac.at/

Organizations/Commercial

Alliance for Marine Remote Sensing	http://www.amrs.org
Am. Soc. Photogrammetry	http://www.asprs.org
Analytical Imaging	http://www.aigllc.com
ARIES Project	http://www.cossa.csiro.au/ARIES
Australian Geological And Remote Sensing Services Pty. Ltd.	http://www.agarss.com.au/MAIN.HTM
Borstad Associates Ltd.	http://www.borstad.com/homepage.html
Canadian National Atlas Information	http://www-nais.ccm.emr.ca/
Earth Satellite Corp.	http://www.earthsat.com/ http://www/geocover.com
Earth Science Organizations	http://www-vl-es-orgs-by-location.html
Earth Watch Inc.	http://www.digitalglobe.com/ewhome.html
Environmental Res. Inst. Michigan (ERIM)	http://www.erim.org/
Environmental Resources Info. Network	http://kaos.erin.gov.au/erin.html
EOM, Inc.; GEMI, Corp.	http://www.eomonline.com
Eurimage	http://www.eurimage.com
European Center Earth Observation	http://ceo-www.jrc.it/CEO_home.html
European Space Agency (ESA)	http://www.esa.int/hq/
Geophysical Environmental Res. (GER)	http://ger.com

Table 1.3 (continued)

GlobeXplorer	http://globexplorer.com
HyMap	http://www.aigllc.com
ImageSat International	http://www.imagesatintl.com/
Index of www Remote Sensing Resources	http://ftp.geog.ucl.ac.be/~patrick/geogr/Eteledetec.html
Indian Space Research	http://www.isro.org/
InterMap	http://intermaptechnologies.com
ISPRS	http://www.geod.ethz.ch/isprs/isprs.html
ITRES Research Ltd.	http://www.itres.com/
Marine Remote Sensing	http://www.amrs.org/
Orbimage	http://www.orbimage.com/
Radarsat	http://www.rsi.ca/
Remote Sensing Society (U.K.)	http://www.the-rss.org/
Remote Sensing Virtual Library	http://www.vtt.fi/aut/ava/rs/virtual/
Sensor & Software, Inc.	http://www.sensoft.on.ca
Space Imaging/EOSAT	http://www.spaceimaging.com
SPOT Image Corporation	http://www.spot.com or http://spotimage.fr/welcome.html
Tiger Mapping Service	http://tiger.census.gov/cgi-bin/mapbrowse -tbl/
U.N. Environmental Program	http://www.grida.no/
VTT	http://www.vtt.fi/aut/rs/virtual

ACQUIRING IMAGERY

The second part of any project is obtaining the imagery. There are generally two kinds of sources: government and service companies. A list of service companies can be obtained through technical literature or even the phone book under the heading "Aerial Surveys", or from various photogrammetric societies. A list of governmental agencies can be obtained from the same sources (Table 1.2). The internet is now a leading source of information on image availability (Table 1.3). Agencies generally have archives that can be searched for images that meet the user's requirements. Service companies can fly a survey to the user's specifications. This usually requires a greater investment of both time and money. If a survey is being planned, consult your local photogrammetric society for standard survey parameters, or refer to the U.S. Bureau of Land Management *Aerial Photography Specifications* [7].

Upon receiving the imagery it is necessary to check the quality and confirm that the area you expected to be covered is correct. One is then ready to begin interpretation.

REFERENCES

1. F. Dekker, *Earth Observation Magazine* **2** (1993), 28–35.
2. C.E. Junge, *Air Chemistry and Radioactivity* (Academic Press, 1963).
3. R.G. Reeves, A. Anson, and D. Landen, *Manual of Remote Sensing* (American Society of Photogrammetry, Falls Church, 1975), 1st ed., Chap. 6.
4. M.J. Abrams, J.E. Conel, and H.R. Lang, *The Joint NASA/Geosat Test Case Project* (Jet Propulsion Lab/AAPG Bookstore, Tulsa, 1984) Sections 11 and 12.
5. B.H. Purser, *The Persian Gulf* (Springer-Verlag, New York, 1973), p. 1–9.
6. B.S. Siegal and A.R. Gillespie, *Remote Sensing in Geology* (John Wiley and Sons, New York, 1980), 702 p.

7. U.S. Bureau of Land Management, *Aerial Photography Specifications* (U.S. Gov't. Printing Office, Denver, 1983), p. 15.

ADDITIONAL READING

Z. Berger, *Satellite Hydrocarbon Exploration* (Springer-Verlag, Berlin, 1994), Chap. 1, pp. 3–34.
C. Lawrance, R. Byard, and P. Beaven, *Terrain Evaluation Manual* (Transport Research Lab, Department of Transport State of the Art Review 7, HMSO Publications Centre, London, 1993), Chaps. 1–2, pp. 1–12.
S.A. Morain and A.M. Budge, eds., Earth Observing Platforms and Sensors, **1**, Manual of Remote Sensing, Third Ed. (John Wiley & Sons, Inc., New York, 1999), CD-ROM.
R.K. Vincent, *Fundamentals of Geological and Environmental Remote Sensing* (Prentice Hall, New York, 1997), 480 p.

QUESTIONS TO CONSIDER

1. You have been asked to determine if there are active seeps in the Natuna Sea, Indonesia. This area frequently has cloud cover. What sensor systems should be considered?
2. It is important to know whether the stratigraphy in Somalia is similar to that in Yemen. Would color or black/white images be more useful? Would satellite images or airphotos be better? Give examples of what information each would provide to help make a usable photogeologic map.
3. How would you go about finding and obtaining airphotos over Kenya, Siberia, or Peru?
4. What scale imagery would you use for a regional gold exploration program in Kazakstan? Why? What scale would you use to locate sources of groundwater in one township in Wyoming?
5. What platform, sensor, and wavelength of imagery would you use to evaluate a mining claim in northern Mexico? In southern Mexico? Would they be different, and if so, why?
6. You need an image of Mongolia next week to tell management what roads and other infrastructure exists. Cost is no object. What would you do? Would you handle it differently if you are on a tight budget?

2
Photointerpretation Tools and Techniques

CHAPTER OVERVIEW

In this section we discuss equipment needed to start interpreting imagery and the different approaches and techniques commonly practiced. There are several tools of the trade that make photointerpretation possible, and others that make it easier (Table 2.1). Among those items generally considered indispensable are stereoscopes, magnifying glasses, and light tables. Other items include various types of marking pens, clear overlays, filters, and machines for changing scales from photos to maps (commonly called transferscopes). For a more complete discussion of devices for measuring vertical and horizontal distances, refer to Ray [1] and Miller and Miller [2].

Image interpretation is, by definition, a subjective process based largely on the individual's experience and training. There are, however, techniques such as image enlargement, filtering, and foreshortening that allow an interpreter to "see" and extract more information from photos and images. Some of these interpretation methods are discussed.

TOOLS

Light Tables

Light tables are used to illuminate positive transparencies, a common format of airphotos or space imagery. When used to backlight paper or duratrans prints, it is

Table 2.1 List of Some Basic Equipment Needed to Begin Photointerpretation

---
Large work surface or light table
Colored pens that write on film
Colored pencils that write on photos
Clear film for overlays
Polymer erasers (soft enough that they do not scratch emulsion)
Magnifying glasses (usually between 3× and 10×)
Stereoscopes (pocket or mirror)
Rulers (metric or English)
Scale rulers (with distances given for various scales)
Planimeter (for measuring areas)
Topographic or planimetric maps as paper or film (to use as a base for the interpretation)
Weights and tape (to secure images and maps)
---

easier to see detail otherwise hidden in dark areas of the image, and colors also appear more vibrant. Backlighting also minimizes eye fatigue.

Light tables are available in many sizes and with varying types of illumination. It is necessary to get the illumination as even as possible in order to avoid distracting light bands due to fluorescent tubes. This is best achieved by having an "opal" glass or "frosted" glass cover on the light box to diffuse the light. If the table is also used for cutting and splicing, it is wise to have a second sheet of common glass over the frosted sheet to avoid scratching the diffuser, which can be expensive to replace.

Magnifying Glasses

Magnifying glasses come in many sizes and powers, some illuminated. Illuminated glasses are particularly convenient when a light table is not available to backlight prints. The most common glasses magnify 3 to 10 times. Prolonged use of magnifying lenses can produce eye strain.

Stereoscopes

Stereoscopes come in pocket or table models, have various magnifications available, and make the interpretation of stereo images easier and more accurate. Stereoscopic viewing requires parallax, the apparent displacement of an object as a result of viewing from two different locations. This is achieved in aerial photography by overlapping successive photos acquired as the airplane moves along its flight path. Satellite parallax is acquired by imaging an object from adjacent orbital paths for a vertical scanner, by different viewing angles on a nonvertical scanner, or by overlapping photos in the case of orbital photography. Stereo viewing generally relies on 60% endlap of photos along a flight line, and 10–20% sidelap of photos from adjacent flight lines.

Stereoscopes allow three-dimensional viewing of the earth's surface, and provide the interpreter with the best remote way to estimate the dip of units or the relief in an area (if topographic maps are not available). With vertical exaggeration (see Recognizing Dip, Chapter 5) one can recognize dips as low as 1–2° that otherwise would not be evident. Without stereo, for example, it could be difficult to determine whether a dark area is dark soil/rock or just an area in shadow. Stereo images also allow a photogrametrist to generate a digital elevation model and topographic maps.

Although stereo viewing is possible for some people without a stereoscope, the scope helps an interpreter focus on the photographs. Pocket stereoscopes are convenient for field work, but may require bending of one photo to see the entire area of overlap. Mirror stereoscopes are more bulky, but allow the viewer to see the entire overlap area without moving the photos. Combined with magnifying lenses, these tabletop scopes can provide a maximum of detail with a minimum of eye strain.

Transferscopes

The purpose of these scale changing devices is to transfer an interpretation from the image to a planimetric or topographic map base. This is accomplished several ways, but commonly this is by projection or split optics.

Projectors illuminate the image and the image is then projected through a lens onto a glass surface (screen). The base map is on clear or frosted film, and is positioned on the glass screen. The scale is matched by varying the size of the projection by moving the original photo or an intermediate mirror closer or farther from the lens. One must have at least two landmarks or "ground control points" on both the image and base map in order to match scales. Distortion on these instruments is radial, increasing outward from the center of the projected image, so that only the central portion can be used with any degree of accuracy. Examples of such instruments include the double reflecting projector (P.V. Kail Associates) and reflecting projector (Kargl).

Scale-changing instruments that use split optics are built by Bausch and Lomb, among others. These zoom transferscopes, both monoscopic and stereoscopic, use varying illumination intensity on the photographs and base map to allow the viewer to see either the photo, base map, or both simultaneously. The images are viewed through binocular lenses that can enlarge the photos, and other lenses can be used to enlarge the base map. These also impose a radial distortion on the transferred image.

All scale-changing equipment is more accurate and less time-consuming than "eyeballing" an interpretation from an image to map base. Problems arise out of the radial distortion inherent in airphotos, the differences in cartographic projection between satellite images and base maps, the radial distortions in projecting lenses, and the amount of relief. Flat or gently rolling topography is less affected by the distortion of lenses than high relief areas. Differences of map projection affect all terrains equally. The best technique to minimize distortion is to superimpose a grid on the image and match scales and landmarks within each grid cell. Many of these problems are eliminated by using orthophotos or geometrically rectified imagery.

Scales and Planimeters

Engineering supply stores carry not only metric/English rulers, but often carry "scale" rules that give measurements in meters/kilometers, feet/miles, chains, rods, etc., for a given map or photo scale. These can be convenient for reading distances directly from the images when the scale is known.

Planimeters are devices, either mechanical or electric, that measure the circumference of an object and provide the area of that object. This is often faster and more accurate than overlaying a grid of known size and counting squares.

Parallax Bars

Parallax bars, or stereometers, are used to determine the height of features on photographic stereo pairs, generally for 23 × 23 cm (9 by 9 in) frames. Consisting of a bar with two small glass plates, the instrument is positioned such that dots inscribed on the plates fall over identical ground points on the two overlapping photos. These dots are fused stereoscopically into a single dot that appears to float in the three dimensional stereo image. By placing the dot on the top of a feature (e.g., a hill or strike ridge), then at the base of the feature, one can determine its height, or differential parallax. This technique is described in Ray [1] and Miller and Miller [2]. Of particular interest, this instrument can be used to help calculate stratigraphic thickness, determine dip magnitude, or generate topographic profiles.

Overlays

Many photointerpreters have historically used soft lead or grease pencils to mark directly onto photographs. This has the disadvantage of making corrections difficult (erasing can remove the emulsion) and often covering the feature being mapped. These drawbacks can be overcome by using clear plastic film overlays. Overlays can be cut to fit the photo or image, taped along one edge, and then annotated with special pens that write on film. The marks can be erased without damaging the photo, and the film can be raised to see the original image beneath. Overlays can be removed and the photo reinterpreted if a second opinion is required.

Base Maps

Interpretations are easiest to plot and compile on topographic maps because of the large number of corresponding features on the image and map. Planimetric maps generally have only roads and waterways to use as landmarks. Regardless of which type map is available, they should be in the form of a frosted film, or mylar, so that images can be projected on them, or so they can be directly placed over an image and the annotation traced or transferred to the map. Frosted film takes drafting ink better than clear film and maintains true scale better than paper. In areas with great relief or on cluttered maps it is often helpful to have the map reproduced as a halftone, or screened print. This makes the base map lighter than the inked lines of the interpretation, and makes annotation easier to see and read.

Filters

Filters can be used to help view images and map lineaments. A coarse (79 line pairs per cm) diffraction grating on film known as a Ronchi filter can be used by viewing an image through the film and slowly rotating the filter [3]. Lineaments perpendicular to the grating are sharp, whereas lineaments at angles to the ruling appear diffuse. This filter is essentially an analog edge-enhancement technique.

A low-tech low-pass filter that allows the interpreter to concentrate on gradual tonal changes consists of simply removing one's glasses (or contact lenses). This is only effective when the interpreter requires corrective lenses, and works because the high frequency information in an image becomes unfocused.

IMAGE CHARACTERISTICS

Photo Finish

Most photos have a semigloss or matte finish to reduce glare and allow one to annotate on the photo using colored pencils. Some workers, however, prefer a glossy finish because details appear sharper. Photos carried into the field are often laminated to protect the emulsion from scratches and moisture and still allow annotation.

Paper vs. Film

Paper prints are generally more convenient to carry into the field and annotate, but must be protected from moisture and rough handling. Positive transparencies are often easier on the eyes when used with clear film overlays on a light table. They are, however, difficult to use and easily damaged in the field.

INTERPRETATION MECHANICS

Influence of Scale

The scale of the image has a great influence on the amount of detail observed and mapped. Landsat MSS images, for example, in the early 1970s were often interpreted at scales of 1:1,000,000. Considering the amount of detail visible at 1:250,000, let alone the ease of mapping, it is not surprising that new interpretations of the same images at an enlarged scale can provide a wealth of new information. In the author's experience, larger scales yield better (more detailed) maps. If only regional features are required, it is always possible to move back and view the image from a distance.

Often one has a choice whether to use satellite images or more detailed airphotos. Not only do most airphotos offer stereo viewing, they provide a wealth of detail generally not available from orbital altitudes. The main drawback when dealing with airphotos is that large numbers of photos may be required to cover an area that only one or a few small-scale satellite images would cover. For example, it takes four photos at 1:25,000 to cover the same area covered by one photo at 1:50,000 (see Airphotos, Chapter 3). The interpreter must then consider the trade-off between detail versus time to interpret and the added cost for more photos.

Influence of Color

A similar argument can be made when considering the use of color versus black/white imagery. Many engineers claim that an experienced interpreter can glean as much information from black/white photos as from color; many geological interpreters disagree. Color photography not only makes interpretation quicker and easier, it also makes it possible to recognize lithologic units of interest, trace them laterally with greater confidence, and recognize a repeated section or missing section in faulted and folded terrains. Estimates of additional information over black/white images range from 10–25%, but that information can be critical in the final analysis. Again, the main trade-off is cost. Color photography can cost two to three times as much as black/white photography.

Black/white satellite imagery has several advantages over color. It is often quicker and less expensive to generate one band of data than the three required for a color composite. Black/white (especially of MSS band 7, TM band 4, or SPOT P) is an excellent choice for a "quick look." Most interpreters will see different features on different bands (wavelengths). In some cases a black/white image will highlight features lost in the background of a color scene. As each project has unique objectives, each case must be considered on its merits.

False color images, particularly color infrared, have specific advantages over true color. Whereas it is easy to recognize units using true color photos, color infrared images will present geology quite well, eliminate most atmospheric scattering (haze) by eliminating the blue band, and emphasize variations in vegetation (species, growth stage, seasonal variations, stress). Different false color combinations can emphasize other features, from mineralogical changes to moisture changes, etc. The primary disadvantage is that the interpreter must become familiar with each new color combination and what the colors represent under the unique conditions of image acquisition and processing involved. It may be necessary to obtain control points in the field to improve the accuracy of the interpretation.

Technique

Every interpreter has a preferred technique for working through a photo or image. Generally, one begins in a known area, an area where a complete section is exposed and there are good outcrops. One then works outward into lesser known or poorly exposed areas. This is the process of extrapolating the known into the unknown.

Often there are large areas of poor or monotonous exposure, or none of the area is well known. In such cases one can impose a grid over the image and work in one grid cell at a time. This has the effect of concentrating one's attention on small areas rather than the whole scene, which can have more information than is readily absorbed at a glance.

An image often yields different information depending on the direction it is viewed. For this reason many workers will rotate an image and observe it from multiple angles. The direction that shadows fall often has an effect on the perception of topography. A stream will appear either low or high (as a ridge) depending on how the shadows fall. Generally, if an image is placed so shadows fall toward the viewer the relief will appear correctly [4].

Another technique that is often employed by seismic interpreters is viewing the image at a glancing angle. The effect, known as foreshortening, enhances slight changes in shapes and is especially effective for joining discrete segments of a fault zone into a regional lineament.

Filters such as the Ronchi grating can enhance lineaments without image processing. Viewing black and white images through red film provides the illusion of depth.

Several passes at an image may be required before all the pertinent information is extracted. In the first stage, for example, one might trace all bedding and look for a repeated section, missing section, and other irregularities. The second stage might involve mapping all traces of fracturing. A third phase is looking for tonal or vegetation anomalies, and a final pass would be mapping and interpreting drainage patterns.

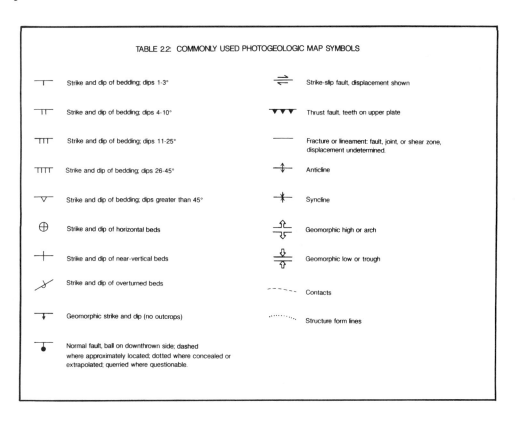

When several images overlap in an area, a comparison of the area on each of the images often reveals new features due to the changes in illumination and plant cover visible on imagery taken at different times. In addition, more details of the surface may be obvious due to the stereo capability that derives from the overlap.

Each problem has a unique set of objectives, and it is worthwhile pausing during the interpretation and reviewing these objectives to ensure that the work stays on track. It never hurts to get a coworker to provide a second opinion: more often than not they will notice something important that was missed.

Some of the more common photointerpretation symbols are shown in Table 2.2. These should be used in conjunction with the standard mapping symbols given in field mapping texts such as those by Compton [5] or Lahee [6].

REFERENCES

1. R.G. Ray, *U.S. Geol. Survey Prof. Paper* **373** (1960): 41–49.
2. V.C. Miller and C.F. Miller, *Photogeology* (McGraw-Hill Book Co., New York, 1961), Chap. 3, pp. 51–58.
3. H.A. Pohn, in D.W. O'Leary and J.L. Earle, eds., *Proc. of the 3rd Intl. Conf. on Basement Tectonics* (Basement Tectonics Committee, Denver, 1978), p. 89.
4. S.A. Drury, *Image Interpretation in Geology* (Allen & Unwin, London, 1987), Chap. 2, pp. 26–27.
5. R.R. Compton, *Manual of Field Geology* (John Wiley & Sons, Inc., New York, 1962), Appendix 4, pp. 334–337.
6. F.H. Lahee, *Field Geology* (McGraw-Hill Book Co., Inc., New York, 1952), pp. 615–620.

ADDITIONAL READING

L. Desjardins, *Bull. of the Am. Assn. Pet. Geol.* **34** (1950): 2284–2317.
C. Lawrance, R. Byard, and P. Beaven, *Terrain Evaluation Manual* (Transport Research Lab, Department of Transport State of the Art Review 7, HMSO Publications Centre, London, 1993), Chap. 3, pp. 1–12.
R.G. Ray, *U.S. Geol. Survey Bull.* **1043-A**, (1956): 1–21.

QUESTIONS TO CONSIDER

1. Your office has a great view, but does not have a light table. The imagery you are using contains dark areas which, you are convinced, contain details important to your project. How might you get around this problem?
2. No imagery or geological mapping exists in your area of interest, and you have been asked to quickly evaluate a broad, low-relief valley for evidence of faults that could localize springs. What other type(s) of data could you use?
3. You have the option of ordering low altitude color imagery or high altitude black/white imagery for a project. What trade-offs are involved? How do you decide which to use?
4. You know your project will take you to the field, where conditions will be harsh. Do you want your imagery on film or paper, and how will you carry and protect it?

3
Remote Sensing Systems

CHAPTER OVERVIEW

Most of us have used cameras to take family or vacation photos, and we are somewhat familiar with cameras and film. Viewers of the evening news have seen weather maps produced from thermal infrared imagery acquired by dedicated satellites. The maps show the distribution of clouds over the earth's surface. Weather radar images indicate areas of light to heavy rainfall. Some camcorders now use thermal sensors to allow recording in the dark. There is a bewildering array of instruments, from cameras to scanners, lasers, and antennas, on satellites and aircraft, which are capable of recording a wide range of the electromagnetic spectrum. These sensors are available for exploration, environmental, and engineering purposes. A summary of some of the better known systems, their useful characteristics, and applications follows. This is not meant to be an exhaustive discussion; references are provided so the interested reader can learn more about these systems.

AIRPHOTOS

Airphotos are acquired by aircraft using cameras specially adapted so that the film advance and shutter exposure is synchronized to flight speed. This is to ensure that each frame has at least the 60% overlap (endlap) required for stereoscopic viewing.

Several terms should become familiar to the user when working with airphotos:

Scale is the ratio of image distance to ground distance and is a function of altitude and camera focal length (in the same units):

$$\text{Scale} = \text{focal length} / \text{camera height}$$

For example: 6 in focal length lens / 24,000 ft altitude = 1:48,000 scale

A 15.25 cm (6 in) focal length is common on most airborne cameras. Larger scales mean more detail can be seen, but less area is included in each frame. For example, standard 22.9 × 22.9 cm (9 by 9 in) frames at a scale of 1:63,360 (flown at 9748 m, or 31,680 ft) each cover an area approximately 210 square km (81 square mi), whereas the same size frame at a scale of 1:24,000 (flown at 3692 m, or 12,000 ft) covers only 30 square km (11.6 square mi). It should be noted that halving the scale from 1:24,000 to 1:48,000 quadruples the area covered, and uses a quarter of the number of frames. This should be considered from a cost standpoint when planning surveys.

Vertical exaggeration occurs when the vertical scale of a stereo image is not the same as the horizontal scale. This can help the interpretation of slopes, dips, fault displacement, and unit thicknesses. Exaggeration can be caused by changes in focal length, camera height, separation of the photo pair, distance between adjacent photo centers, viewing distance, and eye separation. One often chooses to use a 7.6 cm (3 in) focal length lens to increase exaggeration and distinguish dips more clearly in low relief terrain. Exaggeration is increased when the photos are acquired farther apart, or at lower altitudes (lower camera height), with increased separation of the photos being examined, or with increased viewing distance.

Airphotos have radial distortion, meaning that objects appear to lean outward from the center point of a frame, due to the bending of light as it passes through the camera lens. Furthermore, the only area essentially free of distortion on a stereopair is the point halfway between the two photo centers, called the perspective center. In high-relief terrain the distortions can become severe.

Distortions can also be introduced by gradual or abrupt changes in altitude (scale distortion), by tilting of the aircraft (roll and pitch changes that cause oblique distortion), and by crabbing (frames at an angle to flight direction due to flying in a crosswind).

Spectral distortion caused by radial falloff of light intensity outward from the center of the photograph is known as vignetting (pronounced "vin-yetting"). This is again due to the curvature of the lens, and can be compensated for by the use of filters. More information on airphotos and distortion can be found in Miller and Miller [1] and Slater and Colwell [2].

Airphotos are acquired along flight lines, which generally have a minimum of 10% sidelap between adjacent lines. Flight lines are usually laid out north-south or east-west for easy navigation, but can be oriented along strike or structural grain to take advantage of geology, or along pipeline corridors, etc.

Most airphotos are vertical in order to minimize distortion. Under some circumstances, however, it is useful to work with oblique photos, particularly when mapping exposures along cliffs or for orientation purposes.

Low sun-angle photos, sometimes referred to as LSAPs, are acquired early or late in the day, or during the winter when the sun is low in the sky. These have the advantage of exaggerating slight topographic irregularities in seemingly flat landscapes by lengthening shadows and increasing the contrast between shadowed and illuminated slopes. This type of photography is useful for enhancing structures in interior basins, coastal plains, and other low-relief areas.

Airphoto mosaics are also useful. Joining photos into flight lines, and merging lines into mosaics often provides the big picture that cannot be seen or deduced from individual frames. The drawback is that this can remove photos from stereo viewing, and the resulting mosaic often displays a distracting, quilt-like patchiness due to variations in print exposure.

Mosaics can be either controlled or uncontrolled. In controlled mosaics an attempt is made to fit the pieces of the puzzle together in such a way as to minimize distortion. Generally only the center part of each frame is used, and all the frames fit together without obvious offsets along the edges. In uncontrolled mosaics there is no attempt at geometric fidelity. Frames are simply overlapped and pasted together. Typically, uncontrolled mosaics are used only for index maps, to quickly determine which frames cover a specific area. Controlled mosaics are used to see a large area. In many cases this role has been fulfilled by satellite and high altitude aircraft images.

Table 3.1. Film Characteristics

Film Type	Spectral Range	Common Filters	Relative Cost	Ability to Map: Lithology	Structure	Vegetation	Haze Penetration	Water Penetration
Panchromatic (B/W)	0.4–0.7 microns	Wratten 12	low	fair	good	fair	moderate	moderate
Infrared (B/W)	0.4–0.9 microns	Wratten 89B	moderate	fair	good	good	good	poor
Color	0.4–0.7 microns	Wratten 12	moderate	good	good	good	moderate to good	moderate
Color Infrared	0.4–0.9 microns	Wratten 12	high	good	good	excellent	good	poor

Black and White (B/W) Photography

The principal advantage of black and white, or panchromatic photography is its low cost, often a fraction of the cost of color photos (Table 3.1). Most older surveys were routinely flown using B/W; these have been and still are used in exploration programs to great effect (Fig. 3.1). The ability to map structure is usually not compromised. Units are distinguished on the basis of tone (light or dark) and texture (smooth, mottled, etc.). The inability to identify soils, lithologic units, or mineral alteration on the basis of color is the main disadvantage. To the extent that one may not recognize a repeated section due to a lack of color, mapping thrust faults becomes more difficult. Differences in vegetation species and vigor are also difficult to distinguish.

Panchromatic B/W film, such as Kodak 3414 High Definition Aerial film, covers the spectral region from 0.4–0.7 microns, or the visible range. Atmospheric scattering is generally removed using a minus-blue filter (e.g., Wratten 12). In contrast, infrared B/W films such as SO-289 are available that record reflected light in the range from 0.4 to 0.9 microns. Used with a Wratten 89B filter, only wavelengths from 0.7 to 0.9 microns are recorded [2]. This film allows the detection of subtle vegetation changes since the peak reflectance of vegetation is near 0.8 microns. This film also does not record atmospheric haze, and is sensitive to bodies of standing water (black) that absorb infrared light. The cost is less than that of color infrared films.

Color Photography

The chief advantage of true color photography is the ability to recognize units in areas where the stratigraphy is known, and the increased ability to interpret rock units where the geologic section is unknown (Fig. 3.2). Correlation from area to area is improved, particularly if a unique sequence of units can be recognized. Alteration minerals, generally associated with gossans or bleaching, may not be recognizable without color. Color photographs also have an aesthetic quality that makes them particularly appealing for public relations work. The disadvantage of color photography is its cost relative to B/W film. The user must weigh the cost versus benefits to determine the need for color.

Color films such as Ektachrome Aerographic 2448 or SO-242 cover the spectral region from 0.4 to 0.7 microns. Color-infrared film (CIR, "false color," or "camouflage detection film") such as Aerochrome Infrared 2443 covers the range from 0.4 to

30 REMOTE SENSING FOR GEOLOGISTS

Figure 3.1 Black/white U.S. government airphoto of Milner Mountain, near Fort Collins, Colorado. North is up.

0.9 microns and is used with a Wratten 12 filter to remove blue wavelengths [2]. It is particularly sensitive to subtle changes in vegetation and soil moisture and diminishes the effects of atmospheric haze. The cost is generally higher than that of true color film, and there are problems associated with handling the film: it must be stored in a cool or

REMOTE SENSING SYSTEMS 31

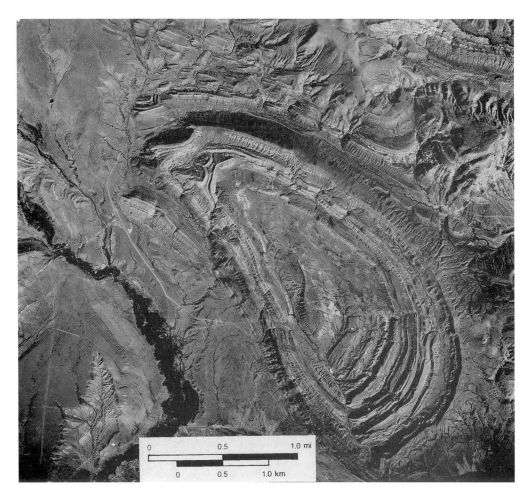

Figure 3.2 Color airphoto of Circle Ridge anticline, Wind River basin, Wyoming. Photo by IntraSearch, Littleton, Colorado. North is at top of photo. (See color plate.)

refrigerated area, and it is sensitive to exposure setting and slight changes in photolab processing chemistry.

The interpretation of CIR photography requires an adjustment in the way the interpreter looks at color on a photo. The sensitivity of the film's pigments has been shifted so that visible blue light is not recorded, and green light is assigned the film pigment blue. Likewise, red light is recorded on film as green, and infrared light is recorded as red. The effect of this shift is to eliminate haze (most scattered light in the atmosphere is blue), to cause deep water to appear black, green rock and shallow or turbid water to appear blue, red units and gossans to appear in shades of yellow to green, and vegetation to appear in shades of red (Fig. 3.3). Most interpreters get accustomed to the color scheme quickly except for the red vegetation.

There is a widespread misconception that CIR photography shows the distribution of heat, and that red areas are warm. In fact, this film records only reflected light in and slightly beyond the visible range (thermal energy, or emitted radiation, has

Figure 3.3 Color infrared NASA airphoto of Milner Mountain, Colorado. North is at the top. (See color plate.)

wavelengths longer than 3 microns and requires special crystal detectors discussed later). The reason that vegetation appears red on CIR photography is that plants are many times brighter in the near-infrared (0.7–0.9 microns) than in the green region (0.5–0.6 microns). Since human eyes are not sensitive to infrared light, we see plants as green, where they are brightest in the visible range (Fig. 3.4).

With a little practice an interpreter will adjust to the color differences between color infrared and true color film, and photointerpretation will be improved by the

REMOTE SENSING SYSTEMS

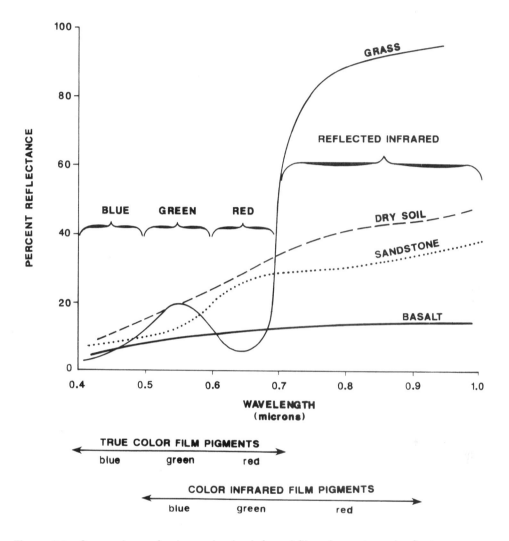

Figure 3.4 Comparison of color and color infrared film pigments and reflectance curves of some typical surface materials. From Condit [3], Moran [4], and Goetz [5].

sharpness, clarity, and vegetation information that is generally not available in standard, or true color photos.

AIRBORNE MULTISPECTRAL/HYPERSPECTRAL SCANNERS

Several airborne multispectral scanners are commercially available. A complete listing of commercial companies involved in remote sensing data acquisition is available from several sources, including the American Society of Photogrammetry (ASPRS). See Table 1.2.

The scanners provide digital information recorded on magnetic tape or CD, although one can request that the vendor provide photographic hardcopy in the form of transparencies or paper prints. The spectral range may be adjustable or set, with the

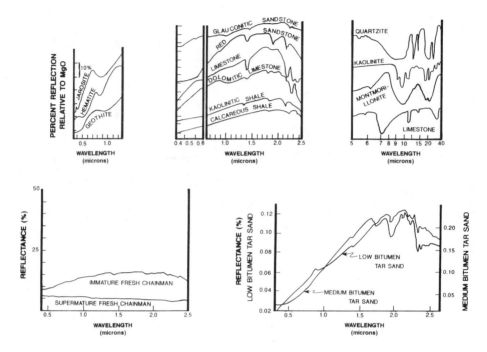

Figure 3.5 Typical spectral reflectance curves of some common geologic materials [6–10].

number of channels (discrete wavelength intervals) and band width (interval range) varying considerably.

One of the advantages of these scanners is the ability to configure the sensor to record spectral features of particular interest, such as clays, carbonates, or silicates (Fig. 3.5). Another advantage is the ability to adjust flight times to avoid adverse weather or atmospheric conditions, or to respond to a crisis. One can also adjust the instruments to vary the resolution, scale, etc. However, if the scanner is not synchronized with flight speed a skewing of the image occurs that must be corrected later during processing.

Flying an airborne scanner survey is considerably more expensive than flying an airphoto survey, both in the cost of the equipment (including mobilization, permitting, etc.) and in the need for computer image processing. On the other hand, working with multiple narrow bands of data and custom processing allows the user to map features not available on standard airphotos.

Imaging scanners are available with spectral bands throughout the ultraviolet (UV), visible, short wave infrared (SWIR, or near IR), and thermal IR range. Besides those developed for research by NASA and the Jet Propulsion Lab, airborne scanners can be purchased or rented from several commercial vendors (contact the ASPRS).

NASA has made available their C-130 for commercial ventures utilizing their NS001 Thematic Mapper Simulator, Zeiss mapping cameras, a thermal infrared multispectral scanner (TIMS), and advanced solid-state array spectroradiometer (ASAS), among others. For more information contact either the Jet Propulsion Lab or the Aircraft Data Facility, NASA-Ames Research Center (see Table 1.2).

The spatial resolution of airborne scanners is a function of the instrument's instantaneous field of view (IFOV) and the aircraft altitude. Generally it is in the range of 1 to 20 meters. The spectral resolution is also an instrument parameter, and can be designed narrow (e.g., 10 nanometers or less) to wide (4 microns or more). Thermal detectors generally have sensitivities from 0.1 to 1.0° C. There generally are three gratings in the instrument that record energy in the visible-near IR (from 0.5 to 0.9 microns), short wave IR (from 2.0 to 2.3 microns), and in the thermal range (from 8.6 to 11.3 microns). The thermal instrument is generally flown at an altitude that provides 15 to 20 meter or better resolution and a 5 to 10 kilometer wide swath. The scanners are gyro-stabilized for roll, pitch, and yaw, so the only distortion is the result of crabbing.

The simultaneous acquisition of many narrow bands allows the analyst to construct spectral reflectance curves (or emittance curves in the thermal range) for the ground area covered by any picture element (pixel). The spectral curve can then be matched against a set of reference curves for various minerals, plants, or soils, in an attempt to identify the surface materials (Fig. 3.6; see also Chapter 6). The greater the spectral resolution (narrower the band width), the more accurate the spectral curve, and the more accurately the material in question can be identified.

NASA has flown several imaging spectrometers (Table 3.2). The imagery produced by these instruments has been given the name "hyperspectral." The *Airborne Imagining Scanner* (AIS), first flown in 1983, had 128 bands in the range 1.2–2.4 microns, each band averaging 9.6 nanometers wide. It had a 320 meter swath and 10 meter pixel. The *Airborne Visible/Infrared Imaging Spectrometer* (AVIRIS) was first flown in 1986 and has 224 bands in the range 0.40–2.50 microns. Each band is 9.6 nm wide. With an IFOV of 1 milliradian, at an altitude of 20 km it has a 10 km swath and 20 m pixel [12, 13]. JPL and NASA Ames are flying the MODIS/ASTER (MASTER) airborne scanning spectrometer. It has 50 channels from 0.4 to 13 microns, a 2.5 mrad IFOV, a 5 to 50 m pixel and 3.5 to 35 km swath, depending on altitude. CASI (*Compact Airborne Spectrographic Imager*) has been flown in both aircraft and helicopters since 1988 and can operate in two modes: spatial mode has 19 bands, and spectrometer mode has 288 bands, each 1.8 nanometers wide. It records images in the range from 0.403 to 1.000 micron. Resolution can range from 0.5 to 10 meters, depending on altitude [14]. A 2 km swath, for example, is composed of 4 m pixels. The *Short Wavelength Infrared Full Spectrum Imager* (SFSI) has been flown by the Canadian Center for Remote Sensing (CCRS) since 1994. The instrument has 120 channels with 0.01 micron band widths in the range 1.2–2.4 microns. IFOV is 0.33 mrad, with ground resolution generally less than 10 m [15, 16]. HYDICE (the *Hyperspectral Digital Imagery Collection Experiment*), flown since 1994, has 210 channels covering 0.4–2.5 microns. The swath is generally 1 km at an altitude of 6 km, with better than 4 m resolution [17]. Other hyperspectral scanners are commercially available, e.g., from Daedalus, Geophysical and Environmental Research (GER), HyVista, and Geoscan. See Chapter 6, Spectral Stratigraphy, and Chapter 7, Exploration Case Histories, for examples of hyperspectral remote sensing applications.

AIRBORNE THERMAL SCANNERS

Thermal scanners can be flown alone or as part of a broader survey. They operate in two spectral regions, from 3.0 to 5.0 microns and from 8 to 14 microns. These instruments do not measure the temperature of objects, but rather the energy radiated from the object. This radiant flux (F) is related to the temperature (T) by the equation:

$$F = EST^4$$

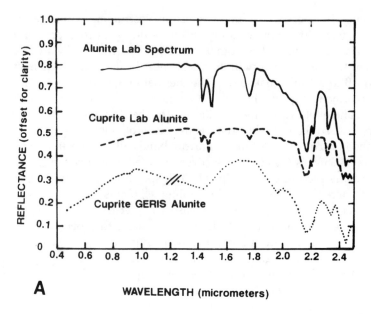

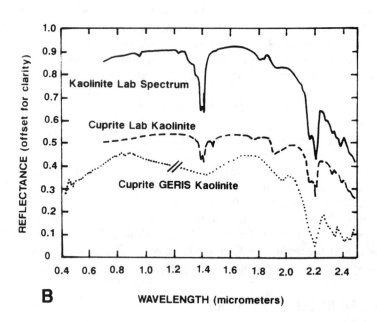

Figure 3.6 Laboratory reference curves and field spectra should be similar, but differences can be caused by variations in grain size, weathering, moisture, soil and plant cover, and atmospheric and illumination conditions. These curves compare pure lab sample reflectance against field measurements of the same mineral and airborne GERIS (a 63-channel imaging spectrometer) reflectance of the mineral. Alunite spectra are shown in (A), whereas kaolinite spectra are shown in (B). From Kruse, et al. [11].

Table 3.2 U.S. Government Operated Aircraft Scanner Systems

System	Spatial Resolution	Swath Width	Instantaneous Field of View (IFOV)	Scanner Field of View (FOV)	Channels
NS001	7.6 m at 3000 m altitude	7.26 km at 3000 m altitude	2.5 mrad	100 degrees	.458–.519 .529–.603 .633–.697 .768–.910 1.13–1.35 1.57–1.71 2.10–2.38 10.9–12.3
TIMS	15 m at 6154 m altitude	9.8 km at 6154 m altitude	2.5 mrad	76.56 degrees	8.2–8.6 8.6–9.0 9.0–9.4 9.4–10.2 10.2–11.2 11.2–12.2
ASAS	2.6 m at 3000 m altitude	1.3 km at 3000 m altitude	0.86 mrad	25 degrees	30 bands from 0.455 to 0.873 (0.014 micron per band)
AVIRIS	20 m at 20 km altitude	10 km at 20 km altitude	1.0 mrad	30 degrees	224 bands from 0.4–2.5 microns (0.0096 micron bandwidth)
MASTER	5 to 50 m, dep. on platform	3.5 to 35 km, dep. on altitude	2.5 mrad	86 degrees	50 bands from 0.4 to 13 microns

where E is the emissivity of the material and S is the Stefan-Boltzman constant, 5.67×10^{12} watts/cm$^{2\circ}$k^4. Emissivity is a function of a material's ability to absorb and reradiate energy. It is defined as the ratio of radiant flux from a body to that from a blackbody at the same kinetic temperature (a blackbody is a perfect absorber and a perfect emitter of radiant energy). Objects that absorb and reradiate large amounts of energy have an emissivity near but always less than 1.0. Objects with low emissivity do not easily absorb and reradiate energy. Pure water has the highest emissivity (0.993); an offshore petroleum slick has an emissivity of 0.972; asphalt has 0.959; granite has 0.815; a polished aluminum surface has an emissivity of 0.06 [18].

When the sun illuminates the surface of the earth, objects facing the sun warm faster than objects in shadows. Therefore daytime thermal imagery generally provides a topographic image showing warm and cool slopes. This can be useful for mapping structures, but other types of images are better for mapping structure and cost less. On flat, homogenous surfaces the thermal images are most sensitive to changes in moisture, since water has a high emissivity. This can help locate sources of water needed in drilling programs, sources of leaks in buried pipelines, or faults that impound groundwater in alluvial areas.

The most useful thermal images for geologic mapping are flown just before sunrise, since the lack of illumination at night gives all slopes a chance to cool to equilibrium. This is when thermal images most accurately show the temperature variations of different materials. Temperatures in predawn images are chiefly a result of rock or soil color, or albedo (darker rocks get and stay warmer than lighter rocks) and density (denser rocks cool slower than less dense rocks). A basalt, for instance, should appear warmer than a granite because of its dark color, and gray sandstone would be warmer than a gray shale in predawn imagery because of the density contrast (Fig. 3.7). In this way thermal imagery can begin to separate lithologic units and indicate structure (Fig. 3.8)

Many materials have unique spectral curves in the thermal part of the spectrum. When several narrow bands of thermal data are acquired, one can match the measured

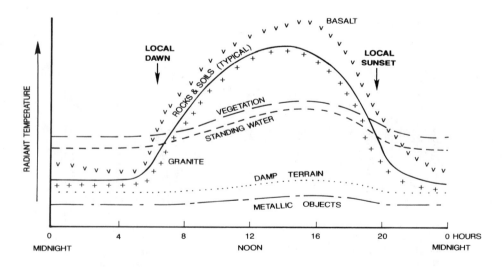

Figure 3.7 Schematic diagram of daily radiant temperature curves for some common materials. From Sabins [18].

Figure 3.8 Pre-dawn thermal image of the Wolverine Creek syncline (arrow) developed in Cretaceous Aspen Formation shales, Caribou Range, eastern Idaho thrust belt. Flown by Mars, Inc., Phoenix.

spectral curve to a reference curve. In the thermal region one can begin to discriminate between silicate minerals on the basis of these spectral curves [19].

The advantage of airborne multispectral scanners is their superior resolution and ability to generate spectral curves and discriminate minerals. They may be used to help identify hydrothermal alteration, oil seeps, source rock, or soil and rock alteration associated with leaking hydrocarbon gases.

The primary disadvantage of airborne scanners is the cost, in time and money, of flying a survey and then processing the large amounts of data generated. Identification of surface materials is complicated by mineral mixing and weathering, and by vegetation cover. Thus, the technique works best in arid or arctic climates where weathering and vegetation are minimal.

In 1982 NASA started flying a six channel *Thermal Infrared Multispectral Scanner* (TIMS) in the range 8.2–12.2 microns. This instrument has a sensitivity of 0.1–0.3° C. The 2.5 milliradian IFOV provides a resolution of 10 m at an altitude of 4000 m. This instrument is particularly well suited to the mapping of silica-rich minerals and silicification due to the silica absorption features located at the instrument wavelengths [19, 20]. The Consiglio Nazionale delle Richerche (CNR) of Italy has flown the MIVIS (*Multispectral Infrared and Visible Imaging Spectrometer*) since 1993. This instrument has a 2.0 mrad IFOV and 102 bands in the range 0.43–0.83 microns, 1.15–1.55 microns, 1.985–2.479 microns, and 8.21–12.70 microns. Ground resolution varies with altitude [21]. NASA is presently evaluating the SEBASS hyperspectral thermal instrument that records 128 channels from 2.1 to 5.2 microns and another 128 channels between 7.8–13.5 microns. Ground resolution varies with altitude.

AIRBORNE RADAR

Radar is an active system in that it illuminates the surface with a beam of microwave radiation. Radar is most sensitive to surface roughness; shorter radar wavelengths are most sensitive to microtopography, whereas long wavelengths are more sensitive to macrotopography [22–24]. The most commonly used wavelengths are designated "K band" (1.1–1.7 cm), "X band" (3–3.2 cm), "C band" (3.8–7.5 cm), and "L band" (25 cm). A useful equation for converting radar frequency to wavelength is:

$$\text{wavelength (cm)} = \frac{30}{\text{frequency (GHz)}}$$

Radar also is extremely sensitive to surface moisture differences (variations in the complex dielectric constant, a measure of the electrical properties of surface materials).

Radar energy is beamed outward from an aircraft perpendicular to the flight direction, this being the "look" direction, e.g., north look (Fig. 3.9). Radar beams can be transmitted and received either "like polarized" or "cross polarized." Like polarized waves are transmitted horizontal and received horizontal (HH) or transmitted and received vertical (VV). Cross polarized waves are either HV or VH.

The radar image consists of three types of radar return: corner reflectors (bright), diffuse reflectors (shades of gray), and specular reflectors and shadows (black; no return). Early imaging radars were called side looking airborne radar, or SLAR. With the development of on-board computers came synthetic aperture radar, or SAR, which synthetically creates a larger antenna, thus achieving greater resolution. Imagery is

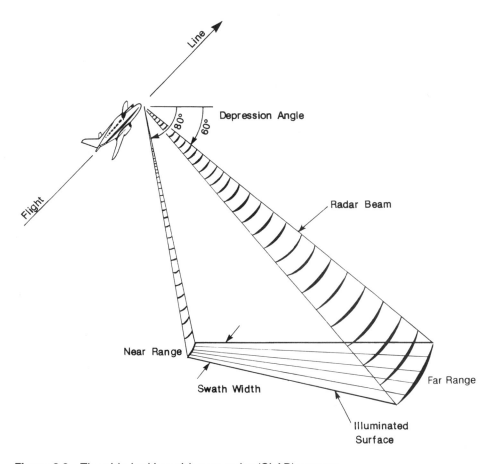

Figure 3.9 The side-looking airborne radar (SLAR) system.

available from a number of sources including the U.S. Geological Survey EROS Data Center, Sioux Falls, South Dakota; the Canadian Center for Remote Sensing in Ottawa; the Earth Remote Sensing Data Analysis Center (ERSDAC) of Japan; and the European Space Agency (ESA), among others. Radar images have been generated by U.S., Canadian, European, Japanese, and Russian satellites.

Advantages of radar imagery over satellite images or air photos include the ability to penetrate clouds (important in the Tropics) and darkness (important in the Arctic and at night). Contrary to common belief most radar systems cannot penetrate a vegetation canopy, yet where the canopy follows topography it generally provides the same information as if there were no ground cover (Fig. 3.10). Radar only penetrates the surface microlayer in soil-covered areas, with the exception of sand cover in hyperarid environments (see Shuttle Imaging Radar). Under rare and unusually arid conditions the microwaves have been shown to penetrate from one to two meters below the surface. Ice is transparent to microwaves, and water reflects them. Thus, radar images of ice-covered areas can reveal the location of water (lakes, rivers) beneath the ice.

Radar enhances topography and therefore structurally controlled landforms. In rugged terrain one can plan a survey with a steep depression angle to minimize

Figure 3.10 Synthetic aperture radar (SAR) image of the El Pilar fault near Arima, Trinidad. This is an example of an area where rainforest canopy mimics topography, yet clear cuts in the forest are also visible as geometric patterns. The brightest areas are towns (buildings act as three-corner reflectors). This is a south look flown by Intera, (now InterMap) Calgary.

shadows; in flat terrain a shallow depression angle highlights subtle topographic changes such as gentle folds or small faults.

Flight lines can be laid out so that structural trends are accentuated. The best results are obtained when flying at about 30° to the strike of structures, so that the beam hits the trend at 60° to strike.

Some companies have experimented with multispectral color radar, but it is not yet common. Experiments by workers at NASA and the Jet Propulsion Lab have demonstrated that multipolarization radar images are sensitive to differences in vegetation size, density, and distribution as well as differences in soil moisture and, in some cases, composition [25]. These tests used an L-band radar over several sites and combined like- and cross-polarized images in color to generate surface feature maps. They found that like-polarized waves are most sensitive to topographic effects and surface water, whereas cross-polarized waves respond to changes in vegetation cover and soil moisture.

Disadvantages of radar include the high cost relative to air photo and satellite surveys, and the general lack of color to assist in lithologic mapping. Layover and foreshortening are unique geometric distortions that make it difficult to estimate dip. These distortions are caused by radar beams returning from elevated objects before beams reflected off low objects, and cause hills, for example, to appear displaced toward the sensor. This distortion is minimized in flat terrain. Most vendors attempt to rectify the imagery to minimize geometric distortions.

In most cases stereo overlap is not available. A kind of stereo can be obtained by keeping all parameters the same but flying again at another attitude, or by flying with two opposing look directions. The first of these options provides low relief because of the low base-to-height ratio that controls vertical exaggeration. The second technique generates shadows on both sides of features, which makes viewing difficult. Unlike photography, radar has no useful information in shadowed areas, since these are true "no data" zones. This may necessitate flying a survey in moderate to high relief areas with two opposing look directions to fill in the shadows.

Radar Interferometry

It is possible to generate digital elevation models from radar imagery, which then allow the creation of topographic maps. Synthetic aperture radar records distance information as well as radar brightness. This technique uses two SAR images to measure small variations in surface elevation (approximately 1 cm) over large ground swaths. The two images are generally flown at different times, or simultaneously with two antennas, so that they have slightly different viewing angles. The images are registered to an accuracy of about 1/8 pixel, the phases of the corresponding pixels are subtracted, and the result is an interferogram. Phase differences in an interferogram have cycles of 360° and must be unwrapped to derive absolute phase. After unwrapping, the absolute phase is used to derive altitude information. The closer the images are acquired in time and space, the more accurate the interferogram will be. Surface variations from one image acquisition to the next result in a phase change known as temporal decorrelation. There should be no phase changes unless there has been a change in surface elevation that changed the distance from the surface pixel to the antenna [26, 27] or a horizontal displacement along the line-of-sight. Using a C-band radar interferometer formed by a pair of antennas displaced across track, Zebker [28] reports that he could measure the topographic height of each point on the ground with accuracies between 2 and 12 m. This error could be improved by removing systematic noise due to the aircraft motion and increasing the signal-to-noise ratio. Some companies have marketed topographic maps generated from radar images

44 REMOTE SENSING FOR GEOLOGISTS

Figure 3.11 Color elevation image of an open pit coal mine at Gillette, Wyoming, created from radar interferometry. Image processing by ERIM. (See color plate.)

(Fig. 3.11). This technique can be applied to satellite radar as well as airborne radar. Radar interferometry has been used to measure creep along faults, subsidence due to pumping of ground water and oil, inflation of volcanos, earthquake deformations, and to make digital elevation models. See Chapter 12, Environmental Hazards, for examples of the use of interferometric radar images.

SATELLITE PHOTOGRAPHY

Satellite photography began with military reconnaissance missions in the late 1950s and early 60s (Table 1.1). The first publicly available photography comprised hand-held, generally oblique true color photos from the Mercury, Gemini, Apollo, and then Shuttle programs (catalogs may be obtained from NASA Johnson Space Center, Houston, or from EDAC, University of New Mexico, Albuquerque).

Skylab

The first serious effort at civilian earth resources photography from space was during the Skylab program from May 1973 to February 1974. Vertical true color and multispectral B/W color infrared photos were acquired from an altitude of 435 km (261 mi) using two camera systems, the S190A and S190B, both of which provided stereo

Figure 3.12 Skylab S190B true color photo of the Mobile, Alabama. Note the blue color due to atmospheric scattering. NASA photo. (See color plate.)

coverage (Fig. 3.12). With a 152 mm (6 in) focal length the S190A covered 163 × 163 km (98 × 98 mi) and had an effective resolution between 60 and 150 meters (195–488 ft). The S190B had a 45.7 cm (18 in) focal length, covered 109 × 109 km (65 × 65 mi), and had a resolution between 15 and 30 m (50–100 ft). Repeat coverage was every 5 days between 50° north and 50° south latitude. Coverage is incomplete because of scheduling and cloud problems. This data was indexed by EDAC, University of New Mexico, and is available from there and the EROS Data Center in Sioux Falls, South Dakota. Much of the photography has changed color or faded with age, but the vertical and oblique images are still useful for a synoptic view of terrain.

Shuttle

The Space Shuttle program has acquired vertical and oblique black/white, color and color-infrared photographs using hand-held cameras and the large format camera (LFC). This instrument has a 30.5 cm (13 in) focal length and was used at altitudes from 239 to 370 km (143–222 mi). Some stereo coverage is available. Contact the EROS Data Center or Johnson Space Center (Houston).

Russian Data

Russia is marketing satellite photography acquired by Soyuzkarta, the earth resources exploration branch of their space program. Imagery is available from several sensors, including the KFA-1000, MK-4, KATE-200, DD-5, KVR-1000, and TK-350 camera systems (Fig. 3.13, 3.14). The KFA-1000 has a resolution of 5 meters (16 ft) and each frame covers 100 × 100 km (60 × 60 mi). Both panchromatic and dual emulsion films (sensitive to 570–670 and 670–810 nm) are used. The MK-4 multispectral camera has 6 meter (19.5 ft) resolution and records 6 channels (Table 1.1). It can achieve 60% stereo overlap using two available cameras, and covers an area 126 × 126 km (75 × 75 mi). The KATE-200 system has 15 to 30 m (49–98 ft) resolution and covers an area 180 × 180 km (108 × 108 mi). Photographs are acquired at an altitude of 200 km (120 mi) with a 200 mm (8 in) lens in three channels: 0.5–0.6, 0.6–0.7, and 0.7–0.85 microns. Stereo overlap from 10–80% is available. The DD-5 system was originally high-resolution photography, but has been degraded to two meter data by digitizing. This data is available on computer-compatible tapes and each scene varies in area between 104 and 200 square km. The KVR-1000 camera, available since 1984, produces black/white photographs with 2–3 m resolution. It is flown on the Kosmos spacecraft at an approximate orbital altitude of 220 km. Image size is about 40 by 50 km, depending on altitude. There is one spectral band in the range 0.51–0.76 microns. The TK-350 camera, available since 1982, produces panchromatic stereo images (0.51–0.76 microns) with 60–80% overlap and 5 to 10 m resolution. Images cover roughly 200 by 300 km, depending on orbital altitude.

Advantages of satellite photography over most digital images include stereo capability and the ability to analyze data without costly image processing. Resolution is generally better than in digital images. Disadvantages include the difficulty involved in making geometric and spectral corrections, the difficulty in filtering high or low frequency data and performing contrast enhancements, and the inability to apply mathematical transformations that enhance lithology, structure, or other objectives of exploration. Provision must also be made to lift film into orbit and later retrieve it. Much of this imagery, particularly the Russian data, is available in digital form.

REMOTE SENSING SYSTEMS 47

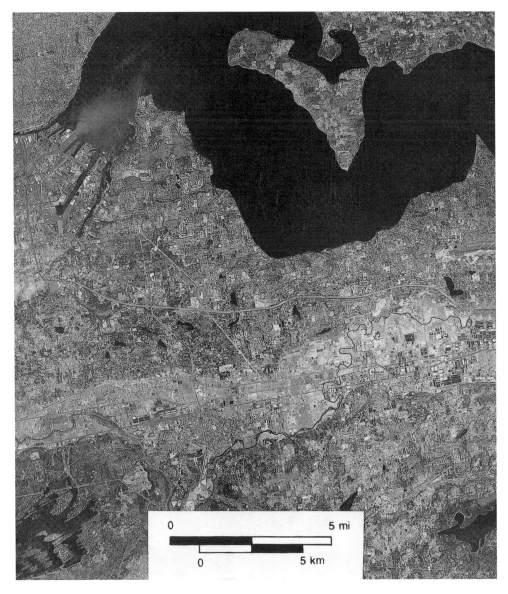

Figure 3.13 Russian KFA-1000 spectrozonal photograph of the Tacoma area and Puget Sound, Washington. (See color plate.)

SATELLITE DIGITAL IMAGERY

The advantages of digital satellite imagery over photography and aircraft data include:

1. Long term repetitive coverage can provide cloud-free images of various seasons (high and low sun elevation; varying plant cover, snow cover, moisture conditions, etc.).
2. Synoptic view at relatively low cost per unit area.

48 REMOTE SENSING FOR GEOLOGISTS

Figure 3.14 Russian KVR-1000 digitized black/white photo of Washington, D.C. EOSAT photo. (Landsat TM images courtesy of Space Imaging, LLC. Denver, Colorado.)

3. Access to all countries without regard to terrain, climate, or politics
4. Confidential exploration and scouting activity.
5. Ability to correct geometric and spectral distortions.
6. Ability to filter, enhance contrast, and apply transformations to maximize geologic information content.

The principal disadvantage is the cost and time required to acquire and operate an image processing system. This can be minimized by purchasing custom-processed imagery from vendors, or by having an efficient in-house system.

Landsat

Civilian satellite imagery in digital form was first made available by the EROS Data Center's Landsat (originally ERTS) program beginning in July 1972 (Table 1.1). Landsat satellites orbit with either a 16-day or 18-day repeat cycle at an altitude of 918 km (Landsats 1,2,3) and 705 km (Landsats 4, 5, 7). The sun-synchronous polar orbit allows the satellite to pass over mid-latitudes at the same mid-morning time every day. Areas north and south of 81° latitude are not covered. Sidelap between orbital paths varies from 7.6% at the equator to about 70% near the poles. Landsats 1,2, and 3 contained a Return Beam Vidicon (RBV) and Multispectral Scanner (MSS); Landsats 4 and 5 contained the MSS and Thematic Mapper (TM) imaging systems. Landsat 6 failed during launch. Landsat 7, launched in 1999, contains the Enhanced Thematic Mapper (ETM) with eight channels.

The RBV on Landsats 1 and 2 consisted of three television cameras that recorded green, red, and near-infrared bands with 40 m (130 ft) ground resolution. Landsat 3 contained two RBV cameras, each with a single panchromatic band in an effort to acquire stereo imagery. Images covered 99 × 99 km (60 × 60 mi). Problems with extreme contrast caused poor-quality images, and the system was not orbited on later Landsats.

The MSS records reflected light in four wavelengths 0.5–0.6 microns ("Band 4," green light), 0.6–0.7 microns ("Band 5," red light), 0.7–0.8 microns ("Band 6," near-infrared), and 0.8–1.1 micron ("Band 7," near-infrared). The reflected light is recorded by crystal detector arrays that convert it to a digital electric signal that is captured by on-board tape recorders. When the satellite is over a receiving station or in range of a data relay satellite the digital data is telemetered to earth and recorded, and can then be copied and sold to the public. The MSS image covers an area 176 × 176 km (110 × 110 mi) with a resolution of 79 m (257 ft). For most geologic applications the MSS image offers the lowest cost per area of any satellite digital imagery, but at the expense of resolution.

The TM system has six reflected light channels ("Band 1": 0.45–0.52 microns, or blue light; "Band 2": 0.51–0.60 microns, or green light; "Band 3": 0.63–0.69 microns, red light; "Band 4": 0.76–0.90 microns, short wave infrared; "Band 5": 1.55–1.75 microns, SWIR; and "Band 7": 2.08–2.35 microns, SWIR), and one thermal channel ("Band 6," 10.4–12.5 microns). Images cover 176 × 170 km (110 × 105 mi), have resolution of 28.5 m (93 ft) in the reflected channels, and 120 m (390 ft) in the thermal band (Fig. 3.15, 3.16). Thermal imagery can be acquired both during daytime (mid-morning) and nighttime (late evening), but reflected imagery is useless during nighttime passes.

Landsat 7 contains the ETM. The first seven bands are the same as those in the TM, whereas an eighth, panchromatic channel, covers 0.52–0.90 microns with 15 m resolution. The satellite has 16-day repeat coverage and a 185 km swath from a 705 km sun-synchronous orbit.

Advantages of the TM imagery include the combination of high resolution and large area coverage, and seven spectral channels that allow better discrimination of surface materials. The blue light band has better water penetration than the green light band available on the MSS, and allows one to generate bathymetric maps and true color images. The thermal band is useful in mapping shallow groundwater, faults that channel groundwater, springs, offshore currents, and color/density contrasts between geologic units. The disadvantage of TM data is that there are four times the data to process in each reflective band, and seven channels compared to four in the MSS. This increases the time and costs involved in computer processing. They also cost more to purchase.

Figure 3.15 Landsat Thematic Mapper image over the Hodna field area, Algeria. Generated using an IHS transform and bands 7-3-1 + 4. Processed by Amoco Production Company, Houston. (See color plate.)

SPOT

Beginning in 1986 the French Space Agency launched the System Probatoire d'Observation de la Terre (SPOT 1) satellites to gather earth resources imagery. The four SPOT satellites orbit at 832 km (500 mi) in a sun-synchronous orbit with a 26-day repeat cycle. Like Landsat, they cross mid-latitudes at mid-morning (around 10 a.m.).

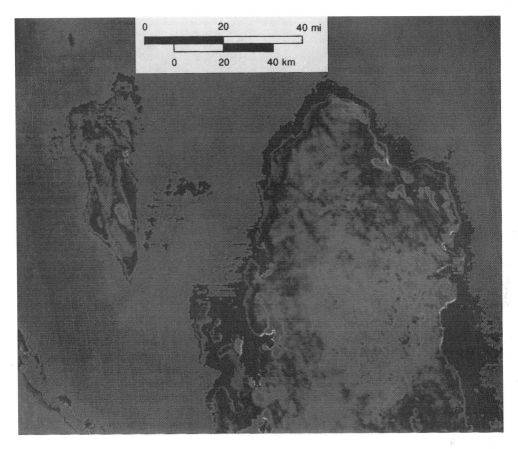

Figure 3.16 Color density-sliced Thematic Mapper thermal image of Qatar and Bahrain. Warm areas are red and yellow; cool areas are light and dark blue. Processed by Amoco Production Company. (See color plate.)

An innovation is the ability to tilt or point the sensor up to 27° from vertical in a cross-track direction, so that the same area can be imaged on successive orbits with the resulting overlap used for stereo viewing. The area covered in vertical mode is 60 × 60 km (36 × 36 mi). Two sensors are available: SPOT P, or panchromatic, and SPOT XS, or multispectral.

SPOT P acquires a single image in the range 0.51–0.73 microns, roughly green and red visible wavelengths (Fig. 3.17). Resolution is 10 m (33 ft). SPOT P trades spectral resolution for spatial resolution. SPOT XS records three channels of data (0.50–0.59 microns, or green light; 0.61–0.68 microns or red light; and 0.79–0.89 microns, or near-infrared). These are commonly combined to create a color-infrared image (Fig. 3.18). The resolution of the XS system is 20 m (65 ft). SPOT XS trades spatial resolution for spectral resolution.

The SPOT system offers high spatial resolution at the expense of large-area coverage. It takes more than nine non-overlapping SPOT images to cover the same area covered by one MSS scene.

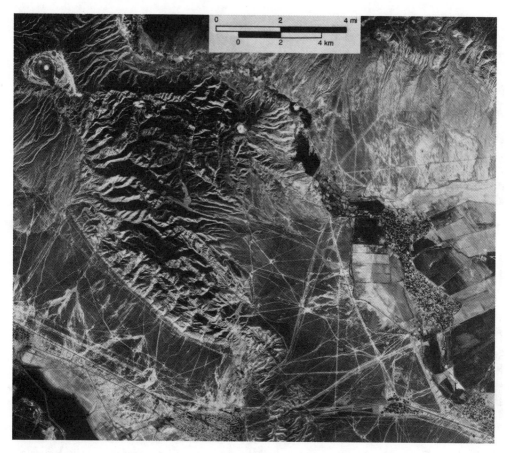

Figure 3.17 Spot Panchromatic image of oil fields south of Baku, Azerbaijan. Processed by Amoco Production Company.

HCMM

The Heat Capacity Mapping Mission (HCMM) was launched in 1978 and generated data until 1980. The objective was to map circulation patterns of marine currents, but large portions of the continents were also covered. Orbiting at an altitude of 620 km (372 mi), this thermal imaging satellite had a ground resolution of 500 m (1625 ft) in the reflected light range (0.50–1.10 microns) and 600 m (1950 ft) resolution in the thermal region (10.5–12.5 microns). Each image covered a 700 × 700 km area (420 × 420 mi), and both day and night images were acquired (Fig. 3.19). Coverage is limited mainly to North America, western Europe, north Africa, and east Australia. Index catalogs and computer tapes are available from the National Space Science Data Center, NASA Goddard, Greenbelt, Maryland.

The advantage of HCMM imagery is the regional coverage: it takes sixteen Landsat images to cover the same area. The disadvantages include limited coverage and low resolution. It is more useful for plate tectonic reconstructions than for detailed exploration.

REMOTE SENSING SYSTEMS 53

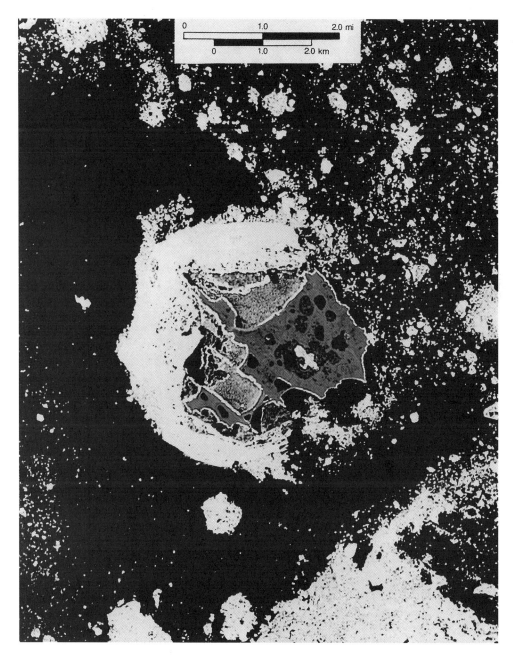

Figure 3.18 Spot Multispectral image of Matveyev Island, Pechora Sea. Processed by Amoco Production Company. (See color plate.)

Seasat

Seasat, designed to monitor the oceans, was launched in June 1978 and failed prematurely in October 1978. This satellite contained an L-band (23.5 cm) imaging radar. Seasat orbited at an altitude of 790 km (474 mi), and had a swath width of

54 REMOTE SENSING FOR GEOLOGISTS

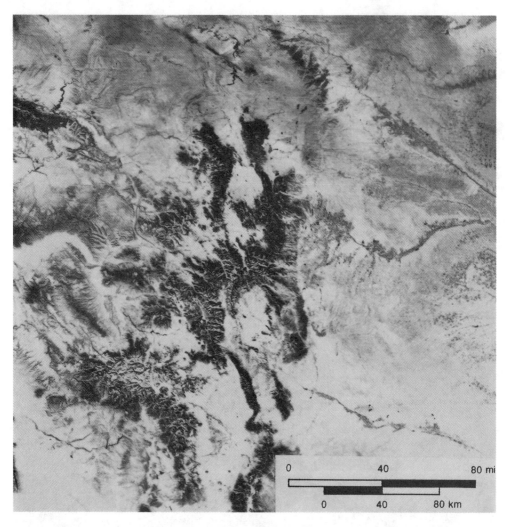

Figure 3.19 Heat Capacity Mapping Mission thermal image of parts of Colorado, Wyoming, and Utah. Daytime image from 6 October 1979. NASA image.

100 km (60 mi), with depression angles from 67° to 73°. Ground resolution was 25 m (81 ft). Coverage was limited mainly to North and Central America, Europe, and part of north Africa. Digital tapes can be obtained from the Environmental Data and Information Service of the National Oceanographic and Atmospheric Administration (NOAA), Asheville, North Carolina. Catalogs are also available from the Jet Propulsion Lab, Pasadena, California.

The advantages of Seasat radar are the cloud-penetrating ability of radar, especially in Central America; the sensitivity to surface roughness changes that indicate changes in soil or lithology, and the ability to map structure by accentuating topographic relief. Disadvantages include the limited coverage and extreme layover caused by the steep depression angle (angle between the radar beam and a horizontal plane). Layover is a geometric distortion caused by the radar beam reflecting off of high points before low,

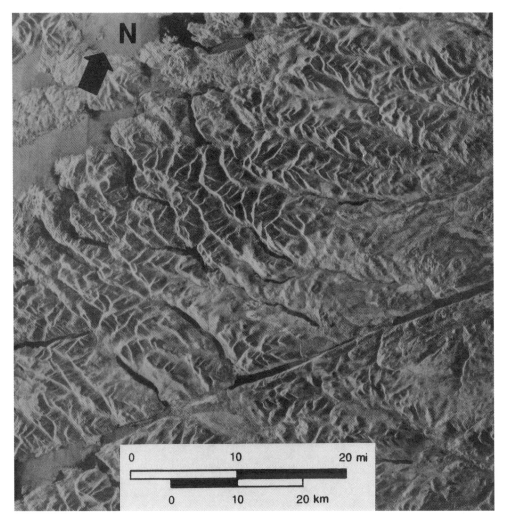

Figure 3.20 Seasat radar image of the Great Glen fault, Scotland. Note the geometric distortion due to "layover." Northeast look direction. Image processed by Jet Propulsion Lab, Pasadena. NASA image.

causing ridge tops to appear closer to the receiver, or to lean toward the radar (Fig. 3.20).

Shuttle Imaging Radar (SIR)

SIR-A was flown on the Space Shuttle in November 1981, and SIR-B in October 1984. Both used an L-band radar, but SIR-A orbited at 250 km (150 mi) and had 38 m (123 ft) resolution, whereas SIR-B orbited at 225 km (135 mi) and had 25 m (81 ft) resolution. SIR-A had a depression angle of 37–43° (50 km swath), and SIR-B had a variable depression angle of 30–75° (40 km swath width). SIR-A, B, and C films are available from the National Space Science Data Center, NASA Goddard. SIR-B data, which also are available in digital form, can be obtained from the Jet Propulsion Lab, Pasadena.

56 REMOTE SENSING FOR GEOLOGISTS

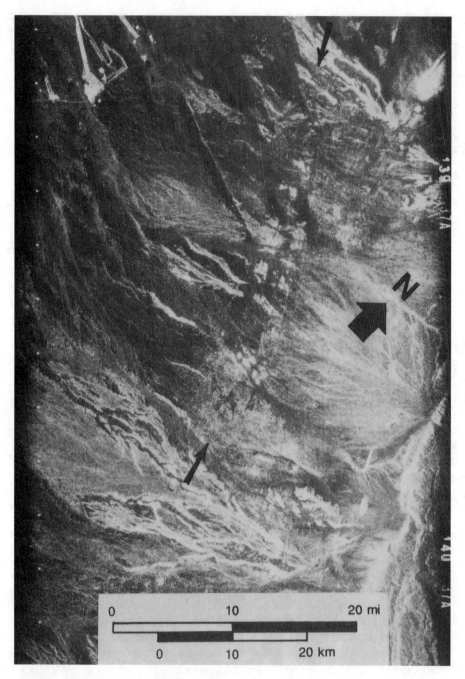

Figure 3.21 Shuttle-Imaging Radar image of a previously unmapped fold (arrow) in a sand-covered area south of Jebel Hafit, Abu Dhabi. NASA image.

Shuttle imaging radar is best known for the phenomena, reported in *Science* in 1982 [29], of penetration through as much as two meters of sand in the hyperarid deserts of Egypt and the Sudan (Fig. 9.2). These workers reported mapping paleodrainage channels not evident at the surface. In other areas, such as the Oman-Abu Dhabi border, anticlines have been mapped on SIR-A imagery through a veneer of windblown sand (Fig. 3.21).

The Shuttle Imaging Radar-C (SIR-C), first launched in 1994, is a joint effort between NASA, the German Space Agency, and the Italian Space Agency. There are two instruments on board: a dual-frequency radar with L-band (23 cm) and C-band (6 cm) wavelengths, and an X-SAR with an X-band (3 cm) wavelength. The ground swath is variable, from 15–90 km (9–56 mi), and resolution varies, depending on the instrument and configuration, from 10 to 200 m (33–656 ft). Incidence angles range between 15 and 55° off nadir. Flight altitude is 225 km (135 mi), resulting in a one-day repeat cycle.

During 2000 the shuttle radar mission imaged the entire earth between ± 60° using C and X band radar. This imagery will be provided as two image sets (one ascending and one descending orbit) with perpendicular look directions. In addition, this mission will use radar interferometry to generate a global digital elevation model with 30 m postings and 10–15 m vertical accuracy. World-wide topography will be provided for $1° \times 1°$ areas. Contact Jet Propulsion Lab for more information.

JERS

The JERS-1 satellite was launched into a sun-synchronous 568 km (340 mi) orbit in February 1992. It has a 44-day repeat cycle and two onboard systems which can be co-registered: The Optical Sensor (OPS) and a Synthetic Aperture Radar (SAR). The OPS has eight bands between 0.52 and 2.40 microns. Ground resolution is 18.3 m (59.5 ft) cross track and 24.2 m (78.7 ft) along track, with a 75 km (46 mi) wide swath (Fig. 3.22). This instrument can produce stereoscopic images using band 3 (views nadir) and band 4 (looks 15.3° forward in the orbit plane). Both of these bands image the 0.76–0.86 micron, or near-infrared region.

The L-band (23.5 cm) SAR also has a 75 km swath and 18×18 m (58.5×58.5 ft) ground resolution. Viewing angle is 35° to the right off nadir (Fig. 3.23).

Data are available to the public from the national archives at the Remote Sensing Technology Center (RESTEC), an agency sponsored by the National Space Development Agency (NASDA) of Japan.

ERS-1 and -2

ERS-1 and ERS-2 were launched in 1991 and 1995, respectively, by the European Space Agency (ESA) into sun-synchronous orbits at a 777 km (466 mi) altitude. They have repeat cycles of 3, 35, and 176 days. The instrument of interest to earth scientists is the Synthetic Aperture Radar (Fig. 3.24). This C-band (5.6 cm) radar covers a swath 80–100 km (48–60 mi) wide with a ground resolution of 30×30 m (98×98 ft). It has a dual incidence angle of either 23° or 35° to the right off nadir. The North American distributor is Radarsat Corp.

RADARSAT

RADARSAT-1 is a project by the Canadian Space Agency to provide the world with a source of radar satellite imagery. Launched in late 1995, the C-band (5.6 cm) sensor has seven synthetic aperture radar modes: standard, wide, fine, ScanSAR wide, ScanSAR narrow, extended high, and extended low. The Standard beam mode nominally provides 100 km (60 mi) wide swaths with 20×27 m (81×91 ft) resolution. The Wide beam mode nominally provides a 150 km (90 mi) swath, with 21×34 m (69×112 ft)

Figure 3.22 JERS-1 OPS image of part of the Neuquen Basin, Argentina. Image processed by ERSDAC, Tokyo. (See color plate.)

resolution. The Fine beam mode nominally provides swaths 50 km (30 mi) wide with approximately 7 × 8 m (29 × 36 ft) ground resolution. The ScanSAR Wide (nominally 86 × 118-m resolution) and ScanSAR Narrow (nominally 73 × 48-m resolution) beam modes provide-wide swaths (500 km [300 mi] and 300 km [180 mi], respectively) with moderate resolutions. The Extended High beam mode provides (nominally) 75 km (45 mi) swath widths with 17 × 27 m (56 × 89 ft) ground resolution. The Extended Low beam mode provides 170 km (102 mi)-wide swaths with 39 × 27 m (128 × 89 ft) ground resolution. The satelite orbits the Earth fourteen times per day at an altitude of 792 km (475 mi) in a sun-synchronous orbit. RADARSAT's selection of beam modes and positions allows it to image a location on the Earth every one to 24-days. The radar is capable of both right and left look directions. The incidence angle varies from 10–59°. The Canadian Data Processing Facility (CDPF) in Gatineau, Quebec, produces up to 44 images per day with a turnaround of four hours or less for rush orders.

REMOTE SENSING SYSTEMS 59

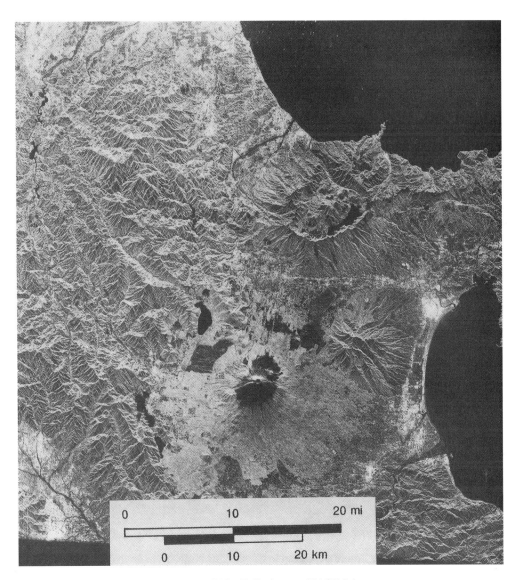

Figure 3.23 JERS-1 radar image of Mt. Fuji, Japan. NASDA image.

ALMAZ-1

The Russian Almaz radar satellite was launched during March 1992 into a 300 km (180 mi) orbit. This is an S-band (10 cm) radar with 13–30 m (40–98 ft) resolution and a one to three-day repeat cycle. The swath width is variable from 200–350 km (120–210 mi) on either side of the orbit. Incidence angle can also be varied from 30 to 60° off nadir.

Resurs-01

Digital products are available from the Russian Resurs-01 series of satellites. The principle sensor is the MSU-SK, which contains five channels in the VNIR and SWIR:

60 REMOTE SENSING FOR GEOLOGISTS

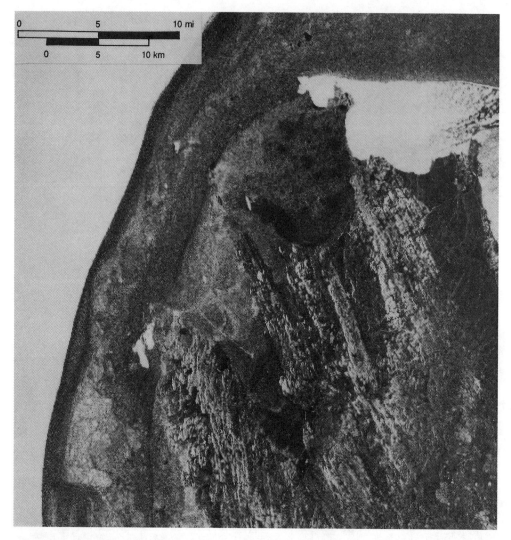

Figure 3.24 ERS-1 radar image of sea ice in the Pechora Sea. The sea ice (left) has a bright radar return whereas outcrops (bottom right) have a mottled texture. Processed by Amoco Production Company.

channel 1 from 0.5 to 0.6 microns, channel 2 at 0.6–0.7 microns, channel 3 from 0.7 to 0.8 microns, channel 4 from 0.8 to 1.1 micron, and channel 5 from 10.4 to 12.6 microns. Resolution in the VNIR is 70 m, although this is routinely resampled to 160 by 160 m; in channel 5 it is 600 m. Swath width is 600 km, and revisit time is 4 days.

IRS

The Indian Remote Sensing Satellites (IRS-1A, -1B, and -1C), launched in 1988 through 1995, contain several sensors: the LISS-1, LISS-2, and LISS-3. The satellites are in a 900–904 km (540–542 mi) sun-synchronous orbit with 22-day repeat coverage. LISS-I has a ground swath of 148 km (89 mi) with 72 × 72 m (234 × 234 ft) resolution. The LISS-II

REMOTE SENSING SYSTEMS   61

Figure 3.25 RADARSAT fine beam mode image of Santafé de Bogotá, Colombia, acquired January 8, 1996. This image has a 46° incidence angle and 8 m pixel. Image size is 37 by 49 km.

has a 74 km (44 mi) swath and 36 × 36 m (117 ft) ground resolution. Each sensor has four spectral bands in the range 0.45 to 0.86 microns. LISS-3, launched on the IRS-1C satellite in 1995, has a 24-day repeat cycle with 23.5 m resolution in the VNIR (visible and near-infrared) range and 70.5 m resolution in the SWIR (short-wave infrared) region (Fig. 3.26). The VNIR swath is 142 km, while the SWIR swath is 148 km wide. It is in a polar sun-synchronous orbit at 817 km.

Figure 3.26 Indian Remote Sensing1-C LISS-3 color infrared image of the Strait of Gibraltar acquired August 30, 1996. Gibraltar is visible in the northeast, and Tangiers in the southwest. (IRS images courtesy of Space Imaging, LLC, Denver, Colorado.) (See color plate.)

The Indian Remote Sensing satellite also contains the WiFS (wide field sensor) and Pan (panchromatic) sensors. WiFS has 188 m resolution over a 774 km swath and 24 day nadir and 5 day off-track repeat coverage. IRS-1C has two channels, 0.62–0.68 and 0.77–0.86 microns. IRS-P3 has, in addition to these two, a channel from 1.55–1.70 microns. The Pan sensor records a single channel from 0.5–0.75 microns with 5 m resolution over a 70 km swath (Fig. 3.27). The repeat cycle is the same as for WiFS.

Figure 3.27 IRS panchromatic 1-C image of Almaty, Kazakhstan, and the Tian Shan foothills, acquired October 8, 1996. (IRS images courtesy of Space Imaging, LLC, Denver, Colorado.)

MOMS-1

The Modular Optico-electronic Multispectral Scanner (MOMS-1) was built by the German Aero Space Research Establishment (DFVLR) and flown on Shuttle Missions STS-7 and STS-11. It has two bands in the range 0.57 to 0.92 microns. Swath width is 140 km (84 mi) with a ground resolution of 20 × 20 m (65 × 65 ft). This instrument acquired a limited number of images, and they are available only to workers chosen by the DFVLR.

ASTER

The Advanced Spaceborne Thermal Emission and Reflection radiometer, launched in December, 1999, carries three instruments designed to obtain detailed maps of surface temperature, emissivity, reflectance, and elevation. ASTER acquires data in three visible bands with 15 m resolution (0.52–0.86 microns), an additional visible band (0.76–0.86 microns) for stereo, six near infrared bands with 30 m resolution (from 1.6–2.43 microns), and five thermal infrared channels with 90 m resolution (ranging from 8.125–11.65 microns). It has a 60 km wide ground swath from a 705 km orbit, with a 4 to 16-day repeat cycle (Fig. 3.28). This program is run for NASA by the Jet Propulsion Lab.

MISR

The Multiple Imaging SpectroRadiometer was launched in December, 1999, to image the Earth at nine different incidence angles simultaneously in each of four channels. The purpose is to study how atmospheric aerosols, clouds, and the surface reflect and scatter light. The angles are 0 (vertical), ±26.1, ±45.6, ±60, and ±70.5 degrees. Repeat coverage is every nine days with a 360 km swath width and a 705 km orbit. Resolution is 250 m (nadir) and 275 m for all other angles. The four channels are in the blue (0.443 microns), green (0.555 microns), red (0.67 microns), and near-infrared (0.865 microns) regions.

In addition to government sponsored systems, a number of commercial satellites are planned. These companies plan to market imagery to national and local governments, the agriculture industry and news media as well as earth scientists and others interested in natural resources [31, 32].

IKONOS-1

Space Imaging-EOSAT of Thornton, Colorado, launched IKONOS-1 in September 1999, a satellite with a one meter resolution panchromatic channel (0.45–0.90 microns) and four sensors with 4 m resolution: blue, green, red, and near-infrared bands. The swath is 11 km wide, and the ability to tilt the camera allows for stereo imagery. Repeat coverage is every 11 days (Fig. 3.29).

EO-1 Hyperion

NASA launched the Earth Observing-1 (EO-1) Hyperion satellite on November 22, 2000. Hyperion is in a 705 km sun-synchronous orbit that collects images 7.5 × 100 km with 30 m resolution. This hyperspectral instrument records 220 channels of data between 0.4 and 2.5 microns.

EROS A1

ImageSat International launched the EROS A1 satellite into a 480 km sun-synchronous orbit on December 5, 2000. The imagery has 1.8 m resolution and a 12 km wide swath. The ability to point the camera provides both stereo coverage and 1.8 to 4 day repeat coverage.

Other Satellites

Most other satellite systems were designed for weather monitoring and have resolution too coarse for most resource exploration. Several, however, are worth mentioning briefly.

Figure 3.28 ASTER false color image of San Francisco, acquired 3 March 2000. Level 1B image processed by MIT/ERSDAC, Tokyo. (See color plate.)

66 REMOTE SENSING FOR GEOLOGISTS

Figure 3.29 Sydney, Australia, as seen on 1 m-sharpened IKONOS image acquired 1 December 1999 by Space Imaging, Thornton, Colorado. (See color plate.)

Orbview-1 and -2

Orbital Imaging launched Orbview-1 in 1995 to monitor the atmosphere for weather forecasting. Spatial resolution is 10 km and swath width is 1300 km from an altitude of 740 km. There is one channel centered at 0.777 microns. Revisit time is 2 days. Orbview-2 was launched in 1997 with the purpose of multispectral earth observation. Spatial resolution is 1 km from an altitude of 705 km, with a 2800 km swath. There are eight channels in the visible and near-infrared: 0.402–0.422 microns, 0.433–0.453 microns, 0.48–0.50 microns, 0.50–0.52 microns, 0.545–0.565 microns, 0.67–0.68 microns, 0.745–0.785 microns, and 0.845–0.885 microns. Repeat cycle is one day.

GOES

GOES (Geostationary Operational Environmental Satellite) are stationary with respect to their position over the earth, orbiting at an altitude of 35,000 km (22,000 mi), and recording weather patterns over most of North and South America. GOES has a visible band (0.55–0.70 microns) with 1 km (0.6 mi) resolution, and a thermal band (10.5–12.6 microns) with 8 km (5 mi) resolution. Data is acquired every 30 minutes and is distributed by the National Weather Service, NOAA, and other federal agencies. These images can be used to monitor not only weather, but also currents, winds, and snow cover, helping in the operation of shipping, offshore platforms, and flood prediction.

AVHRR

The AVHRR (Advanced Very High Resolution Radiometer) system is mounted on NOAA weather satellites in a polar orbit at an altitude of 850 km (510 mi). Each satellite covers the entire earth each day. The ground swath is 2700 km wide (1620 mi), with a 1.1 km (0.7 mi) resolution in five spectral bands: 0.55–0.68 microns (visible), 0.73–1.10 microns (near-infrared), 3.55–3.93 microns (thermal), 10.5–11.5 microns (thermal), and 11.5–12.5 microns (thermal). This data is also available from NOAA.

MOS-1

MOS was designed for remote sensing of oceans and coastal zones. Japan launched MOS-1 in 1987 into a 909 km sun-synchronous orbit with a 17-day repeat cycle. It contains three instruments that sense the visible, near and thermal infrared, and microwave regions at resolutions from 50 m (visible and near-IR) to 23 km (thermal and microwave). The Multispectral Electronic Self-Scanning Radiometer (MESSR) has four bands between 0.51 and 1.1 microns. The swath varies from 100 to 185 km (60–111 mi). There are now two MOS satellites in orbit.

GMS-4

The GMS-4 is a Japanese meteorological satellite with visible and infrared radiometers that have a ground resolution of 1.25 km to 5 km (0.75–3.0 mi). The primary advantage of these systems is their broad area coverage. Their main disadvantage is the poor resolution of the sensors. Contact NASDA about availability (Table 1.2).

ADEOS

The Japanese Advanced Earth Observing Satellite was launched in August 1996 and carries two instruments of interest: the Ocean Color and Temperature Scanner (OCTS) and

Advanced Visible and Near Infrared Radiometer (AVNIR). The OCTS has 12 bands, whereas the AVNIR has 5 more, each with ground resolution of better than 20 m (65 ft). This satellite was cosponsored by France and the United States, and was being used primarily for global environmental change monitoring. It failed during June, 1997.

Early Bird

Launched in 1997, this private satellite launched by EarthWatch of Longmont, Colorado, failed only a few days after launch. Still, it demonstrates that commercial systems can be built and launched. Several more commercial satellites are planned (see below).

Quick Bird

EarthWatch launched its second commercial satellite, Quick Bird, in November of 2000, but the launch failed. It would have carried a panchromatic band (0.45–0.90 micron) with 0.82 m resolution and a four channel multispectral sensor with 3.28 m resolution. The multispectral sensor had a blue band (0.45–0.52 microns), a green band (0.52–0.6 microns), a red band (0.63–0.69 microns), and a near-infrared band (0.76–0.90 microns). The swath was to be 22 × 22 km, extendable to 44 × 44 km, with a three-day repeat cycle. The company is deciding how to proceed.

FUTURE SATELLITE SYSTEMS

In addition to U.S. spy satellite images that continue to become declassified on a regular basis, there are several systems being worked on at this time.

HIRIS

The High Resolution Imaging Spectrometer (HIRIS) is expected to be put in a 705 km orbit by NASA at the start of the next decade [30]. It is designed to acquire a 24 km wide swath with 30 m pixels, and will image in 192 spectral bands in the range from 0.4 to 2.45 microns. With the use of mirrors that can point either along track (+56 to 30°) or across track (+45 to 45°), the instrument will be able to sample any area on the earth, except the poles, every two days. The improved spectral resolution and intermediate spatial resolution should allow both identification of many geologic-scale features as well as global environmental monitoring applications.

MODIS

The purpose of the Moderate-resolution Imaging Spectroradiometer is to measure a variety of biological and physical processes, over both land and oceans and in the atmosphere, every two days in 36 channels. It will be in a 705 km sun-synchronous orbit with a 2330 km wide swath. There will be 2 bands in the region 0.62–0.876 microns with 250 m ground resolution; 5 bands from 0.459–2.155 microns with 500 m resolution; and 29 bands from 0.405–14.385 microns with 1000 m ground resolution. Thermal resolution will be 0.3–0.5° C over water and 1° C over land. A proto-flight model (PFM) was launched in December, 1999.

Orbview

Orbital Imaging of Dulles, Virginia, is planning to launch two satellites. Orbview-3 will contain a 1 meter resolution panchromatic channel (0.45–0.90 microns) and 4 meter multispectral scanner with blue (0.45–0.52 microns), green (0.52–0.62 microns), red

(0.63–0.69 microns), and near-infrared (0.77–0.90 microns) channels. Swath width will be 8 km. Orbview-4 will have the same sensors as Orbview-3 and also a 280 channel hyperspectral scanner with 8 m pixels and a 5 km swath. The 470 km orbit will allow a 3 day revisit time. No launch date has been set.

RADARSAT 2

The second RADARSAT is scheduled for launch in 2003. This instrument will have a swath that varies between 20 and 500 km wide, and will have between 3 and 100 m resolution, depending on beam mode. The C-band radar will orbit at 798 km, and will have a 3 day revisit time.

EROS A2 and B1-B6

ImageSat International plans to launch up to seven satellites through 2005. The EROS A2 will have 1.8 m resolution and a 12 km swath from a 480 km orbit. The EROS B series will have 0.82 m resolution and a 16 km swath from a 600 km orbit.

SIDE-SCAN SONAR

Imaging sonar presents a display similar to radar imagery, i.e., an oblique view perpendicular to the ship's track. The sonar instrument is carried in a "fish" that is towed at varying depths below and behind the survey vessel, and can have adjustable depression angles and swath widths. Depths as shallow as 15 m (50 ft) or as deep as 1,000 m (3000 ft) have been surveyed. Generally speaking, resolution becomes coarser as depths increase. Unlike radar, this system uses acoustic waves instead of microwaves. Sonar often acquires

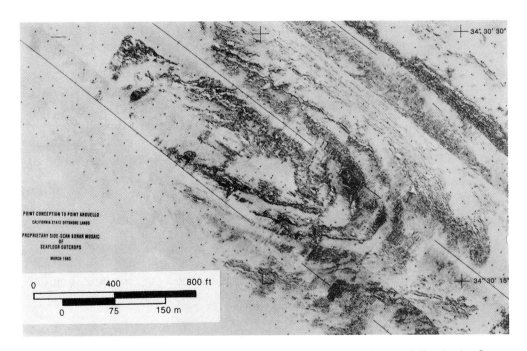

Figure 3.30 This side-scan sonar image of a submerged, breached anticline in the Santa Maria basin, offshore California, was acquired by GeoCubic, Inc., Ventura.

opposite look directions simultaneously, with a no data zone directly below the towed fish. Overlapping swaths are required to properly mosaic images, prevent shadows from falling in opposite directions, and provide stereo viewing.

Sonar imagery can be used at any depth, and can locate not only shipwrecks but has been used to map slope stability for oil platforms and pipeline and cable routes, pinpoint pipeline damage, and map geologic structures such as anticlines and faults [33, 34, 35]. In depths up to 100 meters, where surfaces were exposed to erosion during lowered Pleistocene sea levels, sonar is particularly good at mapping geomorphic features such as breached anticlines and fault scarps (Fig. 3.30). Mudlumps, salt domes, and tar mounds have been identified, and seeps are evident because gas bubbles cause the acoustic signal to break up in a vertical, upward-widening pattern. Gas leaking from the seabed sometimes leaves craters that can be identified. Deep sea manganese nodule fields and nearshore placer deposits can also be mapped.

Several companies provide commercial sonar surveys. They are generally listed under Engineers or Engineering Surveys in the phone directory. The National Institute of Oceanography of Great Britain, and the Institute of Oceanographic Sciences, Wormley, U.K., have been involved in the GLORIA (Geological Long Range Inclined Asdic) program since the early 1970s and have mapped large areas of the seafloor around the world [36]. Their Mark II system uses a 100 Hz linear FM pulse, usually 2 seconds long and repeated every 30 seconds. A 5° wide transmit beam provides ranges of 7 to 30 km (900 m swath at maximum range) and can be acquired in two opposite look directions simultaneously at speeds up to 18 km/h (10 knots). Resolution is 45×120 m (146×390 ft) at these parameters [37].

The U.S. Geological Survey has been using the GLORIA Mark III system to map the Exclusive Economic Zone (EEZ) of the United States and its territories starting in 1984. More than 7.5 million square kilometers (2.8 million mi^2) have been surveyed as of 1993. This system uses either a 20-, 30-, or 40-second pulse repetition rate and images a swath on both sides of the fish 15, 22, and 30 km (9, 13.2, and 18 mi) wide, respectively. The beam is 2.7° wide and 10° vertical. The GLORIA Mark II uses a frequency of 7 KHz or less with a 100-Hz bandwidth to image in depths up to 4,000 m (13,000 ft) [38].

Interferometric sonar, like interferometric radar, can be used to measure water depth and even for classification of sea-bottom types [39]. It is possible to determine the acoustic depth and amplitude and analyze these attributes for sediment properties, then assign them to broad classes such as bedrock, gravel, sand, and silt. Applications exist in slope stability studies, for marine minerals mining, cable or pipeline surveys, and 4D marine seismic surveys.

LASER ALTIMETERS

Many airborne surveys include lasers altimeters linked to a global positioning system. Altimeters may image profiles along the flight path or may be scanning lasers that image a swath. Altitude data can be provided as a topographic profile or as a geometrically-corrected image of topography. These images are especially useful for facilities planning, pipeline routing, and other engineering applications.

REFERENCES

1. V.C. Miller and C.F. Miller, *Photogeology* (McGraw-Hill Book Co., New York, 1961), Chaps. 1 and 2, pp. 8–50.

2. P.N. Slater in R.N. Colwell, ed., *Manual of Remote Sensing* (American Society of Photogrammetry, Falls Church, 1983), 2nd. ed., pp. 231–291.
3. H.R. Condit, *Photogram. Engr. and Rem. Sens.* **36** (1970): 955–966.
4. D.E. Moran, *Geology, Seismicity, and Environmental Impact* (Assn. Engineering Geologists, 1973), pp. 141–155.
5. A.F.H. Goetz, *Remote Sensing Geology, Landsat and Beyond* **SP43–30**, (Jet Propulsion Lab, Pasadena, 1976), pp. 8–1 to 8–8.
6. G.R. Hunt and R.P. Ashley, *Econ. Geol.* **74** (1979): 1613–1629.
7. G.R. Hunt and J. Salisbury, *Modern Geology* **5** (1976): 211–217.
8. A.B. Khale and L.C. Rowan, *Geology* **8** (1980): 234–239.
9. L.C. Rowan, M.J. Pawlewicz, and O.D. Jones, *Bull. Am. Assn. Pet. Geol.* **76** (1992): 1008–1023.
10. E.A. Cloutis, *Science* **245** (1989): 165–168.
11. F.A. Kruse, K.S. Kierein-Young, and J.W. Boardman, *Photogram. Engr. and Rem. Sens.* **56** (1990): 83–92.
12. W.T. Jansen, *Geobyte* (Oct. 1992): 46–49.
13. G. Vane, R.O. Green, T.G. Chrien, H.T. Enmark, E.G. Hansen, and W.M. Porter, *Remote Sensing Environ.* **44** (1993): 127–143.
14. M.J. Wagner, *Earth Observation Mag.* (June 1996): 51–54.
15. N.A. Rowlands, and R.A. Neville, *Proc. of SPIE Infrared Technology XX Conf.* **2269** (SPIE, 1994).
16. P.L. Hauff, P. Kowalczyk, M. Ehling, G. Borstad, G. Edmundo, R. Kern, R. Neville, R. Marios, S. Perry, R. Bedell, C. Sabine, A. Crosta, T. Miura, G. Lipton, V. Sopuck, R. Chapman, C. Tilkov, K. O'Sullivan, M. Hornibrook, D. Coulter, and S. Bennett, *Proc. of the 11th Thematic Conf. on Geologic Rem. Sens.* **1** (ERIM, Ann Arbor, 1996), pp. I-38–I-47.
17. R.G. Resmini, M.E. Kappus, W.S. Aldrich, J.C. Harsanyi, and M. Anderson, *Proc. of the 11th Thematic Conf. on Geologic Rem. Sens.* **1** (ERIM, Ann Arbor, 1996), pp. I-48–I-65.
18. F.F. Sabins, Jr., *Remote Sensing Principles and Interpretation* (W.H. Freeman and Co., New York, 1987), 2nd ed., Chap. 5, pp. 126–131.
19. A.B. Khale and A.F.H. Goetz, *Science* **222** (1983): 24–27.
20. K. Watson, F.A. Kruse, and S. Hummler-Miller, *Geophysics* **55** (1990): 70–79.
21. F.A. Kruse, *Visible-Infrared Sensors and Case Studies, Manual of Remote Sensing* **3** (John Wiley and Sons, 1999), 3rd ed., Chap. 11.
22. F.F. Sabins, Jr. *Remote Sensing Principles and Interpretation* (W.H. Freeman and Co., New York, 1987), 2nd. ed., Chap. 6, pp. 177–233.
23. S.A. Drury, *Image Interpretation in Geology* (Allen & Unwin, London, 1987), Chap. 7, pp. 165–194.
24. R.P. Gupta, *Remote Sensing Geology* (Springer-Verlag, Berlin, 1991), Chaps. 10 & 11, pp. 149–179.
25. D.E. Evans, T.G. Farr, J.P. Ford, T.W. Thompson, and C.L. Werner, *IEEE Trans. on Geoscience and Rem. Sens.* **GE-24** (1986): 246–257
26. A.K. Gabriel, R.M. Goldstein, and H.A. Zebker, *Jour. Geophys. Res.* **94 B7** (1989): 9183–9191.
27. J. Ehrismann, B. Armour, M. van der Kooij, and H. Schwichow, *Earth Observation Mag.* (April, 1996): 26–29.
28. H.A. Zebker, in J.J. van Zyl, ed., *Proc. of 3rd Airborne Synthetic Aperture Radar Workshop*, **JPL Pub. 91–30** (Jet Propulsion Lab, Pasadena, 1991): 230–233.
29. J.F. McCauley, G.G. Schaber, C.S. Breed, M.J. Grolier, C.V. Haines, B. Issawi, C. Elachi, and R. Blom, *Science* **218** (1982): 1004–1020.
30. A.F.H. Goetz, *Episodes* **15** (1992): 7–14.
31. K. Corbley, *Earth Observation Mag.* (Oct. 1996): 26–28.
32. F.A. Kruse, *Visible-Infrared Sensors and Case Studies, Manual of Remote Sensing* **3** (John Wiley and Sons, 1999), 3rd ed., Chap. 11.
33. W.D. Jenkinson, *Oil and Gas Journal* **75** (11/7/1977): 97–102.
34. D.B. Prior, J.M. Coleman, and H.H. Roberts, *Offshore* (4/1981): 151–161.
35. J.E. Clausner and J. Pope, *Offshore Technology Conference* (1988): 329–336.

36. A.S. Laughton, *Jour. Geophysical Research* **86** (1981): 11,511–11,534.
37. R.C. Searle, T.P. Le Bas, N.C. Mitchell, M.L. Somers, L.M. Parson, and P. Patriat, *Marine Geophysical Researches* **12** (1990): 21–39.
38. P.S. Chavez, *Photogram. Engr. and Rem. Sens.* **52** (1986): 1133–1145.
39. S. Hadden and C.D. Green, *Bull. Am. Assn. Pet. Geol.* **83** (1999): 1315.

ADDITIONAL READING

R. Agar, *Earth Observation Mag.* **7** (1998): 36–38.

R. Bianchi and C.M. Marino, *Environ. Eng. Geophys. Soc. Appl. of Geophys. To Eng. & Environ. Probl. Symp. Proc.* **2** (1997): 811–816.

Z. Chen, D. Ahl, J. Albasini, B. Davis, J. Moody, and M. Oxley, *Proc. of the 11th Thematic Conf. on Geologic Rem. Sens.* **1** (ERIM, Ann Arbor, 1996): I-584.

EARSEL Newsletter, **22**: June 1995, 2–10.

EARSEL Newsletter, **26**: June 1996, 2–5.

EOSAT Notes, **11**: Spring-Summer 1996, 4–7.

S. Fujiwara, and P.A. Rosen, *Jour. Geophys. Res.* **103 B2** (1998): 2411–2426.

F.M. Henderson and A.J. Lewis, eds., *Principles and Applications of Imaging Radar, Manual of Remote Sensing* **2** (American Society of Photogrammetry, Falls Church, 1999).

S.J. Hook, E.A. Abbott, C. Grove, A.B. Kahle, and F. Palluconi, *Use of Multispectral Thermal Infrared Data in Geological Studies, Manual of Remote Sensing* **3** (John Wiley and Sons, 1999), 3rd ed., Chap. 2.

F.A. Kruse, *Int. Jour. Remote Sensing* **17** (1996): 1623–1632.

F.A. Kruse, *Visible-Infrared Sensors and Case Studies, Manual of Remote Sensing* **3** (John Wiley and Sons, 1999), 3rd ed., Chap. 11.

F.A. Kruse and K.S. Kierein-Young, *Proc. of the 2nd Thermal Infrared Multispectral Scanner (TIMS) workshop*, **JPL Pub. 90–55** (Jet Propulsion Lab, Pasadena, 1990): 75–81.

C. Lawrance, R. Byard, and P. Beaven, *Terrain Evaluation Manual* (Transport Research Lab, Department of Transport State of the Art Review 7, HMSO Publications Centre, London, 1993), Chap. 5, pp. 64–86.

S.A. Morain, and A.M. Budge, eds., *Earth Observing Platforms & Sensors, Manual of Remote Sensing* **1** (American Society of Photogrammetry, Falls Church, 1999).

S. Parsons, *Earth Observation Mag.* (Sept. 1994): 36–39.

P.A. Rosen, S. Hensley, H.A. Zebker, F.H. Webb, and E.J. Fielding, *Jour. Geophys. Res.* **101 E10** (1996): 23,109–23,125.

D.M. Uhlir, *Earth Observation Mag.* (Feb. 1995).

C. Wicks, Jr., W. Thatcher, and D. Dzurisin, *Science* **282** (1998): 458–462.

QUESTIONS TO CONSIDER

1. What geologic features are you more likely to see on daytime thermal imagery over Qatar (a low-relief peninsula with surface carbonates in the Persian Gulf) versus night thermal imagery?
2. Which would be more useful for detailed geologic mapping in the humid countryside north of Rangoon: SPOT panchromatic 10 m data or SPOT multispectral 20 m data? Why?
3. What do you want to know prior to planning the optimum airborne radar survey over Guatemala?

4. What are the advantages and disadvantages of using Landsat TM vs. Seasat radar to map oil slicks offshore Ivory Coast?
5. What are the advantages and disadvantages of using airborne multispectral/hyperspectral scanners vs. color airphotos in mineral exploration in Australia?
6. How is near-infrared imagery different from thermal infrared imagery?
7. Which wavelengths penetrate water farther, blue or red?
8. Which wavelengths penetrate the atmosphere with the least scattering, blue or red?
9. What is the relationship between microwave wavelength and penetration of surface materials and vegetation?

Section 2
Exploration Remote Sensing

INTRODUCTION

Exploration for minerals and oil and gas is taking the geoscientist into increasingly remote, complex, and poorly understood areas. The days of locating obvious gossans or surface anticlines and drilling out an economic deposit or successful wells lie in the past. Today geophysical and geological surveys are using advanced technology to map subtle indications of mineral alteration, ground water, and hydrocarbon traps. Remote sensing is one high tech answer to surface mapping. Through the acquisition and analysis of airphotos; multispectral/hyperspectral aircraft and satellite imagery; thermal, radar, and sonar images; the explorationist is able to preview and start mapping areas that are remote and structurally complex. In a quick and relatively inexpensive manner one can evaluate the potential of a region and pinpoint targets where field work is necessary to answer remaining questions. Imagery carried into the field, especially with a global positioning system (GPS), allows rapid and accurate location of a geologist's position, and serves as an accurate base for plotting faults, contacts, etc. This section provides a guide to the analysis of imagery to map geologic units, alteration, and geologic structures that can serve as a guide to oil and gas traps or conduits for mineralization.

The uses of remote sensing are as varied as the explorationist's imagination. From regional tectonic analysis to site-specific problems such as stepout well site selection, there is imagery for every need. Generally, the problem that needs answering and the detail on maps already available will determine the imagery to be used.

FRONTIER AREA ANALYSIS

In frontier areas remote sensing data, usually satellite imagery, is used prior to leasing or staking claims to determine which areas contain favorable host rocks, intrusives, or alteration, or which basins, or parts of a basin, contain structures large enough to be worth pursuing for oil and gas. Areas to be leased can be delineated and prioritized.

In hydrocarbon exploration the size and geometry of structures can be measured and the origin deduced, thus providing clues to the subsurface configuration and timing of fold development. When combined with subsidence modeling this suggests which structures are pre-, syn-, or post-generation. Evidence of seals can be obtained by observing shales or evaporites above reservoir units along the basin margin. Source rock distribution can be deduced from outcrops around the basin. The thickness of source units can be estimated and, together with information on the areal extent and total organic carbon (TOC), can be used to calculate the amount of hydrocarbons generated. The presence of intrusive or extrusive units can influence a play concept with regard to thermal maturity of source rocks (and the distribution of volcanics is good to know when planning seismic programs). Sediment distribution networks may provide evidence as to the location of paleodepocenters and the lateral variation of clastic mineralogy and grain size. Such remote sensing-based basin analysis can contribute significantly, for example, to Technical Evaluation Agreements (TEA's) between a company and foreign government.

In mineral exploration one can attempt to map the extent and types of alteration, the nature and extent of a host rock, the location of highly fractured host, or where major conduits of mineralizing fluids exist or intersect. Imagery combined with magnetics can suggest the location of buried intrusives. Imagery merged with gravity data can suggest the depth to a buried intrusive. Paleochannels may be identified that can lead to paleoplacer deposits. An abrupt decrease in vegetation or change in vegetation type noted on imagery can help identify massive sulfides, ultramafic-associated deposits, or kimberlites. Limonitic or argillic alteration, identified on imagery and used in conjunction with a geochemical survey, should help pinpoint prospects. Proximity to old workings support the interpretation of a nearby deposit.

Images also serve a scouting role: they can reveal tracks, old tailings and abandoned workings, folds that are already producing, precisely where recently drilled wells are located, the extent and layout of a seismic program or the location of pipelines and oil storage tanks. Such information can be critical in planning a bidding strategy.

LEASE EVALUATION

During evaluation of a claim or lease, it is important to make the best possible geologic map of the surface in order to locate prospects, lay out geophysical surveys, and identify surface structures to help interpret and tie into other geophysical and geochemical surveys. At this stage in a program it is important to achieve the proper balance of spatial resolution versus synoptic coverage. The geologic mapping will either replace or supplement existing maps. It is important for an interpreter to understand the objective in each area in order to focus on specific evidence that may not be shown on an existing, generalized geologic map. For example, many maps show the distribution of lithologic units but leave out details of subtle structure or any mention of alteration. In other cases units will be generalized as "Cretaceous marine," for example, with no indication of the lithology present in a particular area. The interpreter, aware that the hydrocarbon play is in fractured carbonates, can look specifically for fracture zones at the surface above carbonate units. Alternatively, surface limestones can be targeted if the objective is a Mississippi-Valley type lead-zinc-silver deposit or a manto or skarn deposit.

Individual structures can be assessed and prioritized on the basis of size alone, or the ratio of area under closure divided by drainage area, or proximity to known mines, production, or seeps, or the proximity to mills, roads, pipelines, mature source rock, good reservoir, etc. Several of the parameters that define a prospect can be derived entirely from the imagery.

4
Recognizing Rock Types

CHAPTER OVERVIEW

Rock types generally can be grouped according to characteristics such as resistance to erosion, color, cross-sectional shape of gullies and valleys, vegetation cover, etc. It is often helpful to know if banding is a result of metamorphism or sedimentary layering, or whether the surface is sandstone or limestone. Although one generally can not identify the surface units, it is usually possible to group them into recognizable categories.

Often one begins working in an area with no clear idea of what the local stratigraphy should be. It then becomes necessary to generate a "photogeologic" stratigraphic column based not on age or field criteria, but on features that can be distinguished on airphotos or satellite images. These photogeologic units will be defined by their tone, texture, color, erosional characteristics, and stratigraphic position. It should be possible to *distinguish* between various lithologies such as sandstones and shales (if the vegetation cover is not too dense), but it may not be possible to *identify* the lithologies in all cases. One can still use these units to evaluate the structure of an area or to presume their engineering characteristics (infiltration, slope stability, resistance to erosion, etc.).

IGNEOUS ROCKS

It is often helpful to recognize and take into account the presence of igneous rocks in an area of interest. Many intrusives, such as porphyries, are host to ore deposits. Others provide the hydrothermal solutions necessary to mineralize country rock. Dikes can compartmentalize a petroleum reservoir, but usually do not contribute significant heat flow beyond one or two meters. An intrusive stock, on the other hand, can contribute significant heat and assist in generating hydrocarbons or, worse, overcook them. Surface flows can make it difficult to acquire surveys, to interpret magnetic data, and to get enough seismic energy into the ground for good reflections. Seismic programs generally try to avoid lava flows at or near the surface.

Intrusive rocks usually occur as lobate outcrops covering tens of square kilometers and are fairly homogeneous in texture. Intrusives tend to have uniform plant cover because of their homogeneous composition. These rocks may have arcuate cooling joints, faults, or dikes along their margins (concave inward), and are characterized by

Figure 4.1 Intrusive igneous rocks of the Arabian Shield, Saudi Arabia. Note the multiple, often orthogonal joint sets. Landsat TM band 4. North is at the top of the image. Processed by Amoco Production Company.

multiple, often orthogonal, joint systems (Fig. 4.1). A buried intrusive center may be indicated at the surface only by the conjunction of three sets of fractures at 120º, similar to the pattern formed at an oceanic triple junction. Contact metamorphic halos and skarns around the margins of intrusive centers are often more resistant to erosion than the surrounding country rock. The large crystals found in coarse-grained intrusive rocks tend to erode readily into thick, well-drained soils; fine-grained intrusives are more resistant to erosion. Felsic intrusives tend to be light in color, whereas mafic intrusives are dark colored (Fig. 4.2). Mafic minerals break down into clays and form gentler topography than granitic (arkosic) terrains. Ultramafic, carbonatite, and kimberlite units may have anomalous or sparse vegetation due to exotic soil minerals and clays that cause poor soil drainage [1, 2]. Vein-type mineral deposits are often found in large fracture systems (faults, cooling joints) associated with the margins of intrusives, either concentric to the core (ring faults), radial faults, or at the intersection of these features. Large veins may be recognized by their association with faults, their resistance to erosion, and/or their light color.

Figure 4.2 Light (acidic) and dark (basic) intrusives, Arabian Shield, Saudi Arabia. Landsat TM band 4. North is at the top of the image. Processed by Amoco Production Company.

Extrusive rocks frequently have distinctive landforms. Volcanoes, cinder cones, and flow lavas are all easily recognized. Calderas are often characterized by ring dikes, radial dikes, and craters filled by lakes. Laccoliths, shallow folds caused by volcanic intrusion, may be indistinguishable from normal folds unless aeromagnetic surveys exist in the area and reveal a magnetic core to the fold. Most volcanic rocks are finely crystalline, and these highly resistant layers generally form steep cliffs. Air-fall tuffs, however, may be difficult to distinguish from easily eroded sedimentary rocks unless they are welded. They tend to be very light-colored and erode into a badlands topography. Sulfotaric alteration associated with extrusive volcanics also tends to form rusty colored and/or bleached surface rocks due to breakdown of pyrite and other iron minerals and alteration of feldspars to clays.

Basalts are essentially black and highly fluid and therefore flow through the lowest depressions. Quite often river meanders will be preserved as topographic highs because of basalt flows that filled the stream channel (Fig. 4.3). In other areas flood basalts form dark, resistant layers that cover hundreds of square kilometers. The

Figure 4.3 A basalt flow fills a paleostream channel in the Datil volcanic field, New Mexico. Landsat MSS band 7. North is at the top of the image. Processed by Amoco Production Company.

sequence of flows can often be established by noting increasing vegetation cover with age. Most volcanic terrains have associated reddish soils as a result of rapid weathering of iron minerals [3].

METAMORPHIC ROCKS

Metamorphic rocks can be recognized by disharmonic, ptygmatic, or flexural flow folding, or by pervasive, parallel lineations (foliation or cleavage). Metamorphic units are crystalline and therefore generally resistant to erosion. In contrast to intrusive units, they tend to be heterogeneous, and thus do not have uniform surface texture and color (Fig. 4.4)

It is not always possible to discriminate between compositional or cleavage banding in metamorphics and sedimentary layering unless there is a recognizable foliation or complex folding. Foliation can also be confused with scour or striations in glaciated areas. Low grade metamorphics may be indistinguishable from sedimentary rocks.

SEDIMENTARY ROCKS

Sedimentary rocks are characterized by alternating weak and resistant layers of varying thickness and color. The appearance of the different rock types depends on their grain size and composition and the weathering environment.

82 REMOTE SENSING FOR GEOLOGISTS

Figure 4.4 Metamorphic outcrops near Al Awafin, Yemen. Note layering, faulting, and complex folding. SPOT P image processed by Earth Satellite Corp., Chevy Chase. North is at the top of the image.

Sandstones

Sandstones, composed largely of quartz, feldspar, and lithic fragments, tend to form resistant ridges (Fig. 4.5). This, of course, depends on the degree of cementation and type of cement. Silica cements will be more resistant than calcite cement or iron oxide cement, both of which are more resistant than clay or gypsum cement. The high porosity generally associated with sandstone causes low runoff and widely spaced drainages. High infiltration also favors vegetation cover with deep root systems: every county in the western United States seems to have a "pine ridge" that is usually a sandstone outcrop surrounded by grassy shale valleys. Sandstones display fracturing prominently because they tend to be resistant and brittle. Gullies are generally V-shaped in cross section, becoming U-shaped in siltier soils [4]. Conglomerates and their erosion product, lag gravels, tend to be highly resistant to erosion.

Shales

Shales can be composed either of clay-sized quartz or clay particles. The small grain size makes them very easy to erode, and the product is either a highly dissected

Figure 4.5 The Cretaceous Dakota hogback forms a prominent sandstone ridge south of Durango, Colorado (arrow). Landsat TM band 4. North is at the top of the image. Processed by Amoco Production Company.

badlands (arid climate) or flat to gentle valleys (humid climate). The thin bedding that exists over great thicknesses also contributes to ease of erosion. Drainage is close-spaced because runoff is high (Fig. 4.6). Joints are not well developed because the units tend to be ductile [5]. Gullies range from U-shaped in cross section to gently rounded with increasing clay in the soil [4]. Outcrops can be sparse. Banding is a result more of different colored layers than of varying resistance to erosion. Shales are extremely susceptible to landslides, slumps, and creep.

Carbonates

Carbonates are crystalline, and therefore resist physical weathering. In arid climates, they tend to form ledges and are easily fractured. In humid climates, they are resistant, but more rounded in form. Carbonates are usually light-colored, although they can weather to almost black. The characteristic feature of carbonate rocks in all climates is the karst or sinkhole that yields a mottled, pockmarked terrain (Figs. 4.7, 4.8) and ultimately hums, or buttes, in humid climates such as are seen in Puerto Rico or Southeast Asia. Drainage patterns are considered to be "deranged" as a result of streams suddenly ending in sinkholes. Often there is no recognized drainage pattern, since all precipitation infiltrates through joints. Undisturbed, flat lying carbonates tend to have rhombic joint patterns at all scales. In arid areas carbonates tend to develop little soil, form steep-sided valleys, and have more vegetation in sinkholes where moisture accumulates and some soil is developed (Fig. 4.9). In humid areas the vegetation can be dense. Residual clays associated with carbonates in humid areas often form a terra rosa, or red soil.

Figure 4.6 Badlands are developed in shaly sections of the Tertiary in the northern San Juan basin, Colorado and New Mexico. Landsat TM band 4, processed by Amoco Production Company. North is at the top of the image.

There is no effective way to distinguish between limestones and dolomites except with high spectral resolution imaging spectrometers in areas where there is little or no vegetation cover (see Chapter 6). Dolomites often fracture more readily, and thus may be somewhat less resistant to erosion. Otherwise, if one knows that, for example, a dolomite is black and a limestone is gray in a specific area, one can discriminate them on this basis.

Evaporites

Evaporites such as halite and gypsum or anhydrite exist at the surface only in arid climates because they are water soluble. They are recognized because they are very bright, and usually white. Evaporites accumulate as recent or ancient deposits in dry lake beds and evaporating shallow seas. They occur as layered beds, diapirs, and, rarely, as sand dunes (e.g., White Sands, New Mexico). Subsurface deposits are of greater interest because they act as seals and decollements, and may generate structural traps.

Evaporites are recognized in the subsurface by the land forms they generate. Salt tends to form diapirs (roughly concentric domes) in nearly flat strata (e.g., Gulf Coast, Persian Gulf). In folded terrain evaporites form anticlinal folds over basement faults (e.g., Paradox Basin, Utah) or the mobile core of thrusted folds, occasionally reaching the surface in the core of domes or in rare cases as salt glaciers (e.g., Zagros, Iran, or Atlas, Algeria; Fig. 4.10). Along the south flank of the Salt Ranges in Pakistan, the

RECOGNIZING ROCK TYPES 85

Figure 4.7 Karst topography and deranged drainage is typical of carbonate terrain. This is a tropical area in northern Guatemala. Landsat MSS band 7 processed by Amoco Production Company. North is at the top of the image.

Cambrian salt that acts as a decollement to thrusting is exposed as a bright linear zone at the mountain front. In areas where evaporite layers have been dissolved by ground water they can create a chaotic pattern of irregular synclinal collapse structures (Fig. 4.11) or long, linear synclinal collapse features along fracture zones.

Alteration

Alteration associated with mineral deposits will vary depending on the composition of the host rock and the chemistry of the mineralizing fluids. Alteration generally requires a combination of permeability (e.g., through fractures) and chemical reactivity of the host rock with the hydrothermal solution. It should be noted that not all alteration is associated with mineralization, nor do all mineral deposits have prominent alteration zones. Common alteration types include silicification, argillization, sericitization, propylitization, carbonitization, and hydration. Generally speaking, the more extensive the alteration zone, the more susceptible the country rock has been to ore deposition, since the hydrothermal fluids have obviously penetrated the rocks extensively.

The wall rocks around hypothermal (deep) ore bodies are relatively impermeable and will have thin or no alteration unless there is a strong chemical contrast between the host and country rock. Contact zones around pegmatites may be either narrow and sharp or wide and gradational [6]. Alteration minerals associated with pegmatites include beryl, monazite, sphene, tantalite, phlogopite, topaz, zircon, fluorite, allanite,

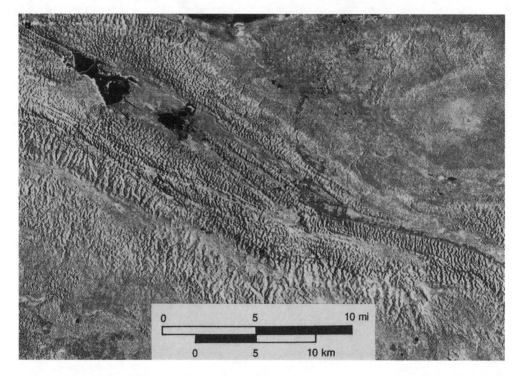

Figure 4.8 Karsted highlands can be recognized by their hummocky, or pockmarked texture in this folded highland area in Chiapas, Mexico. Landsat MSS band 7, processed by Amoco Production Company. North is at the top of the image.

microcline, and biotite. These minerals are important to know when processing hyperspectral imagery for endmember spectra. The alteration zone is generally less than 1 meter wide, but can be up to 7 meters wide. Some pegmatites are characterized by bleaching due to removal of iron and conversion of biotite to muscovite. This bleaching may be visible on imagery.

Alteration in metamorphic wall rocks may contain a conspicuous suite of minerals, including grossularite, andradite, and almandite garnets, wollastonite, epidote, pyroxenes, ilvaite, idocrase, serpentine, spinel, and scapolite, among others. Shales, otherwise impermeable, can become permeable after silicification and fracturing. Shales in a contact zone may develop the sugary texture of hornfels with clusters of epidote, chlorite, andalusite, garnet, and cordierite. Silica-rich fluids injected into a marble host rock generate the minerals wollastonite and idocrase [6]. Intruded carbonates can be metasomatized to form skarns consisting of silicate and oxide minerals. Some limestones and dolomites are simply recrystallized. The process of recrystallization may drive out impurities and lead to a much lighter rock. Gray limestone at Hannover, New Mexico, is coarsened and lightened due to expulsion of carbonaceous material. Silicification is common at igneous-metamorphic contacts, with the silica derived either from the mineralizing fluids or from the country rock. The North Lily deposit at East Tintic, Utah, was found at least partly because of alteration associated with the limestone replacement orebody: hydrothermal dolomite and minor chlorite was followed by dickite, kaolinite, halloysite, beidellite, and rutile, then by allophane, quartz, and barite, and finally by mineralization accompanied by sericite [7].

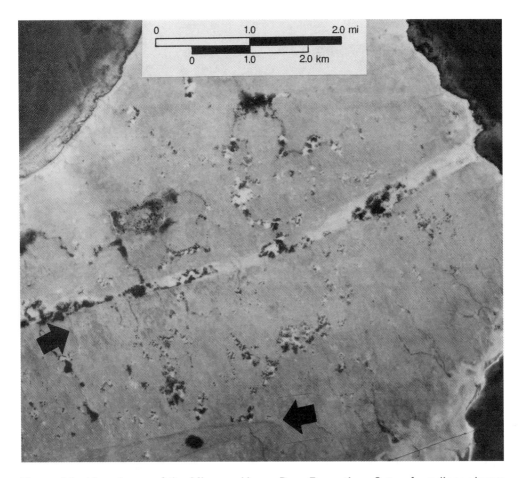

Figure 4.9 Limestones of the Miocene Upper Dam Formation, Qatar, form linear karsts that appear to have coalesced along probable fault zones (arrows). These linear karsts often form superficial "synclines" with inward-dipping beds. Note the concentration of vegetation (dark) within the linear depressions. Airphoto from Hunting Surveys, Ltd. North is at the top of the image.

Sericite is the most extensive alteration mineral in mesothermal (intermediate depth) deposits [6]. Calcite, dolomite, siderite, rhodochrosite, and ankerite are also common. Chlorite is often found paired with sericite around veins. Fine grained silica in the form of jasperoid can be abundant. Pyrite may be conspicuous, as well as feldspars and clay minerals. Porphyry copper deposits are frequently characterized by overlapping or concentric zones of alteration from an inner potassic (quartz-sericite-potassium feldspar), to intermediate phyllic (quartz-sericite-tactite), argillic (quartz-sericite-kaolinite), and outer propylitic (chlorite, calcite, sericite, and montmorillonite) assemblages [8,9]. The deposits at Bingham, Utah, and Miami, Arizona, could have been located purely based on outcrops of altered porphyry containing quartz, sericite, and adularia [7].

Chlorite is perhaps the most abundant epithermal (shallow) alteration mineral. Sericite is common, and other shallow alteration minerals include alunite, zeolites, chalcedony, opal, calcite, dickite, kaolinite, illite, and montmorillonite. The country

88   REMOTE SENSING FOR GEOLOGISTS

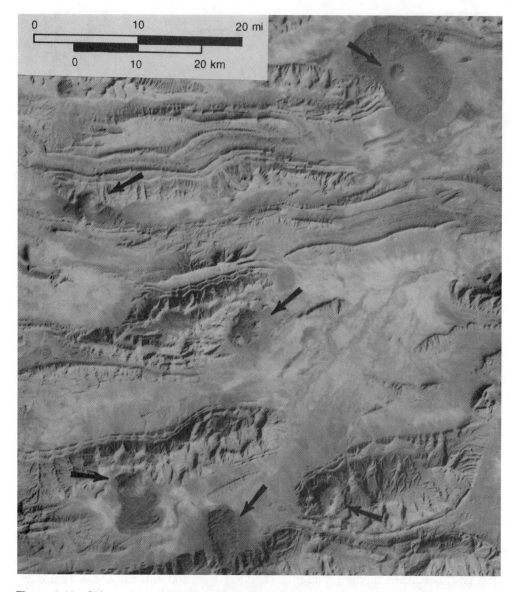

Figure 4.10   Salt-cored anticlines and salt glaciers (arrows) are common in this arid part of the Zagros fold belt, Iran. Landsat MSS band 7. North is at the top of the image.

rock may be extensively altered for several hundred meters from a deposit. Intermediate and mafic volcanics commonly contain propylitic alteration characterized by chlorite, calcite, and epidote.

Mississippi Valley type bedded lead-zinc deposits are characterized by calcite, dolomite, marcasite, chalcedony, and opal. These may be recognized using hyperspectral imagery.

Supergene enrichment occurs in the oxidized zone above deposits where the water table is deep and minerals have a chance to leach. Sulfides decompose to sulfates and

RECOGNIZING ROCK TYPES 89

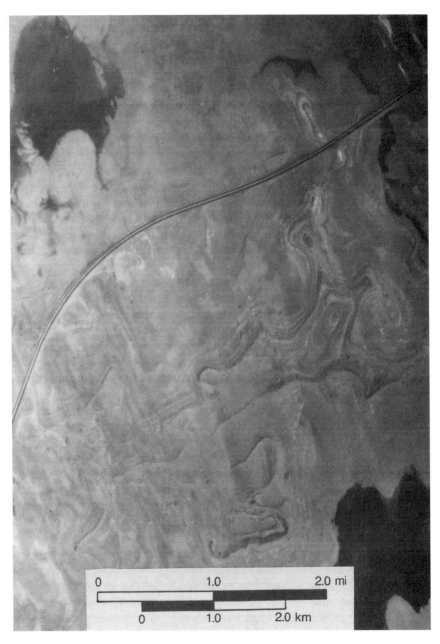

Figure 4.11 Chaotic collapse of Miocene Upper Dam Formation limestone can be seen where gypsum of the Lower Dam Formation has been dissolved in the subsurface near Salwa, Qatar. Airphoto from Hunting Surveys, Ltd., Boreham Wood, U.K. North is at the top of the image.

leave behind native metals, silicates, carbonates, and oxides. Limonite, clay minerals, and silica in the form of jasperoid are characteristic of the alteration assemblage.

Tones and colors associated with alteration include bleaching, darkening, and various colored halos, particularly pastel colors [6]. Clay minerals can be white or

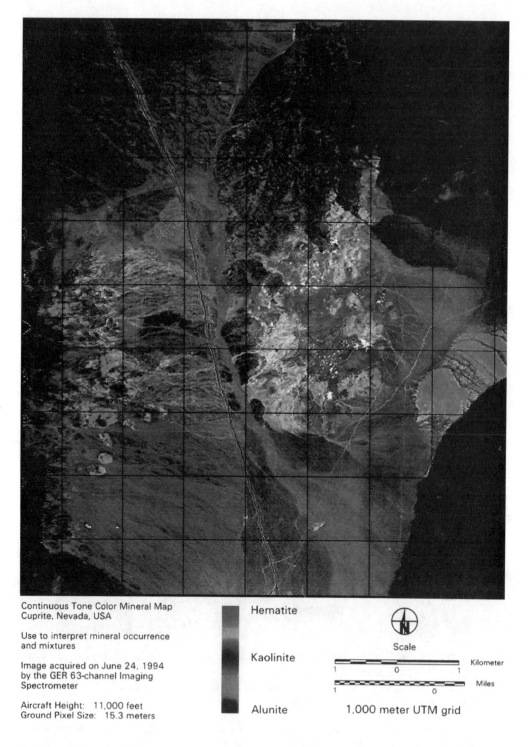

Figure 4.12 Hyperspectral alteration image of Cuprite, Nevada, derived from a GER 63-channel spectrometer. GER Corporation image. (See color plate.)

shades of green or gray. Argillization generally causes a bleached appearance, but the addition of chlorite and epidote causes a green color. Pyritization of red bed sandstones causes a bleached appearance due to reduction of iron. Oxidation of pyrite, on the other hand, produces the conspicuous red to red brown zones known as gossans [10,11].

Whereas bleached or iron oxide stained outcrops are easily mapped using color photography or multispectral imagery [12–16], hyperspectral imagery is required to map suites of alteration minerals (Fig. 4.12; see exploration case histories, Chapter 7).

Alteration associated with mineralizing fluids may have characteristics other than tone, color, and unique spectra. Resistant minerals such as silica in veins or iron-rich gossans regularly form topographic knobs or rises. The outcrop at Broken Hill, Australia, forms a conspicuous ridge visible for miles in the otherwise flat plains [7]. Cerro Negro stands out above the deposit at Parral, Chihuahua, as do the quartz veins at Oatman, Arizona. Veins lacking quartz, and particularly sulfide deposits, often appear as depressions due to the ease of weathering of the soft, altered materials (carbonate gangue and clays). Depressions have also been attributed to the oxidation and attendant shrinkage of deposits. At Bisbee, Arizona, apparent shrinkage cracks extend 200 meters above the deposit to the surface where they outline an irregular, cracked body of ground [7]. Iron ores in the Lake Superior range are associated with hills and ridges.

Vegetation can provide clues to alteration and mineralization. Some plants are poisoned by heavy minerals associated with ore deposits, such as the birch-pine forest over sulfide copper deposits at Karasjok, Norway [17]. Other plants are stunted or deformed, whereas some appear to thrive on these same soils [18–21]. Not all of these plants can be identified on imagery, but areas of stunted or dying vegetation should be recognizable.

Alteration associated with leaking hydrocarbons is considered in a later section (Chapter 5, Color Anomalies Indicative of Folds).

REFERENCES

1. R.G. Ray, *U.S. Geol. Survey Professional Paper* **373**, (1960): 16–19.
2. V.C. Miller and C.F. Miller, *Photogeology* (McGraw-Hill Book Co., New York, 1961), Chap 7, pp. 80–100.
3. S.A. Drury, *Image Interpretation in Geology* (Allen & Unwin, London, 1987), Chap. 4, pp. 72–88.
4. D.S. Way, *Terrain Analysis: A Guide to Site Selection Using Aerial Photographic Interpretation* (Van Nostrand Reinhold, New York, 1978), pp. 31–57.
5. L.H. Lattman and R.G. Ray, *Aerial Photographs in Field Geology* (Holt, Rinehart and Winston, 1965), Chap. 5, pp. 162–188.
6. C.F. Park, Jr., and R.A. MacDiarmid, *Ore Deposits* (W.H. Freeman and Co., San Francisco, 1970), pp. 145–164.
7. H.E. McKinstry, *Mining Geology* (Prentice-Hall, Inc., Englewood Cliffs, 1948), pp. 233–276.
8. M.J. Abrams and D. Brown, Joint NASA/Geosat Test Case Project (Am. Assn. Pet. Geol., 1984), pp. 4–1 to 4–73.
9. L. Lepley, M.J. Abrams, and L. Readdy, Joint NASA/Geosat Test Case Project (Am. Assn. Pet. Geol., Tulsa, 1984), pp. 6–1 to 6–95.
10. T.S. Lovering, *Rock Alteration as a Guide to Ore – East Tintic District, Utah* (Economic Geology Publishing Co., Urbana, 1949), 64 p.

11. R. Blanchard, *Nevada Bur. Mines Bull.* **66** (1968): 196 p.
12. L.C. Rowan, P. H. Wetlaufer, A.F.H. Goetz, F. Billingsley, and J.H. Stewart, *U.S. Geol. Survey Professional Paper* **883** (1974): 35 p.
13. G.L. Prost, *Econ. Geol.* **75** (1980): 894–906.
14. M.J. Abrams, D. Brown, L. Lepley, and R. Sadowski, *Econ. Geol.* **78** (1983): 591–604.
15. M.H. Podwysocki, D.B. Segal, and M.J. Abrams, *Econ. Geol.* **78** (1983): 675–687.
16. D.B. Segal, *Econ. Geol.* **78** (1983): 711–722.
17. J. Läg and B. Bølviken, Norges Geol. Undersökelse **304** (1974): 73–96.
18. D. Carlisle and G.B. Cleveland, *California Div. Mines Spec Rept.* **50** (1958): 31 p.
19. H.L. Cannon, *Science* **132** (1960): 591–598.
20. F.C. Canney, *Second Ann. Earth Res. Aircraft Program Status Review* **1** (NASA, 1969), pp. 7–1 to 7–8.
21. H.L. Cannon, *Taxon* **20** (1971): 227–256.

ADDITIONAL READING

J.L. Berry and G.L. Prost, *Hydrocarbon Exploration, Manual of Remote Sensing* **3** (American Society of Photogrammetry, Falls Church, 1999), pp. 449–508.

C. Sabine, *Remote Sensing Strategies for Mineral Exploration, Manual of Remote Sensing* **3** (American Society of Photogrammetry, Falls Church, 1999), pp. 375–447.

QUESTIONS TO CONSIDER

1. How would you expect limestone to weather, and how would you recognize it in the arctic, in a temperate climate, and in the tropics? A sandstone? A granite? A serpentine?
2. A thrust fault puts a Carboniferous limestone over a Cretaceous Shale in Wyoming. How might one recognize the thrust contact?
3. Although jungle covers the surface, one can make out drainage patterns in the Amazon basin. What is suggested by alternating bands of dense and widely-spaced drainages?
4. While examining Landsat TM imagery over pediment surfaces in the Altiplano of Chile you detect clusters of small yellow-green outcrops on the color-infrared composite image. What could these be telling you?
5. You recognize badlands on an image of the San Juan basin in New Mexico. What lithology does this most likely represent? Outcrops of a dipping unit along the margin of the San Juan basin have pine trees growing on them, whereas the surrounding plains are covered in grass and sagebrush. What is the probable lithology of the ridges?
6. In the same part of the San Juan basin walls of rock radiate outward from a large, almost circular and jagged outcrop of dark rock. Any idea what this could represent?

5
Recognizing Structure

CHAPTER OVERVIEW

In order to recognize structures it is necessary to begin by understanding how the structures form and how they influence surface processes. Simple structures will be examined first, followed by increasingly complex forms. The end result should be the construction of a structure map or structural formline map that integrates what is seen at the surface and what is known or presumed about the subsurface from well data, seismic data, etc. This map should reveal the structural fabric of the region as well as showing individual prospect-scale features. It should assist in preparing for seismic, geochemical, and potential fields programs by providing the location and orientation of folds and faults, and should assist in the interpretation of these surveys and well data.

UNDEFORMED TERRAIN

It may be that there are no truly undeformed terrains. Such a landscape, however, must serve as one end member in the range of deformation from simple to severe. What landforms might be expected in an area with essentially flat-lying, unfaulted, and unfolded strata?

In this ideal case, drainage will develop into a dendritic network on a flat plain (Fig. 5.1) [1]. As erosion progresses toward base level, terraces will develop along the margins of the flood plains. These terraces correspond to the various layers of rock being exposed [2]. As the terraces erode, erosional remnants are left standing as mesas, then as buttes, and finally as small rises. An example of this is expressed in Mesozoic rock of great beauty in Monument Valley, Arizona (Fig. 5.2).

In areas with gently rolling hills, the trace of bedding in flat-lying units will appear as closed polygonal to circular tonal bands on the hillsides (Fig. 5.3).

It is now possible to begin to recognize geologic structure based on deviation from the ideal condition described above.

RECOGNIZING DIP

Dip is the angle at which a rock layer is inclined from the horizontal. Dips may vary from gentle to vertical and overturned, with different topographic expressions of each range.

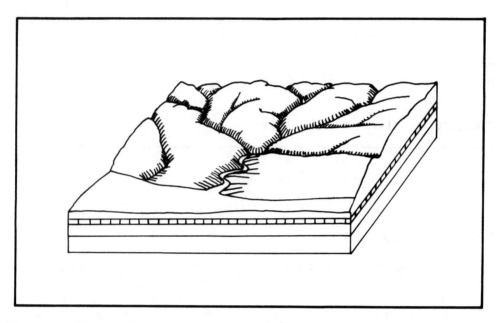

Figure 5.1 Diagram illustrating the development of dendritic drainage on a homogeneous, flat or gently-inclined surface.

Gently inclined layers tend to form long, dip-slope drainages (known as "consequent" drainage) opposed by short tributaries flowing down the bedrock escarpment (known as "obsequent" drainage). Over a large area, this is known as a

Figure 5.2 Buttes and mesas are erosional remnants of flat-lying strata, such as this example from the Monument uplift, northern Arizona.

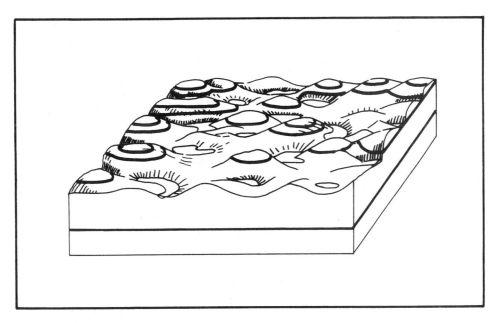

Figure 5.3 Circular or irregular closed tonal bands often represent flat-lying bedding in areas of rolling hills.

trellis drainage pattern (Fig. 5.4). The landform commonly associated with gentle dip is the cuesta or hogback (Fig. 5.5).

All stereo airphotos have some degree of vertical exaggeration that must be considered when estimating dip (see Chapter 3, Airphotos). Vertical exaggeration makes objects appear taller or steeper than they really are, and can be helpful in estimating dips in low-relief terrain (Figs. 5.6, 5.7). Exaggeration is a function of both camera focal length and photobase, the distance between adjacent image centers. A shorter focal length, or longer photo base provides greater apparent relief. On nonstereo imagery such a Landsat it is not too difficult to recognize wide illuminated slopes paired with short, shadowed slopes (or vice versa) and estimate the inclination of strata.

In moderately inclined strata (5–30°) the dip direction and relative magnitude can be estimated using the rule of V's (Fig. 5.8). This rule states that the trace of bedding (either outcrop or color bands) in a valley or water gap forms a V-shape pointing in the direction of dip. This is true in all cases unless topographic relief is greater than the dip of the units (e.g., gently dipping units exposed along a steep cliff). In such cases the bedding trace must be plotted on topographic maps and true dip calculated using a 3-point problem (Fig. 5.9).

As layers become increasingly inclined, the dip slope becomes narrower and the opposing escarpment becomes wider, until in the range from about 45 to 90° the ridge, or hogback, that is formed is almost symmetrical. In near-vertical beds it is difficult to determine dip using the V-shape of the bedding trace. In the case of thinly bedded units, such as sandstone and shale, however, it may be possible to see the trace of multiple beds on the escarpment, and not on the dip slope (Fig. 5.10). It has also been noted [3] that the height of hogbacks decreases with increasing dip (Fig. 5.11). Thinning and thickening of an outcrop along strike may indicate changes in dip

96 REMOTE SENSING FOR GEOLOGISTS

Figure 5.4 Trellis drainage pattern, perhaps modified by jointing, is developed on the Eocene Uinta Formation southwest of Meeker, Colorado, in the Piceance basin. Drainage is developed on gentle dip slopes. Landsat MSS color infrared image processed by Earth Satellite Corp. North is at the top of the image. (See color plate.)

(steepening or flattening, respectively), or may be due to a facies change or true thickness change. If the change occurs in several adjacent units, it is probably a structural change (change in dip or thickening across a fault or hingeline). If a unit appears to dip in one direction and, as one moves along strike the outcrop thins and then thickens and appears to dip in the opposite direction, the bed may have gone to vertical and then overturned. Some anticline-syncline pairs will appear to change along strike to bedding all dipping in the same direction. This may be caused by an overturned common flank where the fold hinge is so tight that it is not recognizable.

A gently inclined surface lacking outcrops is characterized by its "geomorphic dip." Geomorphic dip often coincides with true dip, especially for gentle surfaces. It is possible, however, to have a gently inclined flat surface that is completely unrelated to

RECOGNIZING STRUCTURE 97

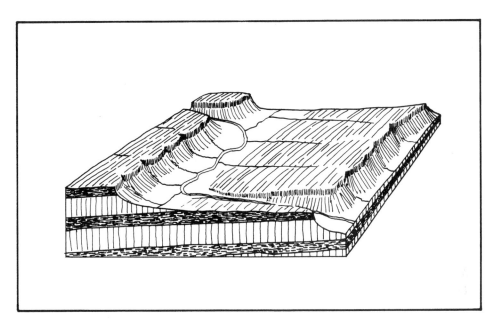

Figure 5.5 The cuesta, a long sloping surface opposed by an escarpment, is generally (but not always) a dipslope.

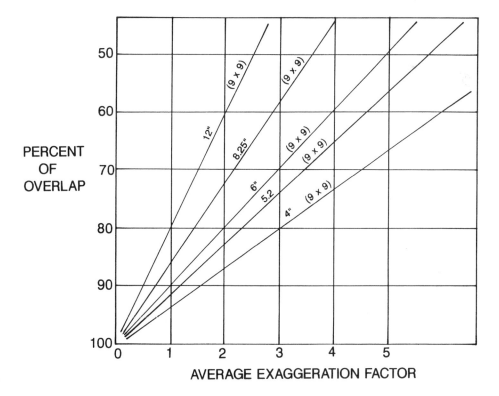

Figure 5.6 Chart relating the average exaggeration factor to the amount of overlap between adjacent stereopairs, print size, and camera focal length. From Thurrell [4].

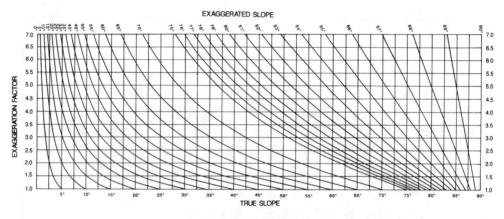

Figure 5.7 Chart showing the relation between true slope (or dip), exaggeration factor, and exaggerated slope (dip). Modified after Ray [5].

the dip of bedding. This is common in the case of pediment surfaces, where rivers or flash floods carve gently inclined surfaces into bedrock between a mountain front and valley bottom (Fig. 5.12). Such surfaces are generally gently inclined toward the valley bottom. Gently inclined smooth surfaces can also be the result of alluvial terraces formed by fluvial processes within basins and preserved as erosional remnants because of a cap of resistant lag gravel (e.g., Meeteetse Rim, Bighorn basin, Wyoming (Fig. 5.13). Similarly, wave-cut terraces can form smooth, gently inclined surfaces along a coastline or around drying inland lakes (e.g., Lake Bonneville, Lop Nor).

ANTICLINES, DOMES, AND HORSTS

Anticlines are, without question, the most important structure in the search for oil and gas. The range of shapes is endless and includes, among others, circular to elliptical patterns with four-way closure, plunging noses, and gentle changes in the strike of bedding (Fig. 5.14). The anticline is referred to as a "positive" structure or "structural high," and commonly is expressed as a topographic high at the surface. Just as common, there is inverted topography associated with this fold, that is, the resistant layers have been breached by erosion and the easier-eroded core is exposed as a valley surrounded by outward-dipping ridges (e.g., Axial Basin anticline, Colorado; Paradox anticline, Utah).

Anticlines generally form as elongated folds in layered strata due to horizontal compression that is maximum at right angles to the fold axis. They may also form by draping of sediments over the edge of a faulted block ("forced folds"), or along the leading edges of thrust faults and over thrust ramps. Domes generally develop as a result of density contrasts between layers, as is the case when a low density layer (salt, gypsum, overpressured shale) lies beneath a higher density layer (sandstone, carbonate). The disequilibrium results in upward movement of the low density layer as a diapir. If the upward movement is such that the overlying layers are pierced, it is referred to as a "piercement" dome. Anticlines have been known to form in this manner, as in the Paradox basin of Utah, or Dukhan anticline in Qatar. In these cases the fold may have been initiated over a basement fault, thus acquiring its elongated

RECOGNIZING STRUCTURE 99

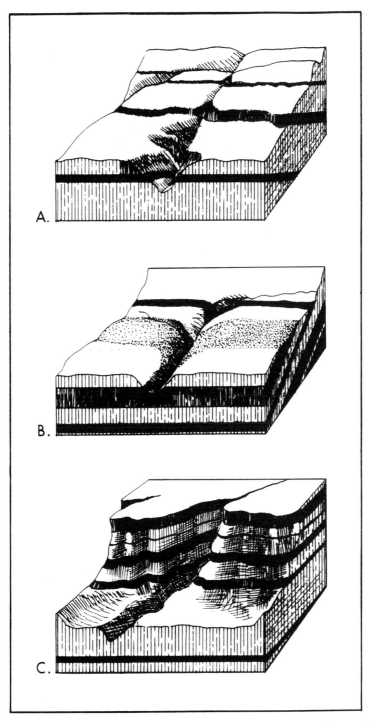

Figure 5.8 The "Rule of V's" A. The trace of beds that dip upstream form a "v" that points upstream. B. Beds that dip downstream form a "v" pointing downstream. C. When topographic relief is steeper than the dip of the layering, the "v" points upstream regardless of the dip. Solving a three-point problem is then required to determine dip.

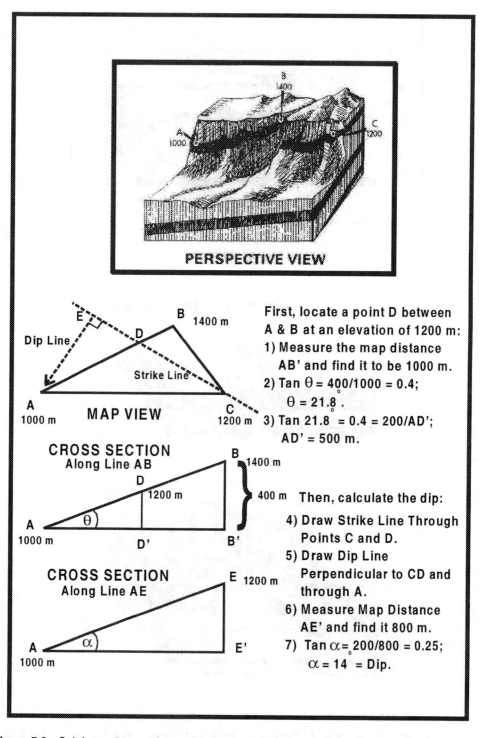

Figure 5.9 Solving a three point problem requires knowing the elevation of at least three points along the trace of a marker bed, and the distances between those points. This can best be obtained when the bedding is traced or transferred onto a topographic map.

RECOGNIZING STRUCTURE 101

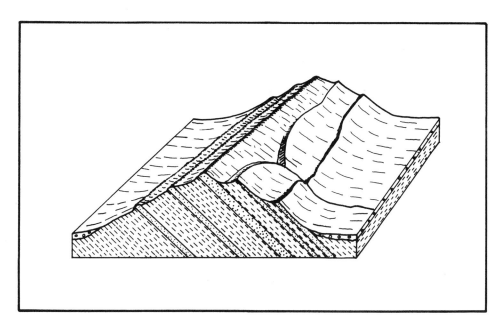

Figure 5.10 On nearly symmetrical ridges the trace of bedding forms a "v" on the dip slope and multiple straight lines or tonal bands on the opposing escarpment.

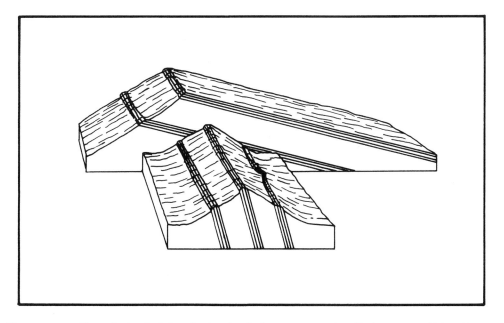

Figure 5.11 The relative height of a hogback decreases as the dip increases. From Miller and Miller [3].

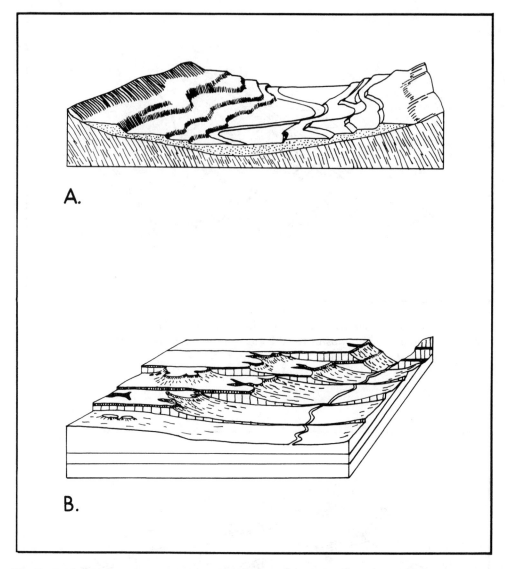

Figure 5.12 Erosional and stratigraphic surfaces. A. Pediment surface carved by a downcutting river. B. Stratigraphic terraces formed by successively exposed bedding surfaces. From Cotton [2].

shape. In the case of diapirism one would not necessarily expect folding to extend beneath the low-density layer (Fig. 5.15). If it is known, for example, that a reservoir unit is capped by an evaporate, and the evaporite has formed piercement domes, there is no reason to drill the flanks or crests of these domes as the reservoir unit is unaffected by the doming. If the seal has been broken by faulting, however, one might expect hydrocarbons to leak upward and migrate into the flanks and possibly the crest of the dome.

Horsts are uplifted blocks bounded on one or more sides by faults. They tend to be rectangular in shape, and accumulate hydrocarbons in sediments draped over the

RECOGNIZING STRUCTURE   103

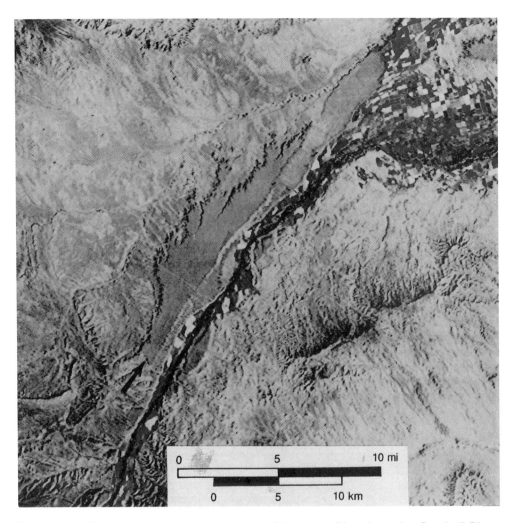

Figure 5.13 Tertiary terrace gravels form the Meeteetse Rim along the Graybull River, Bighorn basin, Wyoming. Landsat MSS color infrared image processed by EROS Data Center, Sioux Falls. North is at the top of the image. (See color plate.)

block, at high corners, or at updip fault truncations along their flanks. Horsts are often associated with forced folds, which are recognized by their flat or gently inclined upper surfaces, near-vertical bedding around their flanks, and abrupt changes in strike at block corners (e.g., Bighorn, Beartooth uplifts, Wyoming, Fig. 5.16).

Surface Folds

Domes and anticlines are expressed at the surface by opposing dips in outcrops on opposite flanks of the structure. In many cases one can follow bedding along a circular to elliptical path, whereas in other circumstances the bedding merely changes strike a few degrees. The bedding may be thought of as structural form lines, similar to structure contours. Ideally bedding will close on itself and a significant exploration

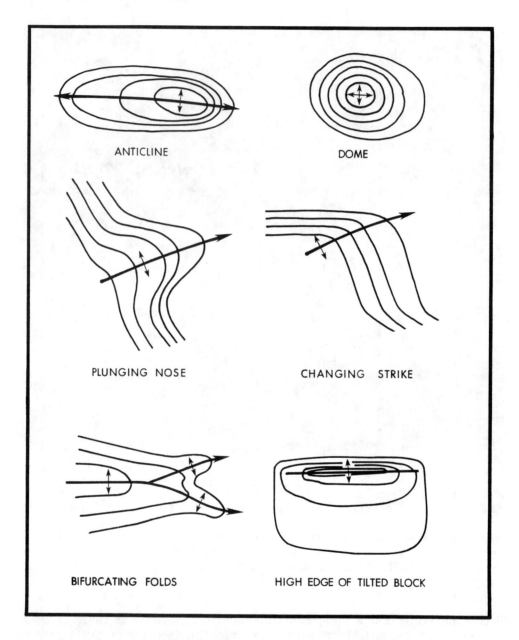

Figure 5.14 Some common fold forms depicted by structure formlines.

target will be revealed. Often the structure is not a closed fold, and an updip facies change or a fault crossing the structure is required for closure (Fig. 5.17).

Buried Folds

Buried or "blind" anticlines occur in areas lacking outcrops at the surface. These require geomorphic evidence in order to be recognized. There are essentially three reasons these structures are expressed geomorphically:

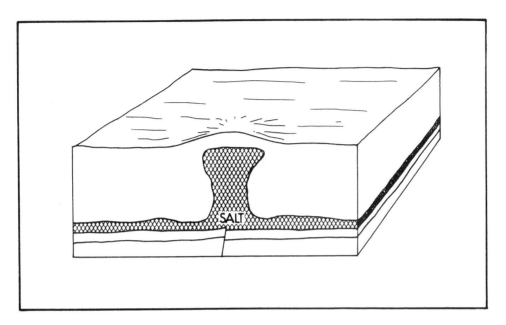

Figure 5.15 Most salt diapirs are expressed as domal topographic highs. Note that there need not be folding beneath the salt.

1. Draping by differential compaction. Differential compaction of recent sediments is generally a result of dewatering of shales [6–9]. The compaction causes overlying material to drape over preexisting structures. The amount of compaction depends in large part on the type of material and stratigraphic thickness of the material over the structure (e.g., shales and gypsum are more readily compressed and dewatered, and are more prone to plastic flow than sandstones and carbonates). The surface expression is also a function of the number of unconformities between the structure and the surface. It should be noted that the surface expression of the buried anticline may be offset slightly from the actual position of the structure at depth.
2. Reactivation. Folds tend to form in areas that are predisposed to deformation, i.e., at points of weakness. They are, therefore, subject to reactivation with renewed tectonic activity [10]. Indeed, isopachs often show repeated episodes of folding in a single structure. In formerly glaciated terrain, old structures can be reactivated by glacial isostatic rebound [11–13].
3. Effect on overlying materials. Buried folds can affect overlying materials by influencing soils, vegetation, and present-day deposition or erosion patterns.

Topographic Expression of Folding

Buried structural highs can be expressed as gentle topographic highs in alluvial cover (Fig. 5.18). This is a result either of compaction over the structure or reactivation of the folding. One must be cautious, however, not to interpret every erosional remnant as a buried structure. In some cases this distinction may not be possible. In permafrost

106 REMOTE SENSING FOR GEOLOGISTS

Figure 5.16 Drape of Upper Cretaceous Wata and Galala limestones over the southeast block corner (arrow) of Gebel Somer, western Sinai, Egypt. Landsat TM band 4, produced by Amoco Production Company. North is at the top of the image.

areas one finds pingos, or frost heaves, that appear to be symmetrical domal uplifts. These are surface effects that may or may not be related to structure (e.g., aligned along faults). Many erosional remnants, and even buried volcanic/intrusive features appear similar to tectonic structures.

In arid environments the zero edge of alluviation is where alluvial fans (distributary drainage) meet the pediment or bajada (tributary drainage). Subtle structure is suggested where the edge of alluviation forms an embayment due to diversion of fan deposition. (Lattman, oral comm., 7/78).

Color Anomalies Indicative of Folds

Color or tonal "anomalies" may be associated with uplifts. These often circular to elliptical tonal changes may be the result of older material being exposed by erosion in the core of an uplift, or by changes in soil type, soil chemistry, moisture differences, or changing vegetation patterns as one proceeds from a low area onto an uplifted area. One word of caution: many of the "hazy" tonal anomalies associated with oil fields that were seen on some of the original satellite imagery turned out to be the result of

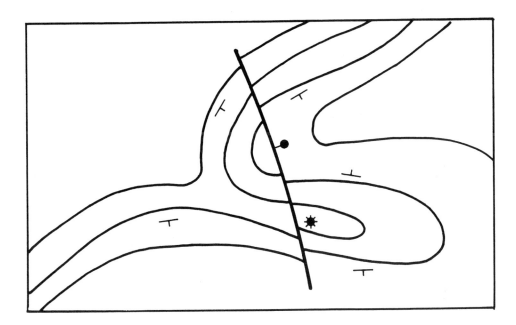

Figure 5.17 Open folding, shown here by structure formlines, requires a fault or other type of up-dip closure to generate a hydrocarbon trap.

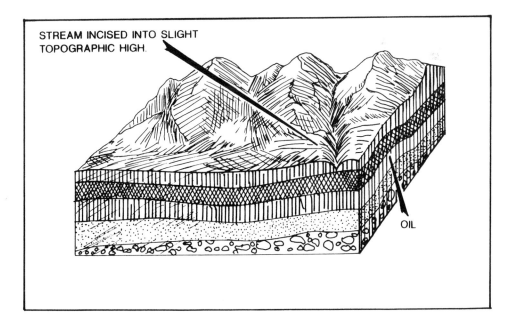

Figure 5.18 Buried structures can be expressed as subtle topographic rises. Here, drainage incised into alluvial cover suggests a buried structure similar to those seen along the Gulf of Suez coastal plain, Egypt.

surface disturbances (tracks, drill pads, mud pits, etc.) that were too small to be resolved on the imagery, yet contributed to a lighter overall tone in the area of production. This problem gave the concept of tonal anomalies a bad reputation that is still difficult to overcome.

Classic examples of soil tonal anomalies revealed on black and white airphotos were presented by Kupsch [11] for the surface expression of Braddock Dome, Nottingham field, and other areas in the Williston Basin of Saskatchewan. These were thought to reflect a change from the background A horizon (lighter) to B soil horizon (darker) along the margins of the topographic high, or a change from high moisture (darker) off the structure to low moisture (lighter) over the structure. Many similar examples of tonal anomalies detected from satellite altitudes have been described (see, for example, Saunders [14]; Morgan, et al. [15]).

Another type of tone or color change may be due to changes in mineralogy related to diagenetic alteration over leaking hydrocarbons (Fig. 5.19). These changes usually occur in the subsurface and are later exhumed and appear at the surface. Donovan [16] noted at Cement field, Oklahoma, that iron-cemented red-bed sandstones are reduced over areas of petroleum microseepage. Thus, iron is mobilized, and moves away from the crest of the producing anticlinal structure. Depending on the pH and Eh conditions, the sequence of alteration under reducing conditions may be:

$$\text{Hydrated ferric oxides} \rightarrow \text{hematite} \rightarrow \text{magnetite}$$

Ferguson [17,18] determined that reduction of ferric iron generates pyrite (2–4% in thin sections) in red bed sandstones with a minimum pH of 8.3 at the Velma, Eola, and

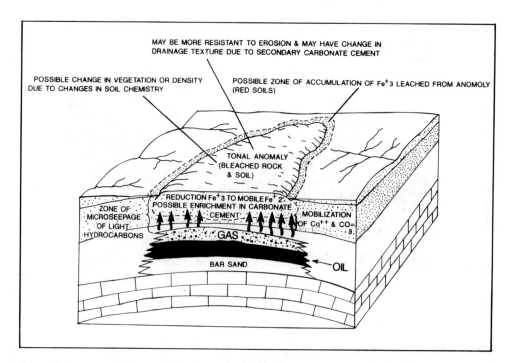

Figure 5.19 Diagenetic soil chemistry changes and the tonal and topographic anomalies that may be associated with them.

Chickasha anticlinal fields, Oklahoma. In all cases this caused a "bleached" appearance in the red beds over production.

More recently Reynolds, et al. [19] noted that ferrimagnetic pyrrhotite exists in well samples above oil and gas reservoirs at Cement field, and does not exist off the field. A major source of sulfide in the iron sulfide minerals may have been microbial sulfate reduction where the sulfate reducing bacteria derive their metabolic energy from leaking hydrocarbons. Machel and Burton [20] also observed that magnetite and pyrrhotite are formed in diagenetic environments containing hydrocarbons, and hematite is generally dissolved or replaced.

Mineral alteration and bleaching have also been documented in red beds of the Permian Cutler and Triassic Wingate Formations over the Lisbon and Little Valley fields, Utah [21–24]. The bleaching coincides with occurrences of near-surface uranium, also believed to have been localized at least in part by seeping hydrocarbons. The light-toned anomalies (Fig. 5.20) result from an absence of pyrite, presence of carbonate cement and feldspars in the sandstone and a relative abundance of clay minerals, primarily kaolinite and chlorite-smectite (corrensite).

Bleaching of the iron-oxide rich Triassic Chugwater sandstone at Spence Dome oil field near Sheep Mountain, Wyoming, is associated with uranium mineralization in the near-surface and oil production at depths less than 1750 m [25]. The authors also noted an absence of kaolinite and decrease in pH in soils developed on the Cretaceous Cloverly shale above production.

Isotopically distinct carbonate cements were reported over the Recluse field, Wyoming [26]. Seeping hydrocarbons were oxidized by microbes in the near surface, and the resulting carbon dioxide went into solution and later precipitated over the field as unique isotopic carbonate cement, depleted in Carbon 13 and either anomalously high or low in Oxygen 18. They reported the same from the Davenport field in Oklahoma.

The manganese content of surface carbonate cements increases dramatically over the Recluse field, Wyoming, and toward the center of a halo-type anomaly over production near Boulder, Colorado (from 120 to 5400 ppm) [26,27]. Apparently Mn++ has a wide range of redox stability compared to iron, and is relatively enriched in soils over areas of microseepage. Also over the Recluse field sage leaves (*Artimesia tridentata*) and pine needles (*Pinus ponderosa*) were found to contain anomalously high concentrations of iron and manganese. This is thought due to reducing soil conditions as a result of seeping hydrocarbons; these conditions allow divalent iron and manganese to be complexed or adsorbed on soil particles, where they are easily taken up by plants.

Lilburn and Zuhair [28] summarized several of these observations as (1) red bed sandstones are bleached as a result of a reducing environment caused by leaking hydrogen sulfide, are relatively enriched in clay minerals, primarily kaolinite and corrensite, and often are associated with uranium deposits, (2) oxidation of hydrocarbons caused caliche or carbonate-cemented sandstones (isotopically distinct and often manganese-rich) in the near-surface and calcite replacement of gypsum, and (3) pyrite and pyrrhotite formed over production as a result of reduction of iron oxide by hydrogen sulfide gas.

It should be noted that the mineralogic changes described here are not restricted to structural traps, and could form just as easily over stratigraphic traps.

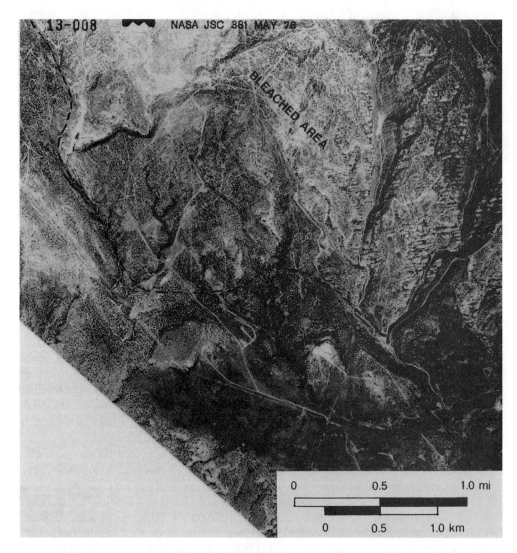

Figure 5.20 NASA airphoto showing bleached areas of red bed sandstone of the Triassic Wingate formation, Lisbon Valley, Utah. Bleaching is believed to be an alteration related to leaking hydrocarbon gases. (See color plate.)

Vegetation Anomalies Suggesting Folds

Topographic highs are often associated with changes in vegetation communities or densities. This may be a result of changes in moisture conditions (as mentioned above) from wetter in the background to drier over the uplift; it may be a result of a change in subcrop or soil type that affects the type or density of plant cover; or it may simply be the result of a change from north- to south-facing slopes across the uplift. In middle latitudes of the northern hemisphere, for example, north-facing slopes tend to be in shadow more, have more moisture and better developed soils, and thus support a more robust plant community than south-facing slopes. DeBlieux [6] described cane stands growing over salt domes in the coastal marshes of Louisiana because the water

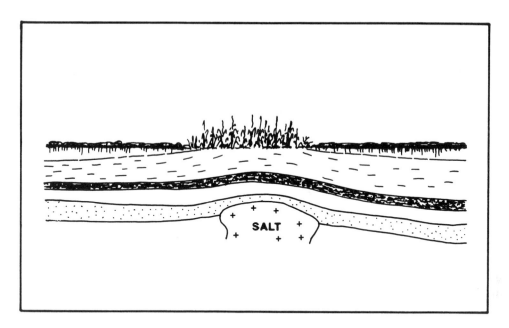

Figure 5.21 Cane stands grow over salt domes in coastal swamps in Louisiana. From DeBlieux [6].

over the domes was shallow enough to support rooted vegetation whereas other areas were too deep (Fig. 5.21).

Vegetation has also been used as an indicator of seeping hydrocarbons, which can occur both over structural traps and stratigraphic traps. Some types of vegetation have been classified as "hydrocarbon indicator" plants, e.g., *Anabasis salsa*, *Salsela* sp. (saltwort) and *Allium* sp. (onion) [29]. In some cases hydrocarbons are detrimental for plant communities over fields. Sagebrush is stunted, and prairie grasses are less dense over the gas cap at Patrick Draw field, Wyoming [30]. High soil pH coincides with high soil gas and causes an inability to absorb micronutrients, especially zinc, that has led to stunting [31]. Also, as mentioned earlier, changing the chemistry of the soil, such as at Cement field, can cause anomalous species to develop.

Many plants are harmed by the presence of hydrocarbons in the soil. Laboratory experiments on marsh plants show that *Spartina patens* suffered increasing damage with increasing amounts of oil and with heavier crudes. It has been noted by pipeline workers that natural gas leaks kill trees and shrubs around leaks. This has been explained [32] as a result of gas driving out soil air and contributing to the growth of methane-oxidizing bacteria that deplete soil oxygen. Flower [33] showed that trees grown above urban landfills were killed when leaking methane decreased the soil oxygen and increased the carbon dioxide in the soil root zone. Rock [34] and Parrish [35] studied vegetation patterns over the Lost River gas field, Virginia. They found that mycorrhizal fungus on plant root hairs (that help with nutrient uptake) are inhibited on Chestnut Oaks in areas of high soil manganese and/or methane concentrations. At the Pico anticline, near San Fernando, California, an oil, tar, and methane seep has harmful effects on orange and walnut groves, as well as on indigenous oak and chaparral.

It appears that other plants do well in the presence of hydrocarbons in soil. Riparian woodland vegetation (including mulefat, royal willow, populus) grow healthy in and around the Pico anticline seep. At Lost River field maples tolerate high soil manganese and methane, apparently because they can tolerate anaerobic root bacteria, and thus replace the natural oaks. The marsh plant *Scirpus olneyi* not only grew well but had enhanced survival in increasing concentrations of crude oil [36].

When looking for vegetation anomalies it is necessary to know what plants are natural in a given area, what environmental factors (moisture, nutrients, soil gasses, soil pH, etc.) they can tolerate, and what factors they cannot tolerate.

Fracture Patterns Indicating Folds

Fracture orientations and intensity have been mentioned as clues to buried folds. Several workers have mapped zones of intense fracturing in order to locate the zones of rapidly changing dip associated with the flanks of folds [37–40]. A tight fold will have a single zone of fracturing along the crest of the fold, whereas a fold with a gentle crest and steep limbs will have a halo of intense jointing above the flanks (Fig. 5.22, 5.23). These joints are a result of flexural folding, and the joint zones extend from the subsurface upward as a result of differential compaction.

An excellent example is the Verde field, San Juan County, New Mexico (Fig. 5.24). The field produces out of zones in the Cretaceous Niobrara (interbedded black shales and thin quartz siltstones) at depths around 1000 m. Matrix porosities and permeabilities are too low to produce oil unless the formation is fractured. The main structure is the hogback monocline, which changes strike and forms both a gentle anticlinal nose and a syncline that cross the Hogback. The best wells produce in the syncline at the position where maximum curvature exists, not at the crest of the anticline [40].

Another aspect of jointing associated with folds is joint orientation (Fig. 5.25). Joint trends expected in a dome-shaped granite hill were first described by Chapman [41]. Jointing in anticlines was described by Price [42] as tensional, sheer, and cross-axis. Clay models reveal surprising detail regarding the orientation and location of joints and faults in circular and elliptical uplifts formed in extensional stress fields, compressive fields, and with no prevailing stress [43]. These fractures are expected in the domed strata, and under favorable conditions they might be perceived through overlying alluvial material.

In alluvial terrain streams may show a marked angularity over buried structural highs. Fractures propagating upsection due to drape or reactivation appear to control stream channel direction more than in other parts of the basin (Fig. 5.26).

Buried Folds in Glacial Terrain

In areas of continental glaciation the abrasive action of ice sheets tends to decrease the relief on the preexisting surface. Yet as the glaciers advance they grind down fractured bedrock faster than unbroken rock, thus deepening fault-controlled drainage. Areas that remain topographically (and perhaps structurally) high as the ice sheet melts will be characterized by "dead ice" moraines (Fig. 5.27). Dead ice moraines are deposited by stagnant ice and are identified by rimmed or donut-shaped kettles which form as the ice remnants melt and deposit debris around the last ice [45]. The last ice melts off

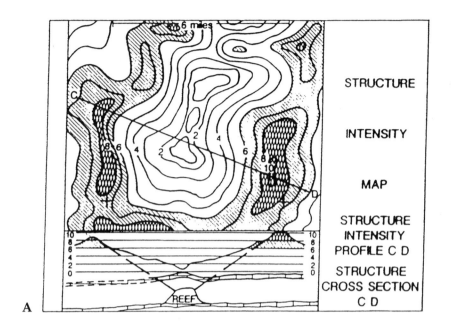

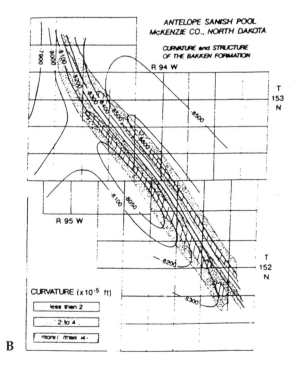

Figure 5.22 Examples of fields that produce out of zones of intense fracturing where the rate of change of dip is greatest. A. Wizard Lake reef, Alberta. From Blanchet [37]. B. Curvature on the Antelope-Sanish pool, North Dakota. From Murray [38]. C. Iso-fracture map of Goose Egg dome, Wyoming. From Harris [39].

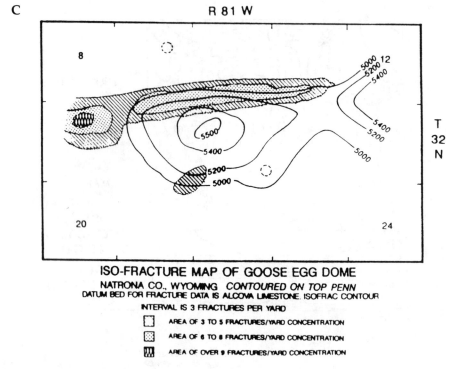

Figure 5.22 (continued)

topographic high points since meltwater flows through and speeds melting in low areas. Dead ice moraines are developed over pre-glacial topographic highs in the Williston Basin of North Dakota along the Missouri Coteau. Kupsch [11] found that 24 of 30 glacial (and subcrop) highs corresponded to seismic highs between Old Wives Lake and Beechy in Saskatchewan. The same correspondence of dead ice moraines to subglacial highs was reported by Mollard [7].

The southern peninsula of Michigan is covered by as much as 330 m of glacial till that is either cultivated or forest-covered. Thus it is a challenge to map geologic structures at the surface. Yet there are reasons surface features should reflect subsurface structure in that area. It has been demonstrated [46] that glacial channels develop above preexisting bedrock channels, which are in turn often eroded into faulted bedrock. In addition, a minimum of 130 m and maximum of 1000 m of uplift has been determined [47] as a result of post-glacial isostatic rebound in Michigan. That amount of ground movement is considered to be enough to reactivate old fault lines and impress their trace on the post-glacial surface [48, 49] (Fig. 5.28).

Folding Beneath Drifting Sand

Where thin, discontinuous sand sheets prevail, topographic (and perhaps structural) highs tend to be windswept and fairly clear of sand cover, whereas great thicknesses of sand accumulate on the leeward side of the high [50]. The high acts as a windbreak, slowing air velocities and causing turbulence, forcing sand to drop (Fig. 5.29). Examples of structures expressed in such an environment are Safir dome, in North Yemen, and El Borma, in Algeria, which appear as wide, smooth, elliptical depressions

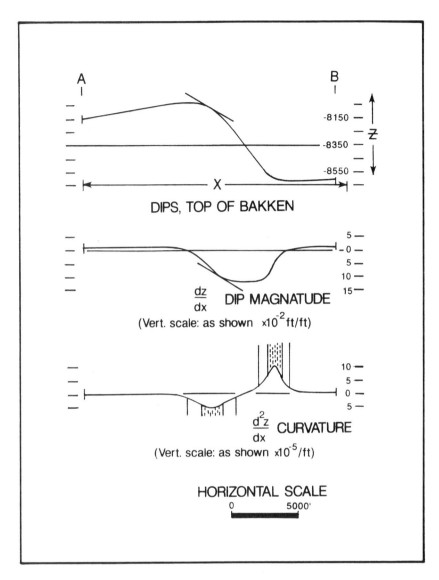

Figure 5.23 Comparison of structural profile, dip magnitude, and structural curvature across the Antelope Sanish field, McKenzie County, North Dakota. From Murray [38].

between sand waves (Fig. 5.30). Thermal imagery may reveal faults that affect groundwater, either by ponding it on one side of the fault or by acting as a conduit beneath a thin sand cover.

Folding Revealed by Drainage Patterns

Stream patterns are especially sensitive indicators of blind structures in flat, alluviated plains. A fractional change in regional gradient as a result of drape or reactivation can affect the course of runoff over structures too subtle to be evident during field mapping.

The following list provides some drainage indicators of folding. One should bear in mind that drainage responds to many factors, and buried structure is but one.

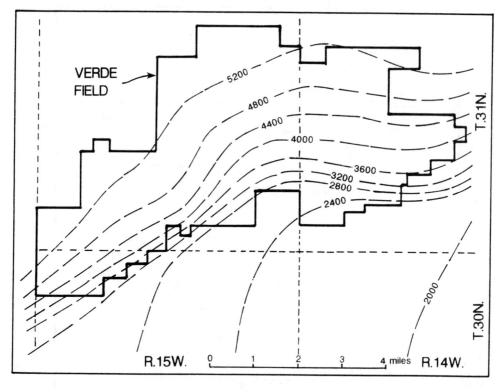

Figure 5.24 Structure contour map on top of the Point Lookout member of the Mesaverde Group, Verde field, San Juan County, New Mexico. The field is located at a syncline plunging across the Hogback monocline. From Gorham, et al. [40].

- Channel Width/Depth Adjustment and Incision
  The width/depth ratio of a stream channel increases as the regional gradient flattens out upstream from the crest of a structure [51]. As the gradient flattens one may observe "alluvial ponding," an anomalous area of alluvium that occurs in the wider channel upstream from a structure. As it passes over the crest of the fold the channel narrows and becomes incised between high cutbanks [52]. Downstream from the fold the stream straightens as a result of increased gradient (Fig. 5.31). Such features are particularly well expressed in sand cover and alluvium over horst blocks, as near Riyadh, Saudi Arabia, and the Tarim Basin, China [53] (Fig. 5.32).
- Compressed Meanders
  Streams may show local compression of meanders (progressive shortening of the wavelength and increase in the amplitude) where the gradient decreases upstream from a fold crest (Figs. 5.33, 5.34). This has been noted at a fold in Kent Co., Texas, [54, 55] and in the Salawati Basin of Indonesia associated with drape over pinnacle reefs [8]. Downstream from the fold crest the channel straightens, then resumes its normal course.
- Drainage Texture
  Lithologic mapping can be based on drainage texture. Coarse clastics tend to have better infiltration and less runoff, causing widely spaced drainage, whereas fine-grained units, particularly shales, have very fine-textured and closely spaced

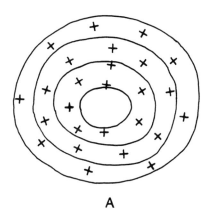

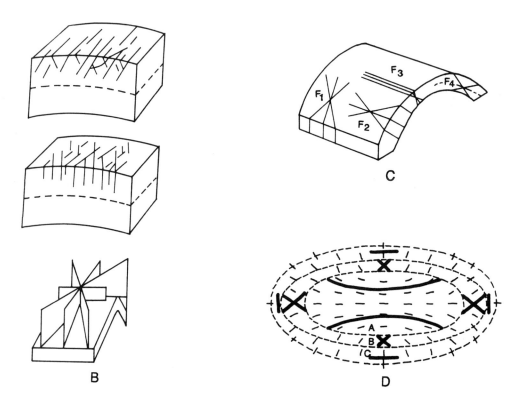

Figure 5.25 Joint orientations with respect to folding. A. Exfoliation joints in a granite dome. From Chapman [41]. B. Joints typically developed in anticlines as described by Price [42]. C. Joint sets defined by Stearns [44]. D. Joints (light) and faults (heavy) observed in clay models of elliptical uplifts. From Withjack and Scheiner [43].

drainage patterns. This criteria has been used in the Ucayali Basin, Peru, to map a fold in a low relief area mantled with alluvium and continuous vegetation cover (Fig. 5.35) [56].

118 REMOTE SENSING FOR GEOLOGISTS

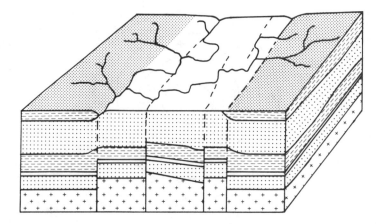

Figure 5.26 Jointing can control drainage over uplifts to a greater extent than in adjacent areas. This is a result of outcrops on the uplifts versus soil in surrounding lowlands, or thinner soils over the uplifts that better reflect shallow jointing because of compaction and drape.

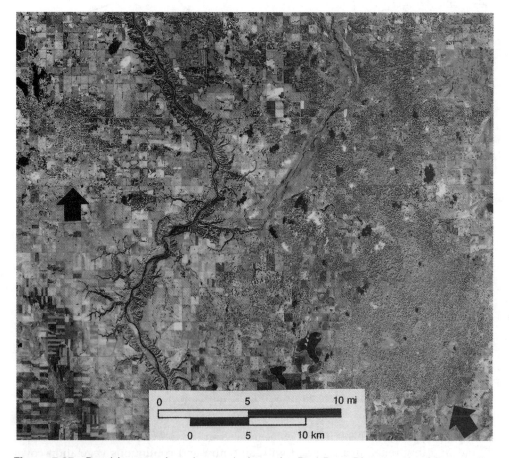

Figure 5.27 Dead ice moraines (arrows) along the Red Deer River north of Drumheller, Alberta. Landsat TM band 5 processed by Amoco Production Company. North is at the top.

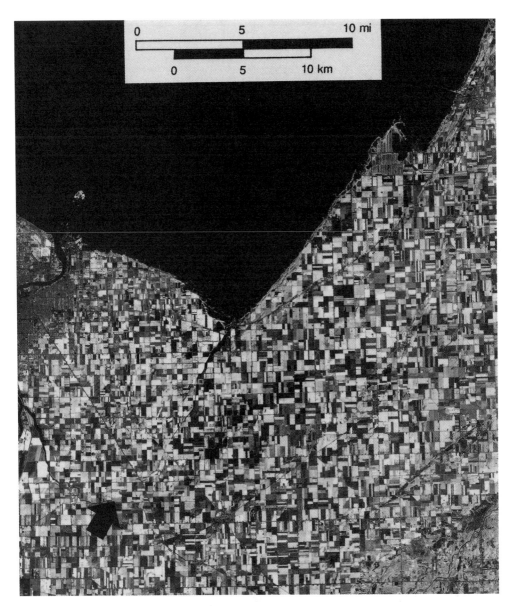

Figure 5.28 A fault bounding the Saginaw Bay graben (arrow) is a result of glacial rebound imposing old faults on recent terrain. Landsat TM band 4 processed by Amoco Production Company. North is at the top.

- Radial and Concentric (Annular) Drainage
  Any topographic high will tend to have drainage that flows away from the crest in a more or less radial pattern. If the pattern is radial in 360°, and has concentric tributaries, then it is likely to represent domed, stratified material (Fig. 5.36). The concentric pattern is a result of arcuate strike valleys. Alternatively, radial inward flowing streams may represent a synclinal structure [3, 51, 56].

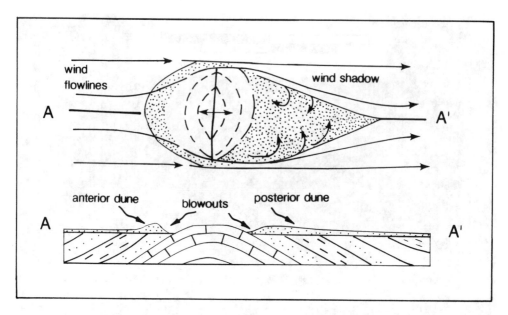

Figure 5.29 Windflow and sand accumulation around topographic (and structural) highs. Modified after Bagnold [50].

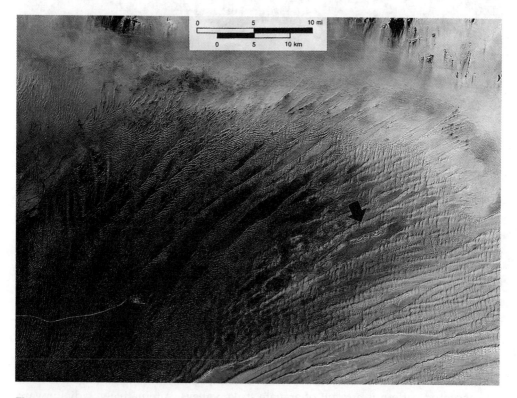

Figure 5.30 Large dunes are developed on the flanks of Safir dome, Yemen, where Jurassic units outcrop (arrow). The Alif field also sits in a sand-free area. TM bands 1-4-7, processed by Earth Satellite Corp. (See color plate.)

RECOGNIZING STRUCTURE 121

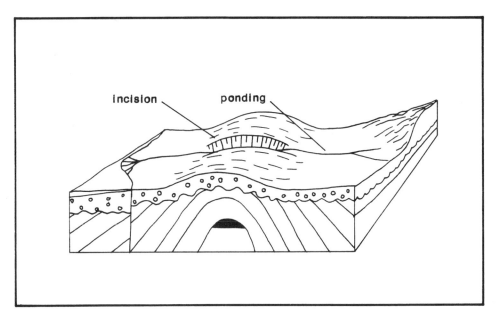

Figure 5.31 Stream channels become narrower and are incised over a structure. Alluvium is "ponded" upstream from the structure.

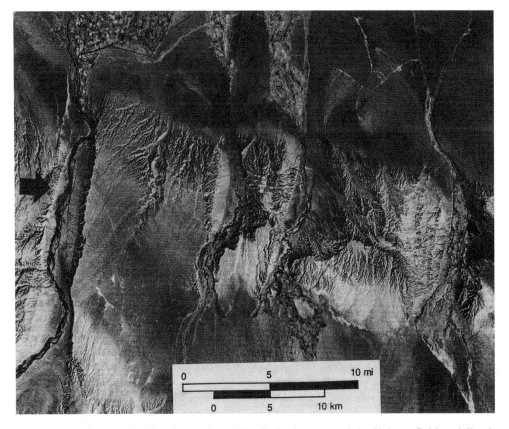

Figure 5.32 Stream incision (arrow) and tilted alluvium reveal the Kekeya field anticline in the southwest Tarim basin, China. Landsat MSS band 7 processed by Amoco Production Company. North is at the top.

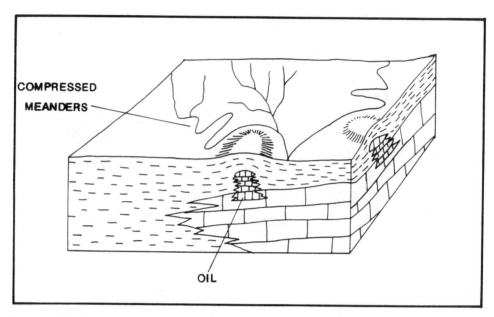

Figure 5.33 Compressed meanders form upstream from pinnacle reefs in the Salawati basin of Indonesia [8].

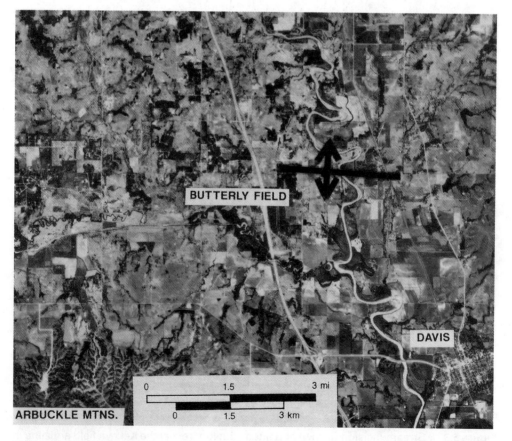

Figure 5.34 Compressed meanders of the Washita River occur upstream from the Butterly field, Oklahoma. U.S. Geol. Survey photo. North is at top of page.

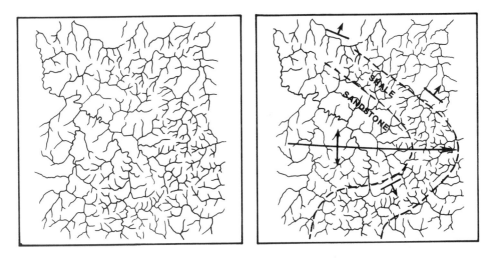

Figure 5.35 Folding is revealed by changes in drainage texture in the Ucayali basin, Peru. From Doeringsfeld and Ivey [56].

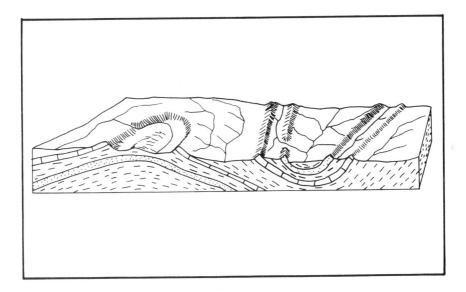

Figure 5.36 Radial and concentric drainage are associated with anticlinal and synclinal folds.

- Braided Stream Segments
  A stream becomes braided whenever the load exceeds the capacity to carry material. One way this can happen is for a local decrease in gradient, such as might be expected over a buried structure.
- Double Drainage Deflection
  The deflection of two streams in opposite directions is generally evidence of a subtle topographic high or buried structure [56]. The streams usually converge on the downstream side of the feature (Fig. 5.37).

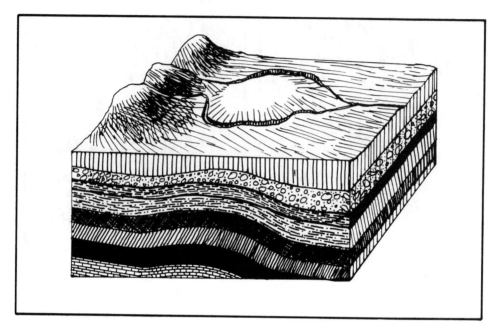

Figure 5.37 A double drainage deflection occurs when streams adjust their courses to account for intervening structure.

- Long Dip Slope Versus Short Backslope Drainage
  Curving patterns of long drainages opposed by short tributaries may indicate changes in strike related to folding [5] (Fig. 5.5).
- Parallel Drainage Deflection
  The deflection of two or more parallel streams in the same direction can be evidence for a plunging anticline (Fig. 5.38). The streams bow around the structure and are increasingly incised as erosion progresses [56].
- Levee Preservation
  In an actively subsiding delta plain, there may be areas where short sections of paired levees remain above water above a fold or dome. An example of such "flying levees" is given by DeBlieux [6] at Cutoff field, La Fourche Parish, Louisiana (Fig. 5.39).

Several examples of drainage indicating deeply buried structures are given by Melton [57], Penney [58], and Trollinger [59].

Topographic Profiles

Drawing topographic profiles across an area of interest may reveal structures not otherwise obvious (Fig. 5.40). The appearance of asymmetric slopes, suggesting cuestas, arranged on opposite sides of a presumed structure, may be an indication of bedding, especially in vegetation-covered areas where outcrops are sparse [56]. This technique has been used to locate production at the Kurten, Fort Trinidad, and Giddings fields, Texas [60,61].

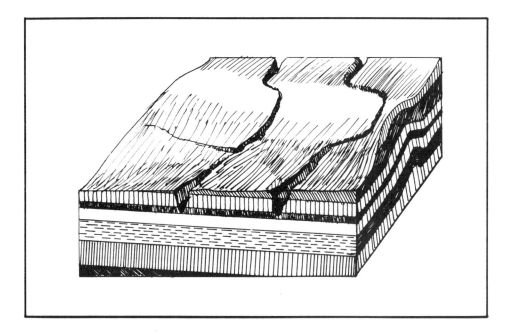

Figure 5.38 Parallel drainage deflections occur when streams shift their courses to accommodate the plunging nose of a fold.

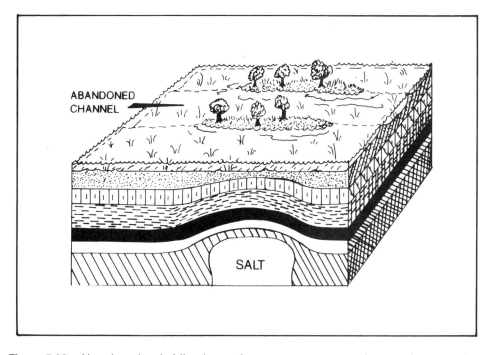

Figure 5.39 Abandoned, subsiding levees in a swamp area are submerged except where they form "flying levees" over a structure. From De Blieux [6].

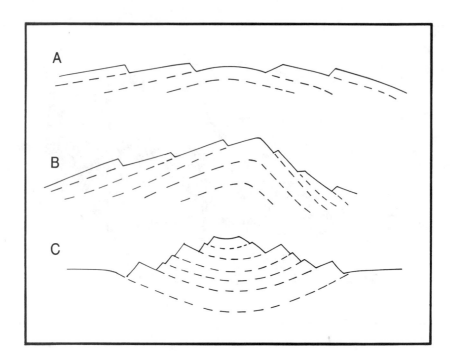

Figure 5.40 Topographic profiles over folds. A. Profile over a symmetric, breached anticline. B. Profile over an asymmetric anticline. C. Profile over a syncline with inverted topography, i.e., the synclinal axis is topographically high because the youngest units are resistant to erosion.

Trend Surface Analyses

In some cases structures as deep as 1830 m have been mapped at the surface through one or more intervening unconformities. For example, the Devonian Leduc reef in the Alberta basin has been mapped by using trend surface analysis on the pre-Cretaceous unconformity [62]. The analysis removes the effect of regional dip, allowing the smaller irregularities due to drape over the reef to become evident. This same type of analysis can be done on surface topography: the smoothing of irregularities may reveal broad, regional highs and lows.

Surface Displacement of Subsurface Structure

It must be recalled that the surface trace of a fold axis is above the subsurface high only when the fold is symmetrical. Many drillers have learned to their dismay that the structural high at depth can be offset a significant amount from the surface crest. The best evidence for an inclined axial plane is an asymmetric fold, where one limb is steeper than the other. The axial plane often (but not always) bisects the interlimb angle, and thus dips away from the steeper limb. In many cases an asymmetric fold is cored by reverse or thrust faults, which further complicate predicting the structural high at depth (Fig. 5.41 A).

When gently inclined strata are draped over a fold (or reef, etc.) at depth, the surface trace of the high will be offset updip, as demonstrated at the McLouth oil field, Kansas [63] (Fig. 5.41 B).

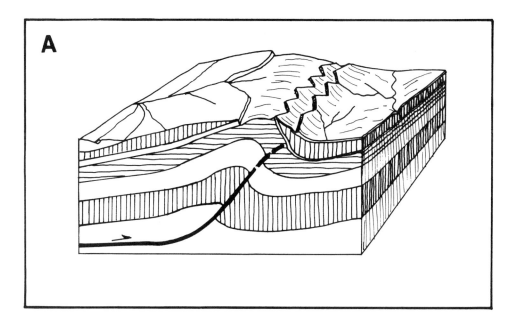

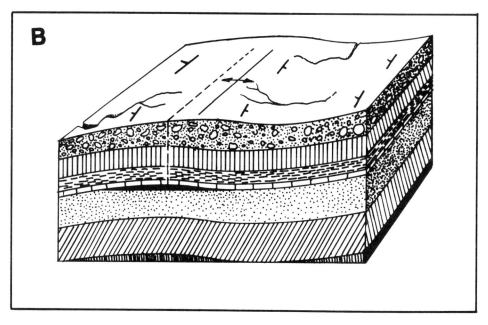

Figure 5.41 Asymmetric folds. A. Thrusting causes an inclined axial plane, so the surface trace of the fold axis is offset from the axis at depth. B. In gently dipping strata the surface trace is offset updip from the subsurface fold crest, as at the McLouth field, Kansas.

CIRCULAR FEATURES

Some interpreters are prone to highlight circles of unknown origin as important to an exploration program. Unless these features can be attributed to some process they are meaningless. Natural circular features of interest in petroleum exploration are usually related to salt diapirs or drape folds over pinnacle reefs (Figs. 5.15, 5.33). Gentle drape folds often appear to be no more than soil tone changes [11,14,15]. Other common circular geologic features, which may be of interest in petroleum or mineral exploration, include meteor impact craters, volcanic craters, circular karst sinkholes, diatremes, and granite exfoliation domes. Oil has been produced from meteor impact craters (Ames Hole, Oklahoma; Little Knife, North Dakota; Steen River, Canada), diatremes (south Texas), or volcanic necks (Dineh b'kiyah field, Arizona). Uranium has been produced from karst collapse breccias in northern Arizona. Diamonds have been produced from kimberlite diatremes (South Africa and elsewhere).

FRACTURES

Fractures are breaks in rocks, and include both joints and faults. Joints have no discernable movement parallel to the fracture surface. Joints can be open, closed, or filled with minerals (veins). Faults have measurable offset in a dip-slip, strike-slip, or combination mode. Faults can be further subdivided into normal, reverse, rotational, strike-slip (or wrench), and thrust displacement (Fig. 5.42).

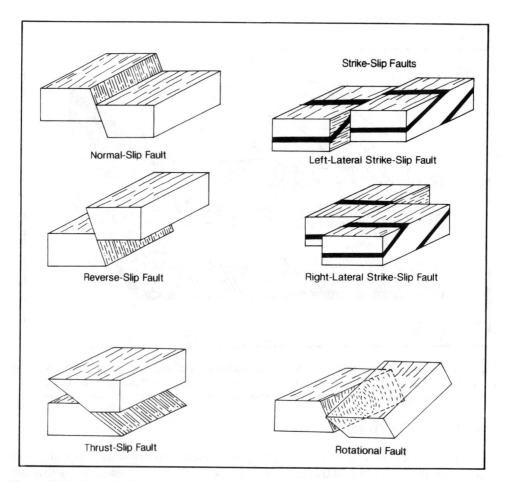

Figure 5.42 Types of faults.

"Lineaments," much discussed in the remote sensing literature, are here assumed to be evidence of fracturing, and may indicate zones of increased porosity, the boundaries of uplifted blocks, fault traps, drape folds, mineralized veins and shear zones, or petroleum reservoir boundaries. Dense fracturing may reveal the zone of maximum curvature on a fold (Fig. 5.22).

Propagation Mechanisms

Some discussion of the mechanisms involved in fracture propagation is required because the relationship between surface structures and those at depth is often controversial and not always evident. Mechanisms invoked for fracture propagation include reactivation, settling, minor readjustments associated with seismicity, earth tides, thermal contraction, uplift, or regional extension.

It has been demonstrated that fractures can extend upward from faults in the basement. The theoretical basis for the upward propagation of faults into flexures and fracture zones is found in the pioneering work of Sanford [64] who used displacement field and stress distribution diagrams to predict the surface expression of vertical displacement at depth. The Bright Angel fault in the Grand Canyon [65, 66] is a classic example of multiple periods of reactivation that have caused a fault to display varying displacements in successively younger units from Precambrian through Permian time. Jamison [67] and Heyman [68] have shown examples of faults extending upward into joints and folds along the Redlands/Monument fault system on the flank of the Uncompahgre Uplift, Colorado. Stein and Wickham used a viscosity-based finite element model to demonstrate that "failure begins near the basement deflection and propagates ... upward" from a vertical step in the basement [69, p. 225].

Hodgson [70] and Alpay [71] believed that a possible mechanism for fracture propagation without fault rejuvenation was high frequency, low amplitude vibrations related to seismic activity. They cited an analogous situation in cracked pavement, where old fractures reappeared in new surfacing after being subjected to traffic vibrations [72]. Free oscillations in the earth caused by large earthquakes have "been compared to the ringing of a bell" [73, p. 108]. Oscillations have been measured with periods from a few minutes to 57 minutes. On the other hand, microseisms are continuous vibrations with periods of 5–10 seconds and maximum amplitudes exceeding 10 microns. Barosh related the pattern of fractures observed in alluvium at Yucca Flats, Nevada, to the joint patterns in the adjacent bedrock and proposed the mechanism "is that of upward propagation of fractures from differential movement between joint-bounded blocks during [underground nuclear] explosions" [74, p. 216]. Brown and Hudson [75] show that it should be possible to cause fatigue failure by cycling almost any level load. The implication is that even loads as small as those caused by earth tides may cause jointing over time [76]. Another means of propagating fractures is by cooling a large region. Coefficients of thermal expansion for minerals and rocks were catalogued by Skinner [77]. The volume of quartz, for example, increases 3.8% as temperature increases from 20 to 570° C; the volume of calcite increases 1.8% from 20 to 600° C. Sandstones have a coefficient of thermal expansion of 0.00001 per degree C in the range 20 to 100° C, and granites have an expansion coefficient of 0.000008 per degree C in the same temperature range. This is equivalent to a change in length (contraction) of 20 meters (65 feet) for a granite layer, and 25 meters (81.3 feet) for a sandstone layer 10 km (6.1 mi) long uplifted and cooled from a depth of 5 km (16,250 feet) with a higher than normal thermal gradient of 5° C/100 m (as measured at Cisco, Utah, by Hallin, [78]). This strain would be accommodated by jointing. In sedimentary rock there is no preferred orientation to thermal contraction joints, but preexisting planar flaws may control fracture trends. Continental crustal extension joints (regional joint sets) due to thermal cooling were proposed by Turcotte and Oxburgh [79].

Regional extension may affect large areas of the earth's surface (e.g., the Basin-Range province). Strain takes the form of normal faults and overlying near-vertical joint patterns parallel to the maximum horizontal compressive stress. In areas with a strong preexisting structural fabric, joints may form above and parallel to or rotated slightly from the preexisting fabric.

Fractures can propagate downward from the surface due to local uplift or volumetric changes due to dewatering of sediments. Price [42] and Nur [80] proposed that joints could form in uplifted strata because of a volume increase (Fig. 5.43). Such joints tend to form parallel to the margins of the uplift, and decrease in frequency toward the center of the uplift. Nur [81] suggested that the depth of penetration was related to the length of the fracture and spacing between parallel fractures.

Arcuate fractures similar to glacial crevasses (and often containing clastic dikes) have been related to down slope creep on a decollement or to collapse of the overlying section as an evaporite layer is slowly dissolved. The surface geometry suggests a listric nature at depth. Large volumes of brittle rock are capable of moving down slope on ductile layers and forming fissures. In the Canyonlands of Utah the Colorado River has cut through 460 m (1500 feet) of strata into the Pennsylvanian Paradox evaporites [82, 83]. The rocks dip toward the river and move slowly down slope forming grabens that are arcuate in plan view (Figs 5.44, 5.45).

Fractures that originate as the result of overpressuring within a specific horizon are generally confined to that horizon and its overlying or underlying seals. Once a seal is breached by fracturing, the pressure returns to normal and there is no need for fractures to continue to propagate. One would not expect, then, and one generally does not see, fracturing at the surface associated with overpressured, fractured reservoirs such as are found, for example, in the Uinta Basin, Utah.

Joints

Joints at the surface can be valuable indicators of folding, as seen in the previous section (Fig. 5.22), and of faults or joint zones at depth. In particular, fractured reservoirs may be revealed at the surface as zones of intense jointing if the fracturing is tectonic (as opposed to overpressured) in origin. For all of these reasons it may be important to identify and locate joints at the surface.

Joints can be recognized by linear ridges or valleys, tonal and vegetation alignments, and orthogonal or parallel streams or lakes (Fig. 5.46). In some basins fracture systems have been injected by dikes, and the dikes clearly reveal the joint trends (for example, in the northeast San Juan basin, New Mexico (Fig. 5.47). Where carbonates lie at the surface, there are often alignments of sinkholes along fracture zones (Fig. 4.9). In Saudi Arabia and Qatar, the Shamal winds have etched joints to the extent that outcrops are serrated and some have been reduced to fins of rock oriented parallel to the wind (Fig. 5.48).

Joints in intrusive rocks tend to form orthogonal patterns, probably reflecting the relatively homogenous nature of the rock. Often the margins of the intrusion will be marked by arcuate and concentric cooling joints. Metamorphic basement is often

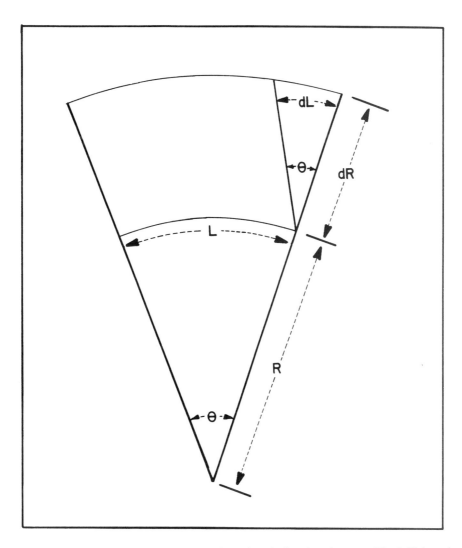

Figure 5.43 The volume increases in a slice of rock that has been uplifted. This volume increase may be accommodated by distributed jointing or normal faults parallel to the edge of the uplift. From Price [42].

characterized by a strong parallelism reflecting foliation and its influence on jointing. A similar strong parallelism of surface lineations exists on fluted till plains in areas previously covered by continental glaciation. These scour features were developed by ice movement, and are unrelated to structure.

Carbonates sometimes have a characteristic pattern of rhombic jointing, as if the crystal structure were expressed on a regional scale. A convincing explanation has not been presented for this observation.

Joints in dipping sedimentary units are most easily recognizable on imagery when the jointing is oriented perpendicular to strike. Joints that parallel strike generally have the same appearance as strike valleys. This can introduce a bias when mapping fracture densities.

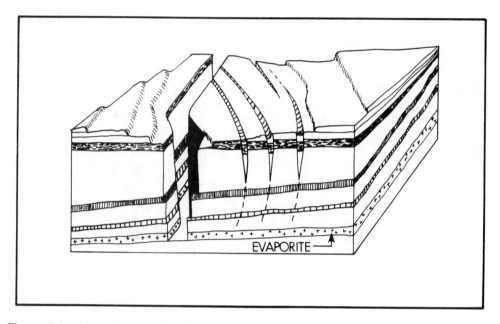

Figure 5.44 Arcuate extension fractures probably go listric into a decollement in the Pennsylvanian salts, Canyonlands, Utah. The section is laterally unconstrained due to erosion by the Colorado River.

Subsurface fracture shape and distribution may be predicted with a knowledge of the stratigraphy of the section. Joints in massive or crossbedded sandstones tend to be linear, wavy, or curved, whereas joints in siltstones and shales are more irregular (Figs. 5.49, 5.50). Fractured reservoirs in essentially undeformed rocks will be best developed in brittle, well-cemented and fine-grained rock, i.e., in units with the greatest tensile strength [84]. Thin brittle units (sandstones, marls) interbedded with weaker units (shales, anhydrite) will generally be highly fractured.

It has been observed that primary joint sets are often perpendicular in adjacent sandstones and shales. This may be a result of a reorientation of stresses in ductile units (shales) following failure in brittle strata (sandstones), and may allow prediction from the surface to the subsurface if the stratigraphy is known.

There is some controversy regarding joint orientations near the edges of cliffs. Are these the result of an unconstrained surface and the consequent expansion perpendicular to that surface, or are they joints that already exist in the area and are obvious only because the rock is spalling along the cliff face? As this issue is unresolved at this time, it is probably best not to assume that the orientation of cliffs is related to regional or local joint trends unless one sees a marked parallelism to joints in adjacent areas of continuous bedrock.

Locating subsurface fractures using surface fractures is inherently risky. Not only can the joint surfaces be inclined, but field measurements have shown that near surface joints rarely exceed 40 m of vertical extent in sedimentary rock [84].

Contour maps of fracture intensity are one way to define fracture zones. Many fracture density maps ignore the bias caused by mapping in areas with cover (low fracture density) versus outcrops. Meaningful mapping of fracture density requires

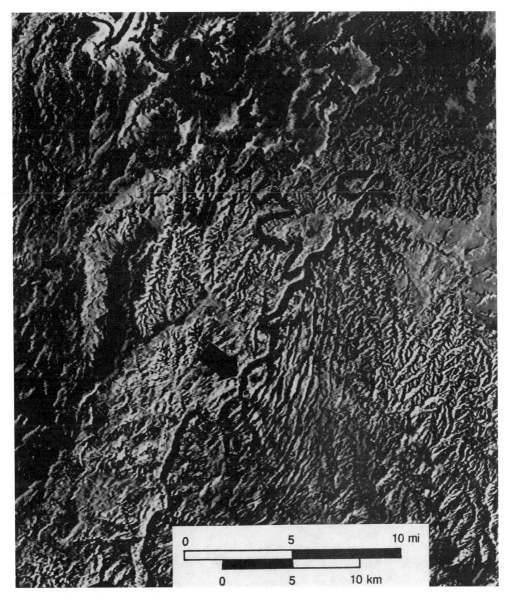

Figure 5.45 Landsat MSS of the Grabens area, Canyonlands, Utah. The Colorado River (arrow) has carved a space that allows the section to slide downdip on arcuate, listric normal faults. Processed by EROS Data Center. North is at the top. (See color plate.)

continuous, homogeneous surface units to avoid the necessity of normalizing units by applying lithologic factors. Even then, drilling in a high density fracture zone only increases the chance of encountering fractures. In producing areas, the large variation in IP (initial production) within such zones implies that hitting an open joint is not guaranteed. The critical factor in a successful well is tapping into a fracture plumbing system. Horizontal drilling perpendicular to the primary open fracture trend should improve the chances of a successful water or oil well under these conditions.

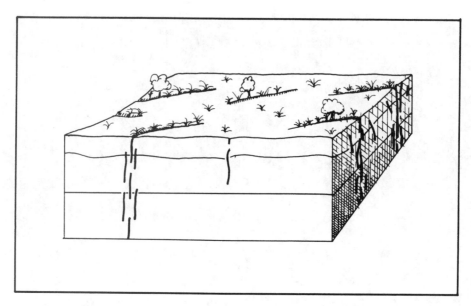

Figure 5.46 Jointing is often recognized by vegetation alignments, because plants obtain moisture that collects in the fractured bedrock.

Figure 5.47 Basaltic dike (arrow) intrudes the Tertiary in the northeastern San Juan basin, New Mexico. Intrusion was contemporaneous with jointing. Processed by Earth Satellite Corp. (See color plate.)

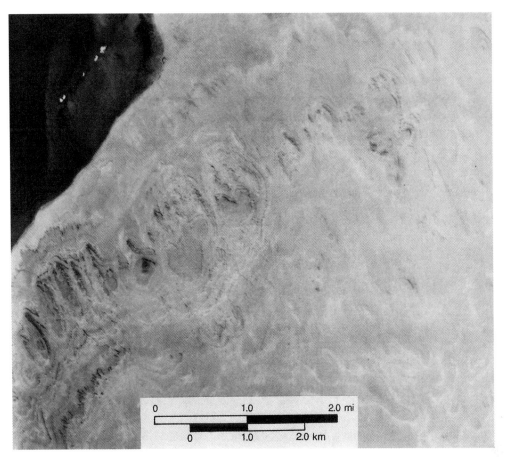

Figure 5.48 Joints in the Miocene Dam Formation near Jebel Al Udayd, Saudi Arabia, are etched by eolian abrasion. Photo provided by Hunting Surveys. North is at the top.

Surface mapping provides information necessary to predict fracture trend and location at depth. Fracture orientation data is most reliable: (1) in basins without complex deformation or several tectonic episodes, (2) where surface units are about the same age as reservoir units, and (3) where there are few or no unconformities. Although surface fracture sets are often the same as those at depth, the dominant set at a given horizon cannot be predicted with confidence.

Faults

Faults can be recognized by topographic or lithologic offsets, offset streams, streams flowing around acute angles, or streams flowing against a regional gradient. Within a flood plain, meander loops that reverse stream flow 180° from the regional gradient indicate that the valley floor has been locally tilted upstream by a fault crossing the valley (Fig. 5.51). A fault trace may be as subtle as an alignment of springs, or as obvious as the juxtaposition of two dramatically different rock types. In glaciated terrain they may simply be a scoured linear depression. Faults are classified by the kind of offset, or slip, which is the actual displacement of a point on the fault surface.

Figure 5.49 Jointing in a 200 m high cliff of the Triassic Wingate Formation sandstone, Ute Canyon, Uncompahgre uplift, Colorado. The longest vertically continuous joints are approximately 42 m long.

Unless the offset is measured on a key bed, the interpreted displacement is apparent and is based on the topographic expression of the fault, which can be misleading (Fig. 5.52).

Movement on inclined fault planes is described in terms of the hanging wall (above the fault) and footwall (beneath the fault). Normal faults are those in which the hanging wall moves down the fault plane. A listric normal fault, or growth fault, is one where the fault plane flattens at depth, generally dying out in bedding. In a reverse fault the hanging wall moves up with respect to the footwall. Thrust faults are reverse faults inclined less than 45°, and are often nearly flat (Fig. 5.42).

Strike-slip (or "wrench") faults tend to be near vertical at depth, and blocks on either side of the fault move horizontally in opposite directions more-or-less parallel to the fault. They are considered right-lateral or left-lateral strike-slip depending on the direction moved by the block across the fault from an observer (Fig. 5.53). Transpressional strike-slip faults separate blocks that are converging; transtensional wrench faults separate blocks that are moving apart. A special category of strike-slip fault, called a tear fault, is confined to the hanging wall of a thrust fault, and allows deformation to vary in intensity across the fault (Fig. 5.54).

Most faults have a combination of dip-slip and strike-slip displacement, and are classified on the basis of the dominant component. Some are rotational, changing from normal to reverse displacement along strike. Others, like some strike-slip faults, split and curve upward and outward from a single fault plane at depth, creating "flower" or "palm tree" geometries that look much like thrusts in cross-section (Fig. 5.55). Each

Figure 5.50 Irregular joints in the Triassic Chinle Formation shales, Unaweep Canyon, Uncompahgre uplift, Colorado.

Figure 5.51 Streams flowing against the regional gradient suggest fault control. A meander loop that reverses the direction of flow 180 degrees (a) suggests that a fault has tilted the alluvium slightly. A barbed tributary (b) also shows fault control.

category will be examined in turn for those characteristics which make the faulting recognizable at the surface.

Normal, Reverse, and Listric Faults

Normal and reverse faults are often near-vertical at the surface, and thus are characterized by a linear fault trace and are nearly or completely indistinguishable from each other. Many of these faults are marked by topographic offset, or an escarpment, with the upthrown side ranging from an abrupt mountain front to a structural terrace with only a few meters relief (Fig. 5.56). For example, the Wasatch Front in Utah has several thousands of meters of dip-slip, whereas many growth faults along the Gulf coastal plain in Texas are only noticeable as slight dips in the surface (Fig. 5.57 A). Uplifted basement blocks are often characterized by triangular facets along the scarp, whereas drag along the fault can cause sedimentary layers to form pronounced flatirons (Fig. 5.58). Faults frequently offset landforms such as coastlines or the crest of a mountain range. Two different lithologies are commonly juxtaposed by the fault, such as crystalline basement and sandstones, carbonates and volcanics, or outcrop and alluvium. Faults are more difficult to recognize if the same unit occurs on both sides. If displacement is significant there can be an abrupt change in strike, or dip,

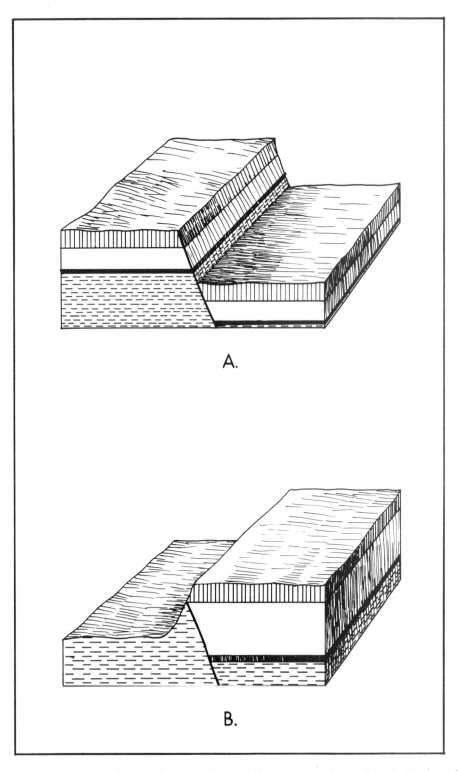

Figure 5.52 Topographic relief along a fault. A. Fault scarps give a direct indication of dip slip. B. Fault line scarps often give a misleading indication of dip slip. As the block shown in (A) erodes, the resistant layer on the downthrown side becomes topographically high (B).

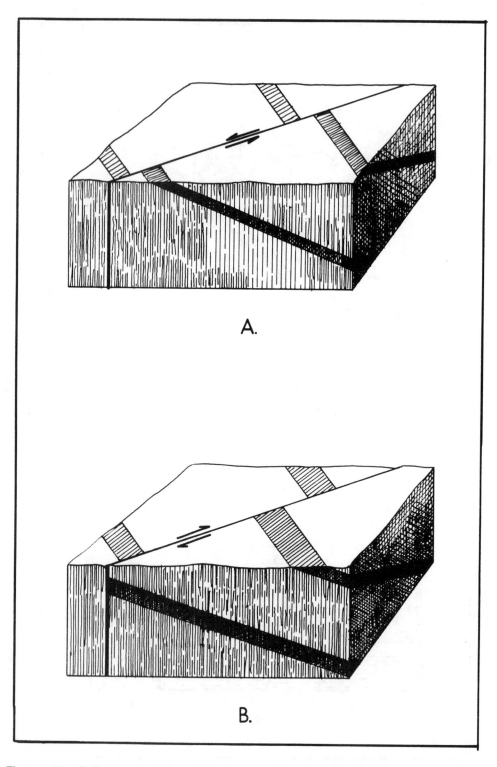

Figure 5.53 Strike-slip, or "wrench" faults. A. Left-lateral offset. B. Right-lateral offset.

RECOGNIZING STRUCTURE 141

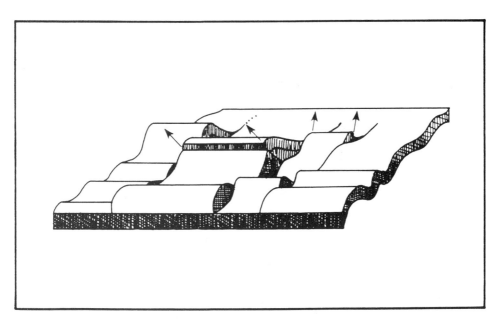

Figure 5.54 Tear faults in thrust sheets allow differential movement across the fault, often resulting in changes in fold vergence. From Davis [85].

Figure 5.55 "Flower" structures are caused by convergent strike-slip faulting, or transpression. From Lowell [86].

or both across the fault trace (Fig. 5.57 B). High angle normal faults are often associated with tilted blocks or horst and graben topography. Intrusion of basaltic dikes along the trace of a fault suggest that these faults penetrate the lower crust. Mineralized faults

Figure 5.56 A normal fault scarp along the east side of the Teton Range, Wyoming, raises the basement several thousand meters.

may appear either as resistant ridges and knobs or linear depressions, depending on the material in the fault zone. Silicified fractures should resist erosion; the presence of iron oxides will make them linear zones of reddish hues. Faults filled with clay gouge will probably weather rapidly and may form sags filled with poorly drained soils and perhaps water, or may show a change in vegetation from surrounding areas.

Where there are no outcrops, or they are covered by vegetation, near-vertical faults may appear as vegetation or soil tone alignments (Fig. 5.59). The story is told of an old water well driller who would climb the hills of North Carolina in the spring to look for the green streaks of early grass in the fields. He knew from experience that these were the areas with the best chance of finding water (Fritz Johnson, oral comm., 1985). We can suspect that these are areas where faults brought groundwater near the surface. Linear stream segments or linear valleys extending for several kilometers suggest faulting, as the broken rock is more readily weathered and eroded into valleys [55]. Tributaries that are aligned on both sides of a major stream, or across drainage divides strongly suggest structural control, as do barbed drainages and offset streams (Fig. 5.60). Springs often indicate groundwater ponding on one side of a fault, and aligned seeps indicate a fault zone acting as a groundwater conduit. On thermal images these faults appear cooler than adjacent rock and soil due to evaporative cooling both during the day and night [87].

Normal faults provide updip fault closure petroleum traps as well as traps at the high corner of tilted blocks. Accommodation or transfer zones are formed by the overlap of normal faults in rifts (Fig. 5.61). These zones serve to channel sedimentation,

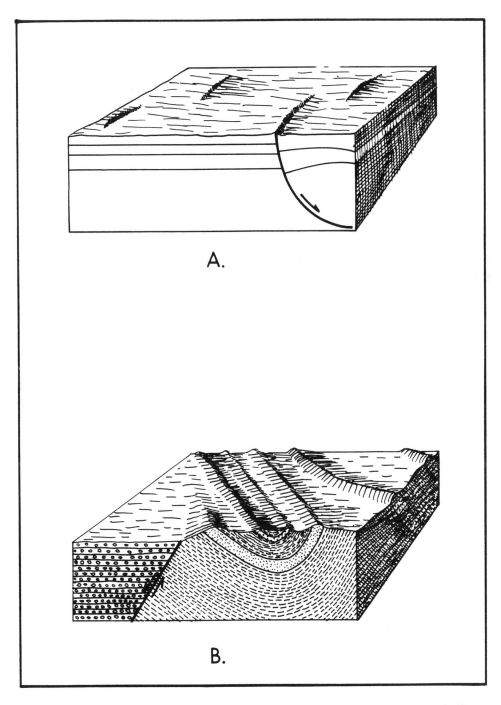

Figure 5.57 Fault scarps. A. Growth faults such as those found along the Gulf coast generally have relief from one to three meters. B. Many high-angle normal faults have abrupt changes in structure and topography across the fault.

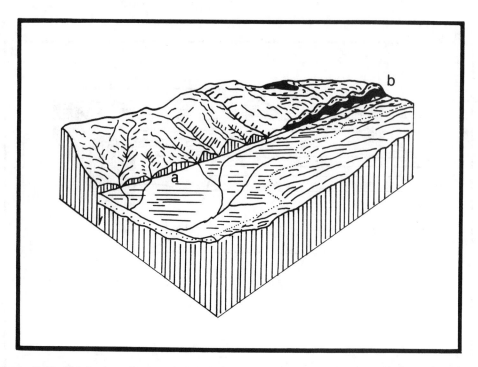

Figure 5.58 Fault scarp forms triangular facets (a) where ridges are cut off, and flatirons (b) where bedding dips steeply into the fault. From Cotton [2].

juxtaposing structural highs and depocenters. Provided they have updip seals, these zones can be the location of prolific oil production (e.g., Morgan field, Gulf of Suez). Faults serve as conduits for hydrothermal and epithermal mineralizing solutions and as the locus of mineral and metal deposition, forming vein deposits along the fault, or manto deposits where the solutions impregnate a carbonate or reactive clastic unit. Faults also commonly help localize hot springs and hydrothermal sources.

Listric faults flatten with depth and die out in bedding. These faults tend to be arcuate, or scoop-shaped in plan view. For purposes of interpretation, faults that are concave toward the downthrown side are more likely to be listric, whereas straight faults are more likely to be planar and high-angle. This relationship of curving trace to shallow dip is based on slope stability work by engineering geologists [89]. The ideal shape of gravity-slide blocks (slumps, or landslides) tends to be concave-up in cross section, and concave toward the downthrown side in plan view (Fig. 5.62). Moore [90, p. 411] proposed that, in the Basin and Range of Nevada "curvature of the fault block ranges reflects the curvature of the main bounding fault in plan; hence, the fault itself is believed to be convex toward the direction of the tilt of the range, or concave toward the downthrown side of the fault."

Rotation of bedding and subsequent continued listric faulting leads to a sequence of steep to shallow dips as one proceeds into the basin (Fig. 5.63).

One type of listric normal fault, the growth fault, is often characterized by dip reversal into the fault, called rollover (Fig. 5.64). This is, in effect, an anticline formed by collapse of the hanging wall strata as it pulls away from the footwall. These folds are economically important in areas of rapid sediment accumulation such as the Gulf of Mexico coastal plain and river deltas.

RECOGNIZING STRUCTURE 145

Figure 5.59 Vegetation (red) is aligned along springs, which in turn are controlled by faulting along the margins of Spring Valley, Nevada. Also note the paleo-shorelines at the north end of the valley. This Landsat MSS color infrared image was processed by Earth Satellite Corp. (See color plate.)

Reverse faults bound many uplifts, such as those in the Colorado Plateau-Southern Rocky Mountains provinces. These uplifts are often characterized by overhangs and zones of fractures parallel to their margins. The uplifted basement and bounding fault

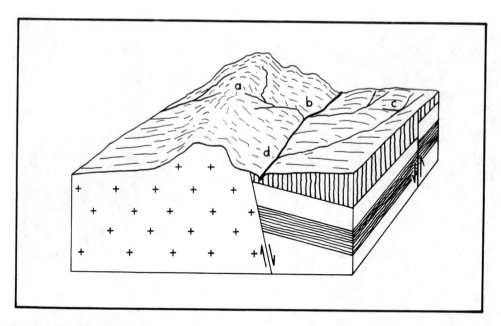

Figure 5.60 Drainage indicators of faulting. Streams aligned across drainage divides (a) or on both sides of a major drainage (b) suggest fault control. Barbed drainage (c) and long, linear valleys (d) also suggest faulting.

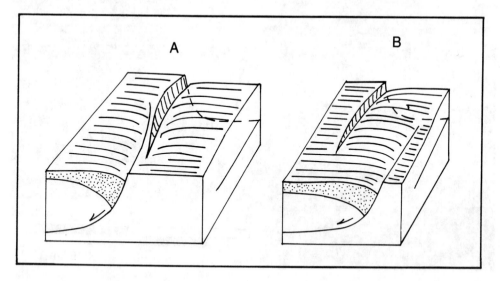

Figure 5.61 Antithetic en echelon transfer zones. A. Back to back. B. Overlapping. Note that the downthrown side is not only a depocenter, but the rollover associated with the faulting causes a structural high that may be a site for the accumulation of hydrocarbons. From Morley, et al. [88].

may form an updip seal for migrating hydrocarbons. Fracture zones can serve as conduits for mineralization, such as at the Schwartzwalder uranium mine in the Rampart Range, Colorado. Overhang plays have been studied and observed on

RECOGNIZING STRUCTURE   147

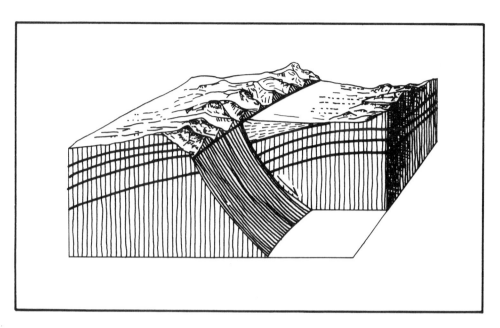

Figure 5.62 Gravity slide and listric normal faults are concave toward the downthrown side in both map view and cross section. Note triangular facets along the fault. From Moore [90].

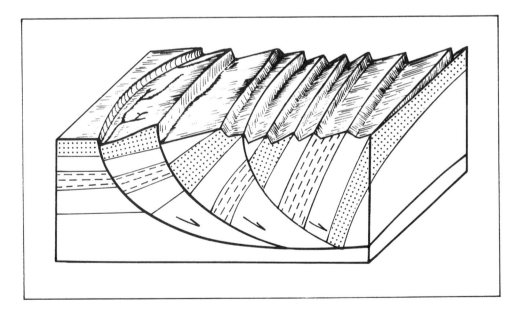

Figure 5.63 Block diagram illustrating the effect of successive listric faulting. Note how the oldest fault block has been rotated so that bedding is almost vertical.

satellite imagery in many areas including the Rampart Range and Uinta and Wind River mountains, Wyoming. Faulting and jointing parallel to the mountain front occurs in each area. It should be noted that the Uintas are composed primarily of bedded

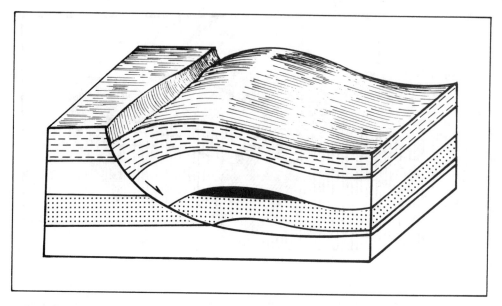

Figure 5.64 Oil can accumulate in an anticline generated by listric growth faulting.

Precambrian sediments, and thus may be fundamentally different in their mechanical response to reverse fault/thrusting than the granite-cored Wind River and Rampart Ranges. Yet the fracture pattern parallel to the mountain-bounding faults as mapped in the Uintas by Ritzma [91] is essentially the same as that mapped in the granites of the Wind River Range by Love, et al. [92–94], and by Bryant, et al. [95] in the Rampart Range. This coincidence goes beyond the tendency for granites to form orthogonal fracture patterns. The patterns mapped on Landsat and airphotos are all more or less parallel to the mountain front (Fig. 5.65).

Faults parallel to the mountain front, diagnostic of overhangs, are probably extensional features that developed following compression (?) and uplift (Figs. 5.43, 5.66) [42]. These normal faults are most likely to penetrate the overhang and extend into the sediments and reservoirs below. These fractures should be targets of geochemical surveys.

Faults that are convex toward the downthrown side tend to be thrusts, and will be discussed in a later section.

Whenever a normal fault cuts inclined strata there will be an *apparent* strike-slip offset, with a key bed on the downthrown side being displaced so that the trace is opposite older strata (Fig. 5.67). It is not always possible to distinguish between such a normal fault and strike-slip faults. Strike-slip faults, however, often cause bedding to turn into the fault as a result of drag. They can also be characterized by numerous features discussed in the next section.

Strike-Slip Faults

Strike-slip, or wrench faults are best recognized by the consistent lateral offset of lithologic units, facies, streams, ridges, alluvial fans, or any feature that crossed the fault (ore bodies, roads, fences, homes). They tend to be near-vertical at the surface,

RECOGNIZING STRUCTURE 149

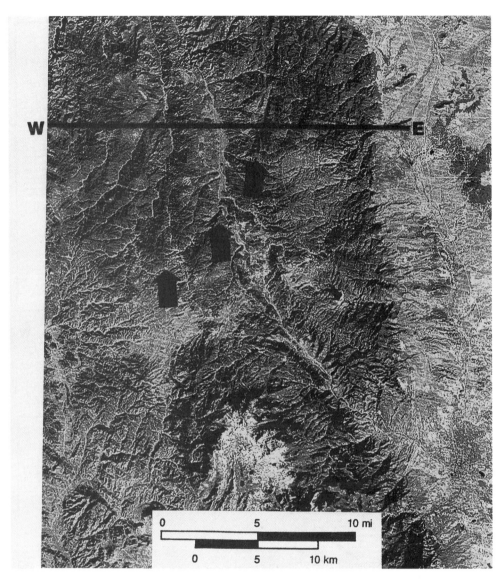

Figure 5.65 Landsat MSS color infrared image of the Rampart Range, Colorado, from Pikes Peak to Castle Rock. Arrows indicate faults parallel to the mountain front. North is at the top. Processed by Earth Satellite Corp. (See color plate.)

and thus have long linear segments (Fig. 5.68). Drag along the fault often causes these features to bend or rotate into the fault. Faceted spurs are ridges that have been abruptly cut off by the fault. When these ridges block an old stream valley this is called a headless valley. There are also a host of accessory features that are developed with these faults and help to identify them. These include en echelon folds oriented 30–45° to the fault and perpendicular to the direction of maximum horizontal compression (Fig. 5.69). These folds begin to develop prior to faulting and are most intense (highest amplitude) at the site of faulting.

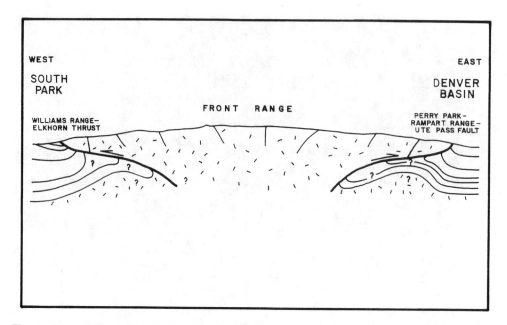

Figure 5.66 Schematic cross section of the Rampart Range, Colorado, along line W-E shown on Figure 5.65. Note the extension faults parallel to the margins of the uplift. From Jacob [96].

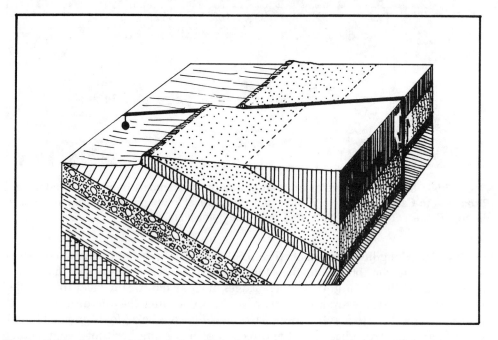

Figure 5.67 A normal fault can offset the contract of inclined units and give the appearance of strike-slip faulting. One must look for other indicators of displacement, such as consistent offsets, en echelon folds, etc.

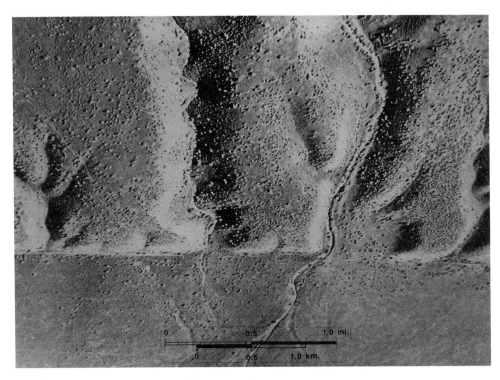

Figure 5.68 Scarp of the San Andreas fault near Frazier Park, 100 km north of Pasadena, California. This is a right-lateral strike-slip fault. Note the linear scarp. North is up. From Anderson [97].

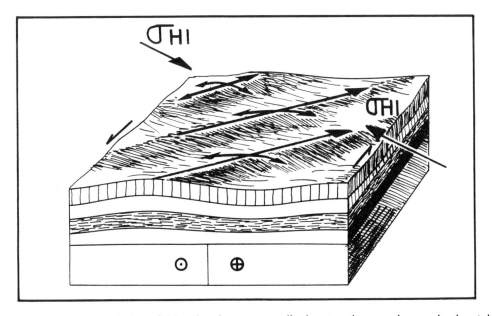

Figure 5.69 En echelon folds develop perpendicular to the maximum horizontal compressive stress, and at 30 to 45° to the master wrench fault.

Figure 5.70 Sag ponds lie along the Kaltag left-lateral strike-slip fault near Unalakleet, Alaska. View east. Photo by M. McGuire.

As the fault develops the original folds are cut and offset, and new en echelon folds form along the flanks of the fault zone [98]. The Newport-Inglewood trend in the Los Angeles basin has produced much oil from such en echelon folds [99]. En echelon normal faults (e.g., Lake Basin and Cat Cr. fault zones, Montana) can also be associated with wrench faulting. Irregularities in the fault trace cause other characteristic features [100]. Where the fault bends one will find either pull-apart zones (oblique rifting, sag ponds; Fig. 5.70) or areas of compression (buckle folds, flower structures; Fig. 5.71). Examples of releasing bends include the Salton Trough, California, and the Dead Sea rift, Israel-Jordan. Examples of restraining bends include the Mecca Hills, California, and Ocatillo badlands, California [101, 102].

Much has been made of using fracture patterns to determine both the fault trend and stress field (and thus offset) around strike-slip faults. Except in the case of en echelon folds or fractures, there are a plethora of wrench-associated fractures. Those parallel to the master fault are "Y" shears; "R" (Reidell) shears are en echelon fractures oriented 15° to the master fault; "R-prime" shears intersect the master wrench at 75°; the "P" shear is symmetric to the R shear and intersects the master fault at 15° (Fig. 5.72). One should take care in the interpretation of a fault pattern as indicative of strike-slip offset: if each set of shears is given a range of ±15° most fractures seen on imagery can be used to justify almost any proposed wrench fault.

Another common approach is to categorize fractures into two primary sets, consider them to be conjugate shears, and interpret the principal horizontal compressive stress as bisecting the acute angle between the shears. The problem with this technique is that most areas have at least two principal joint sets, but they are not necessarily caused by strike-slip faulting.

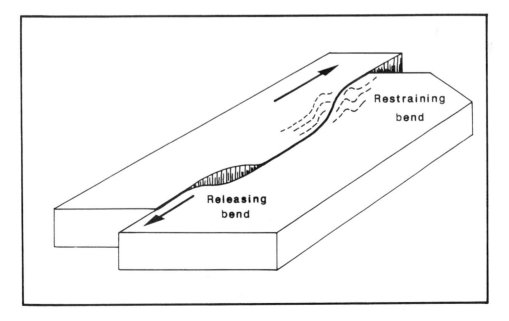

Figure 5.71 Releasing bends tend to form basins; restraining bends cause uplifts along strike-slip faults. From Crowell [100].

154 REMOTE SENSING FOR GEOLOGISTS

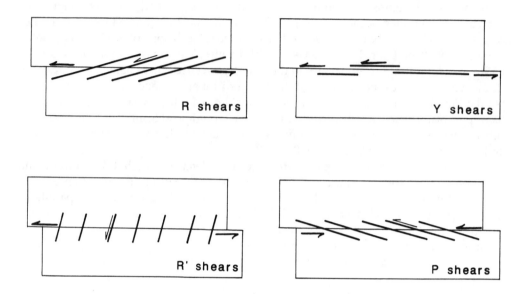

Figure 5.72 Subsidiary faults that have been described associated with strike-slip faulting include R, Y, R¹, and P shears. These are usually not all present in any given area along the master fault.

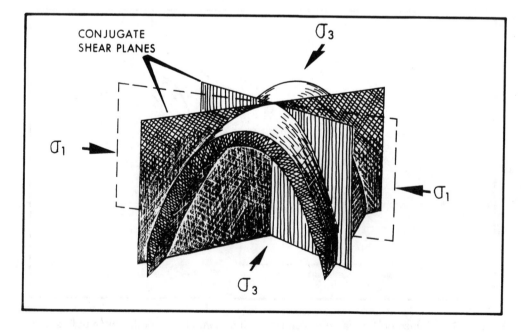

Figure 5.73 Conjugate shears form at 30° on either side of the maximum horizontal compression (sigma 1).

Under ideal conditions of pure shear (no rotation), conjugate fractures will form with an acute angle of 60°, bisected by the direction of maximum compression (Fig. 5.73). Under these circumstances each of the conjugate shear fractures accommodates strike-slip displacement with an opposing sense of offset (one right-lateral, one left-lateral). In most cases wrench deformation involves simple shear and rotation of material within a shear zone.

Shear faults die along strike by transferring their displacement. This can be done through a series of "horsetail" faults, which disperse displacement among many segments. These en echelon horsetail features are sites of local extension and often contain veins and mineral deposits. Whereas a major fault frequently contains gouge, the small auxiliary faults adjacent to it may contain relatively clean breccias with good permeability and serve as excellent sites for mineral deposition. Likewise, clean fault breccias assist in vertical migration of hydrocarbons. Shear faults can also transfer displacement by folding strata (shortening) on one side and extending (down-dropping) strata on the other side of the termination.

Thrust Faults

Most faults are recognized by abrupt changes in strike, dip, or rock type, often associated with an escarpment. The trace of a high-angle fault tends to be linear, hence the term "lineament." The trace of a thrust is generally irregular, following topographic contours, and making identification of these faults difficult. There are, however, both direct and indirect criteria for recognizing thrusts on imagery.

Direct indications of thrusting include abrupt changes in strike or dip, representing different structures in the hanging wall and footwall (Figs. 5.74, 5.75). The hanging wall is often topographically higher than the footwall, and the fault trace runs along a

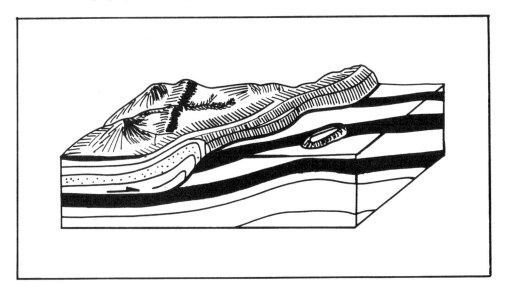

Figure 5.74 Leading-edge anticlines along thrust fronts tend to have irregular to arcuate fault traces, and may have erosional outliers (klippen). They can often be recognized because of the abrupt change in strike and dip of units on either side of the fault. Beds in the hanging wall generally strike parallel to the thrust trace and dip parallel to the fault. From Prost [103].

156 REMOTE SENSING FOR GEOLOGISTS

Figure 5.75 Landsat MSS image of part of the Tsaidam basin, China, illustrating multiple northeast-directed thrust sheets with leading-edge anticlines. North is at the top of the image. Image processed by EROS Data Center.

break in slope at the base of the upper plate. Where there is no topographic relief, the hanging wall can be recognized because the strike of bedding in fault bend folds and fault propagation folds is usually parallel to the thrust front and dips are more-or-less parallel to the fault surface. In many cases the fault trace in plan view is convex in the direction of transport (Fig. 5.76). Part of the irregular fault trace may be a result of erosional outliers, or klippen, in front of the main sheet. Leading edge anticlines characterize thrusts that terminate by ramping upsection; otherwise the hanging wall units generally dip in the same direction as the thrust fault. Imbricate fans, where thrust splays cut the hanging wall into stacked slivers, appear as zones of parallel ridges, and are recognized as thrusts by observing repeated section. Imbricate fans often develop over footwall faults or thrust ramps. Folds carried on a thrust tend to have a common asymmetry, verging toward the foreland. These often form imbricated, or stacked anticlines or synclines, depending on the level of erosion, with each fold separated from the others by a thrust (Fig. 5.77). Detachment folds, formed by flexural slip, tend to be concentric and have steep to overturned limbs, no preferred vergence, and no obvious thrusting at the surface (Figs. 5.78, 5.79, 5.80). The axial length of these folds is generally several times the wavelength, and these folds are frequently arcuate or sinuous. Abrupt changes in structural style, as from tight to open folding, or imbricate ridges to folds, can be observed at the foreland edge of many thrust belts, such as the Brooks Range, Alaska, or the Sawtooth Range, Montana, and often occur across thrust faults (Fig. 5.81).

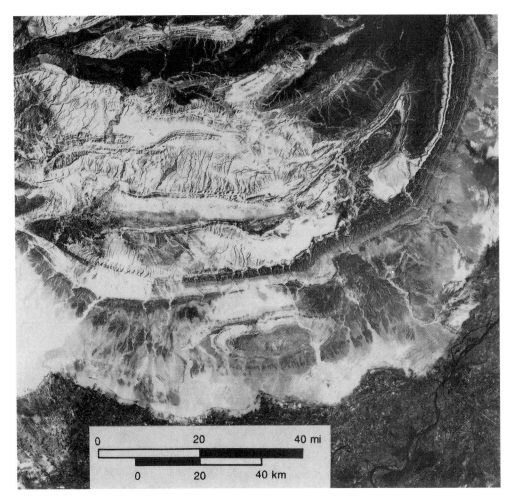

Figure 5.76 The southern Sulaiman Range of Pakistan is a south-directed series of thrusted folds. This illustrates how a thrust front is often convex in the direction of transport. Landsat MSS band 7 image processed by Amoco Production Company. North is at the top.

Indirect clues to thrusting include tear faults, lateral ramp anticlines or monoclines, and relaxation faults. Tear faults tend to be linear and have characteristics of strike-slip faults, but can also have normal offset, or have monoclines developed along trend. They generally end at the thrust front, either abruptly or by curving into the thrust fault (Figs. 5.82, 5.83). Folds on either side of a tear fault may have opposite vergence, or be at different stages of structural development. Monoclines form in the hanging wall where a tear fault lies above a faulted footwall. In other cases lateral ramp anticlines form over a footwall fault (Fig. 5.84). Relaxation faults develop on the backlimb of hanging wall folds or where the thrust ramps upsection. These listric normal faults are scoop-shaped, concave toward the hinterland, and sole in bedding or the thrust. Thrusts terminate along strike by transferring their displacement to folds or overlapping thrust faults. In some cases folds that merge into faults along strike may suggest thrusting [103].

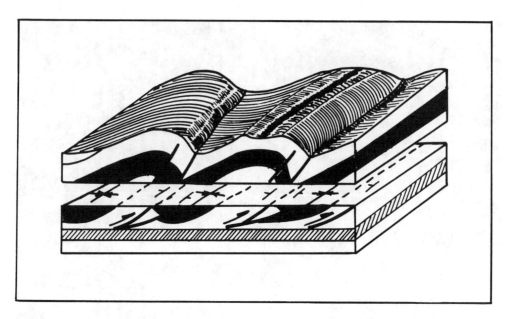

Figure 5.77 Stacked anticlines (or synclines, depending on the level of erosion) imply thrusting. From Prost [103].

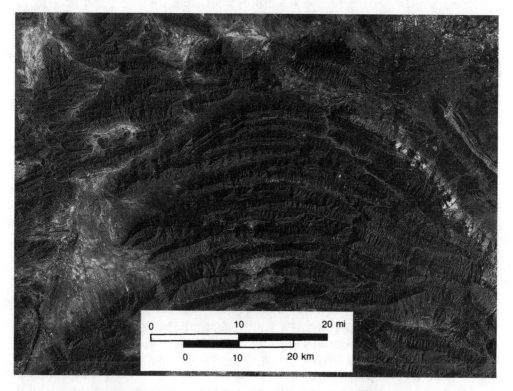

Figure 5.78 Landsat TM image of the Sierra Madre Oriental near Monterrey, Mexico. Folding developed in Cretaceous and Jurassic strata is believed to be detached over an upper Jurassic evaporite decollement. North is at the top of the image.

Figure 5.79 Landsat TM image of the northeast flank of the San Julian uplift near Caopas, Mexico. In this natural down-plunge projection one can see a cross section of Laramide-age detachment folds (arrows) that were rotated by later uplift. These folds clearly show a detachment at the upper Jurassic evaporite level. North is to the left of the image.

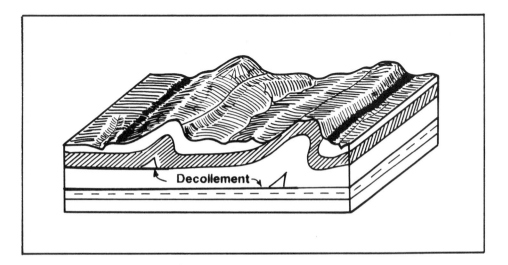

Figure 5.80 Folds with steep limbs often overturned on both sides, no preferred vergence, and no obvious faulting at the surface imply detachment thrusting. From Prost [103].

Figure 5.81 The transition from imbricate ridges to folds (at arrow) marks the frontal thrust of the Brooks Range at the Ivishak River, Alaska. Thrusting here is from south or southeast to north or northwest. Landsat MSS image processed by Earth Satellite Corp. North is at the top of the image. (See color plate.)

Imagery can be used to help determine the direction of thrust transport. This is important, among other things, because balanced sections are required to be in the direction of transport. Thrusted folds are generally perpendicular to the transport direction. Tear faults are usually parallel to the direction of thrusting. The thrust front is often convex toward the foreland. Folds tend to verge toward the foreland (steep forelimb; gentle back limb).

It is difficult to predict depth to the thrust or detachment based on image interpretation alone. A knowledge of stratigraphic thicknesses and the units at the

RECOGNIZING STRUCTURE 161

Figure 5.82 A major tear fault (arrow) cuts thrust sheets consisting of Paleozoic sediments in the Kalpin uplift, Tarim basin, China. Landsat MSS band 7 image processed by Amoco Production Company. North is at the top.

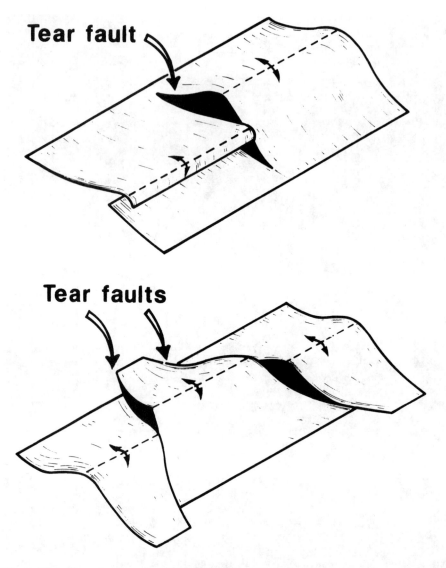

Figure 5.83 Tear faults separate thrust sheets with different amounts of displacement. They often curve into and offset the thrust front. Displacement can appear normal, strike-slip, or have components of both. These faults may be caused by subthrust structures or facies changes in the thrust sheet. Folds often have different vergence or amplitudes, or die out across these faults. From Prost [103].

surface can help. High amplitude (tight), high frequency folds suggest a detachment closer to the surface than broad, open folds [104].

If one can map a thrust along the forelimb of the fold, one may assume it is a leading-edge fold (sometimes called truncation anticlines or fault bend folds [105]. These folds suggest relatively shallow thrust faults. All factors being equal, the amplitude and frequency of thrust-carried folds decreases as the thrust plate thickens. Lithology and stratigraphic thickness also play a part: thick, competent units form

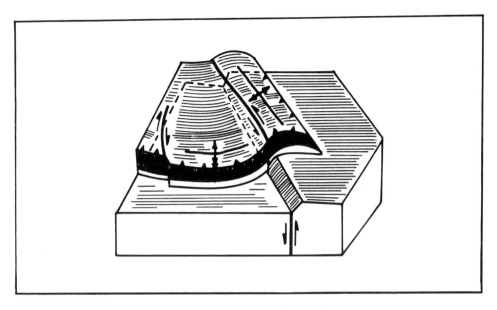

Figure 5.84 Lateral ramp anticlines can form over footwall faults by oblique convergence. From Prost [103].

broader folds than thin, ductile units. Fold length should have no relationship to depth to detachment.

Factors other than fold geometry can be used to estimate thrust plate thickness (Fig. 5.85). If the section contains sequences of shales or evaporites overlain by sandstones or carbonates, the tendency is to form large amplitude detached buckle folds above the decollement [105]. Carbonates are more apt to form imbricate thrusts [104]. Interbedded sandstones and siltstones tend toward fault propagation folding with multiple, leading edge thrust-cored folds [105]. "In shale sequences, faults tend to be blind and die out in complex, hanging-wall folds" [104, p. 339].

If there are blind thrusts (thrusts not exposed at the surface), the deformation above the thrust may be either more intense or less intense than that in the lower thrust sheet. In this case there are only indirect clues to lower sheet deformation. If subthrust folding has longer wavelengths than upper plate folding, the vergence of surface fold axes may change from foreland to hinterland depending on position with respect to underlying folds (Fig. 5.86). If subthrust folding is higher frequency than thrust-carried folds, surface highs may develop over ramps or subthrust folds.

Angular unconformities have some of the same characteristics as thrusts. They are characterized by abrupt changes in strike and/or dip. Folding above an unconformity will be seen below the unconformity, unlike many hanging wall folds. Bedding above an unconformity is, by definition, less deformed than below, whereas the upper plate of a thrust may be more deformed than the lower. Thrusts repeat the section; unconformities do not.

Reverse faults superimpose older rocks on younger, as do thrusts, (except in special circumstances, as might occur when thrusting is out of sequence). The high angle of the fault plane causes a linear fault trace, distinguishing this class of faults from thrusts.

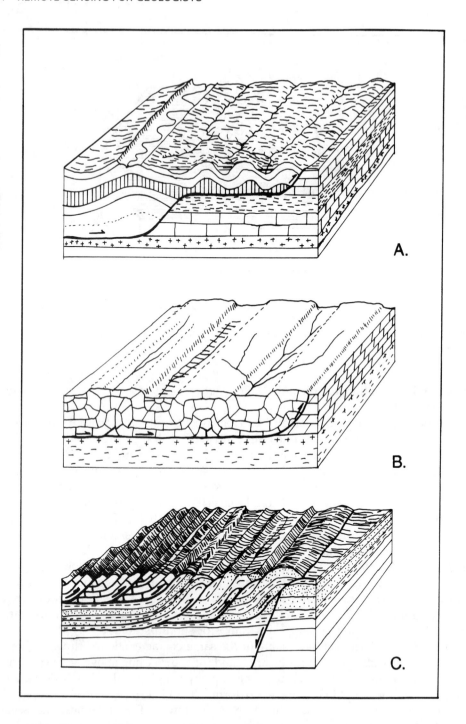

Figure 5.85 Clues to thickness and lithology of the thrust sheet. a. Tight folds imply a thinner thrust sheet than broad, open folds. B. Depth to the detachment is approximately equal to the thickness of a fold flank for box (detachment) folds. C. Imbricate thrusts are more common in carbonates, whereas fault propagation folds are more common in clastic sequences [104].

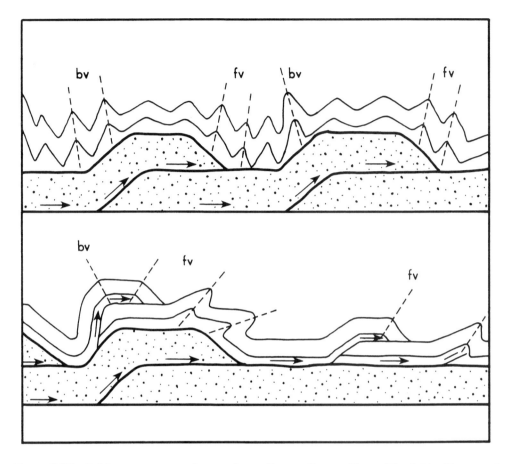

Figure 5.86 Fold vergence can change depending on the position of the fold with respect to sub-thrust structures. fv = forward verging; bv = backward verging. From Dunne and Ferrill [107].

Subthrust plays are receiving increasing attention, and are difficult to recognize. Folding that occurred post-thrusting will be expressed at the surface and will extend below the hanging wall, but the age of folding generally cannot be determined with certainty from remote sensing data alone. Subthrust highs can act as a buttress, causing deeper erosion into the sheet and along tear faults or lateral ramps at the margins of the uplift. The depth to the detachment can be estimated on the basis of the geometry of detached folds. A series of curves developed by Jamison [105] for detachment folds relate fold interlimb angle and backlimb dip to the ratio of fold amplitude (from elevation data) divided by thrust sheet thickness (Fig. 5.87). Ramsay and Huber [106] show how one can calculate the depth to a decollement for fault bend folds by measuring the uplifted area, original line length, and length after deformation (Fig. 5.88). Hanging wall thickness changes can be hung from surface elevations to reveal the shape of the thrust surface. Fold vergence of minor structures can also be used to speculate whether the thrust sheet is riding up onto a structure (hinterland verging) or moving over a subthrust structure (foreland verging; Fig. 5.86) [107].

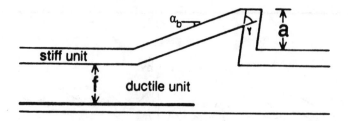

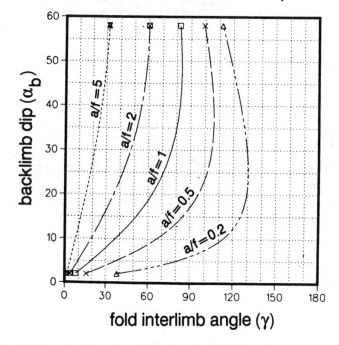

Figure 5.87 Graph for estimating the depth (f) to the detachment surface for detachments folds when the interlimb angle, backlimb dip, and fold amplitude are known. From Jamison [105].

In some cases (e.g., gold at Gold Acres, Nevada) mineralization is localized in breccias associated with thrusting. In general, however, thrust faults are thin and gouge-filled and do not serve as good conduits for mineralizing fluids or hydrocarbon migration. Extension faults in the upper plate, however, may be excellent conduits for the vertical migration of fluids.

INTERPRETING THE TIMING OF STRUCTURAL DEVELOPMENT

Structural timing is important in mineral exploration because mineralization may occur pre- or post-folding and faulting. Early mineralization may be linked to stratigraphic

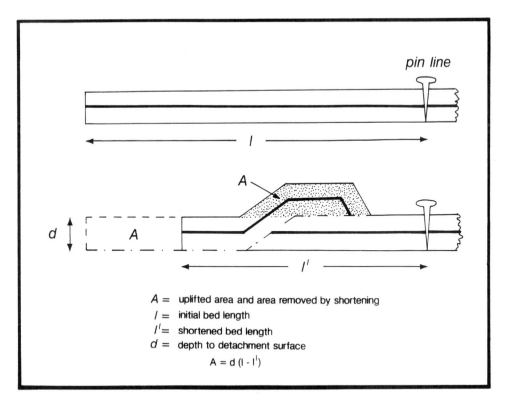

Figure 5.88 Estimates of depth to the detachment for fault bend folding can be obtained by the technique presented by Ramsey and Huber [106].

changes, whereas late mineralization may be related to fractures (veins) or folds that controlled the flow of hydrothermal fluids. Likewise, hydrocarbons may have been generated before folding and moved through the area, or after folding and accumulated in structural traps. Timing from image interpretation will be relative, e.g., a fold developed after thrust faulting, or contemporaneous with normal faulting. For absolute time of development it is necessary to obtain radiometric ages, fossil ages, or some other technique that involves sampling of the rocks themselves. The relative timing of structures can often be interpreted from map patterns seen on imagery (Fig. 5.89).

If a unit is dipping and one can map a thickness change across a fault, it is likely that the fault was contemporaneous with sedimentation and controlled sedimentation either by allowing growth on the downthrown side of a normal fault, or by condensing the section on the upthrown side of a thrust or reverse fault. Strike-slip faults can have depocenters on alternating sides of the main fault, depending on the location of restraining (thinner section) and releasing bends (thicker section). One should be aware that a change in dip can cause an apparent thickening (with decreasing dip) or thinning (with increasing dip) of a section of uniform thickness.

If one can identify an angular unconformity surface on imagery, the units parallel to the unconformity are younger than those that terminate against the surface at an angle. One should be aware, however, that some downlapping units may appear to form

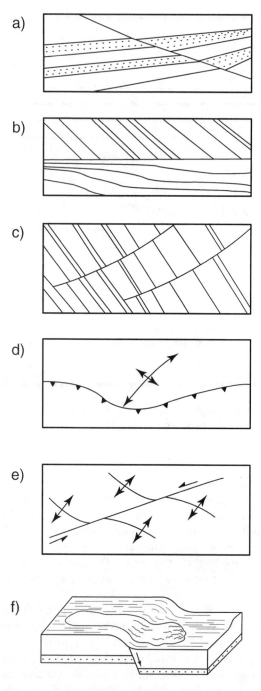

Figure 5.89 Some indicators of relative structural timing. a) Unit increases in thickness across a fault indicating the fault was active during deposition. b) Units parallel to an unconformity are younger than those terminated by it. c) Faults die out in younger rock and coalesce in older rock. d) A thrust embayment may indicate a pre-existing buttress. e) Folds displaced by a fault are older than the fault. f) Undeformed lava flows covering a fault scarp are younger than the scarp.

angular unconformities when in fact they are younger than the surface they are lapping onto. Clinoform bedding is the best clue to downlapping or onlapping units.

When a section is inclined and faults are visible, extensional faults generally die out upward in younger units and coalesce downward, with increasing displacement, in older units. If the fault is listric, movement must have occurred between the age of the decollement horizon and the horizon in which the fault terminates. The same occurs in thrust faults, where displacement decreases in the direction of transport (unless a fault has broken through to the surface, in which case the displacement is constant along the fault). The age of thrusting is younger than the age of the decollement and older than the oldest undeformed unit.

Folds and faults are younger than the units they deform. Cross-cutting relationships imply that the continuous fault is younger than the unit, fold, or fault that it displaces. A folded fault, fold, or unit indicates a period of deformation that post-dates the original fault, fold, or unit. A fault that terminates against another fault is probably younger than the throughgoing fault. The same applies to joints.

Multiple structural events can sometimes be unraveled by careful attention to regional kinematic indicators. For example, if a thrust is clearly offset by extensional faulting, one can infer the thrust is older than the normal faulting. If a thrust front forms an embayment around a large fold, the fold probably existed prior to thrusting and acted as a buttress to forward movement of the thrust sheet. If en echelon folds are cut and offset by a strike-slip fault, the folds probably were formed early in the same stress field that later caused a fault to break through.

A lava flow, alluvial fan, or lake bed deposit that covers a fold or fault scarp with no indication of displacement is younger than the structure it covers.

THE FIELD CHECK

It is not possible to overemphasize the importance of field checking a photointerpretation. Although ground "truth" is as subjective as any photointerpretation, there are certain measurements (e.g., dip of fault plane) and details (lithology, direction of slickensides, age of units) that are just not available on imagery, regardless how detailed. Even dips are, for the most part, estimates. For every feature described here there are numerous other features that can appear the same. Ideally, an interpreter would be able to visit the field briefly prior to each project to determine the types of ground cover and erosional patterns, measure the section, and obtain some strikes/dips for control. In reality it is fortunate if the interpreter gets the opportunity to do field work after a project is completed. But remember: an image interpretation is only an interpretation until one has been in the field and confirmed the mapping.

Bilotti, et al. [108] demonstrated a rapid, low-cost method of delineating prospects by carefully combining satellite stereo imagery and interpretation of contacts with field-measured strike-dips. Interpreted strike-dips agreed within 5° to surface measurements. Surface strike-dip information was then combined with seismic to generate cross sections. Intergration of image interpretation and field work makes for more effective, credible projects.

The purpose of a thorough interpretation is to create a map of identifiable structures, alteration, and units and highlight those areas where critical questions remain. The purpose of field mapping is to focus on those areas and answer the questions. The interpretation thus serves many objectives: providing an inexpensive map in a short period of time, directing field work (thus saving the time and expense of a drawn-out field season), providing exploration access to remote areas, planning logistics of seismic, potential fields, and/or drilling programs, assisting with the interpretation of geophysical or geochemical surveys, and keeping an exploration program confidential prior to ground work. Field work not only completes the interpretation, but adds to the interpreter's experience and knowledge, making the entire process more reliable for future projects.

REFERENCES
1. A.D. Howard, *Bull. Am. Assn. Pet. Geol.* **51** (1967): 2246–2259.
2. C. Cotton, *Geomorphology* (Whitcombe & Tombs Ltd., Christchurch, 1968), Chap. 7, pp. 76–95.
3. V.C. Miller and C.F. Miller, *Photogeology* (McGraw-Hill Book Co., New York, 1961), Chap. 7, pp. 80–100.
4. R.F. Thurrell, Jr., *Photogram. Engr.* **19** (1953): 579–588.
5. R.G. Ray, *U.S. Geol. Survey Professional Paper* **373** (1960): 64–66.
6. C. DeBlieux, *Gulf Coast Assn. Geol. Soc. Trans. 12th Ann. Mtg.* (1962): 231–241.
7. J.D. Mollard, *Canadian Oil and Gas Industries* (1957): 1–8.
8. N.H. Foster and R.A. Soeparjadi, *Am. Assn. Pet. Geol.– Soc. Econ. Paleo. and Min. Ann. Mtg. Abs.* (1974): 35 p.
9. J.F. Conrad, *Wyo. Geol. Assn. Guidebook—29th Ann. Field Conf.* (1977): 391–396.
10. R.J. Weimer, *Colorado Geology* (Rocky Mtn. Assn. Geols., Denver, 1980), pp. 301–313.
11. W.O. Kupsch, *Williston Basin Symp.* (1956): 66–75.
12. R.L. Rieck, Mich. State Univ. PhD Dissertation (1076), 216 p.
13. A. Maslowski, *Northeast Oil World* (Sept. 1985): 19–21.
14. D.F. Saunders, *Unconventional Methods in Exploration Symposium II* (1980): 63–82.
15. K.M. Morgan, D.R. Morris-Jones, and D.G. Koger, *Oil and Gas Journal* (Sept. 1982): 326–327.
16. T.J. Donovan, *Bull. Am. Assn. Pet. Geol.* **58** (1974): 429–446.
17. J.D. Ferguson, MSc Thesis, Oklahoma State Univ., Stillwater, Oklahoma (1975): 88 p.
18. J.D. Ferguson, *Shale Shaker* **29** (1979): 200–208.
19. R.L. Reynolds, N.S. Fishman, R.B. Wanty, and M.B. Goldhaber, *Bull. Geol. Soc. Am.* **102** (1990): 368–380.
20. H.G. Machel and E.A. Burton, *Bull. Am. Assn. Pet. Geol.* **75** (1991): 1864–1876.
21. D.B. Segal, M.D. Ruth, I.S. Merin, H. Watanabe, K. Soda, O. Takano, and M. Sano, *Proc. of the 3rd ERIM Thematic Conf. on Rem. Sens. for Explor. Geol.* (ERIM, Ann Arbor, 1984), pp. 273–292.
22. J.E. Conel and R.E. Alley, *Joint NASA/GEOSAT Test Case Project, Sect. 8* (Am. Assn. Pet. Geol., Tulsa, 1985), pp. 8–1 to 8–158.
23. D.B. Segal, M.D. Ruth, and I.S. Merin, *The Mountain Geologist* **23** (Rocky Mtn. Assn. Geols., April 1986), pp. 51–62.
24. S.J. Morrison and W.T. Parry, *Geol. Soc. Am. Prog. with Abs., Rocky Mtn. Sect. Mtg.* **18** (March 1986), p. 397.
25. R.V. Malhotra, R.W. Birnie, and G.D. Johnson, *Proc. of the 7th Thematic Conf. on Rem. Sens. for Explor. Geol.* (ERIM, Ann Arbor, 1989), 14 p.
26. M.C. Dalziel and T.J. Donovan, *U.S. Geol. Survey Circ.* **837** (1980): 11 p.
27. T.J. Donovan, R.L. Noble, I. Fiedman, and J.D. Gleason, *U.S. Geol. Survey Open File Rept.* **75–47** (1975): 11 p.
28. R.A. Lilburn and A.S. Zuhair, *Shale Shaker* **34** (December 1983): 40–56.
29. F.R. Siegal, *Applied Geochemistry* (John Wiley and Sons, New York, 1974), 353 p.
30. D.M. Richers, R.J. Reed, K.C. Horstman, G.D. Michels, R.N. Baker, L. Lundell and R.W. Marrs, *Bull. Am. Assn. Pet. Geol.* **66** (1982): 903–922.

31. B.N. Rock, *Vegetation at Lost River, Virginia*, in Fourth GEOSAT Workshop (Flagstaff, Arizona, June 1983).
32. S.H. Davis, Jr., *Jour. Arbor.* **3** (1977): 153–154.
33. F.B. Flower, E.F. Gilman, and I.A. Leone, *Jour. Arbor.* **7** (1981): 43–52.
34. B.N. Rock, *Teledetection Appliquee a la Cartographie Geologique* (1984): 299–309.
35. J.B. Parrish and B.N. Rock, *Botan. Soc. Am.* **70** (1983): 52 p.
36. J.L. Gebhart and R.H. Chabreck, *Proc. 29th Ann. Conf. SE Assoc. Game and Fish Comm*, St. Louis (1975), pp. 567–577.
37. P.H. Blanchet, *Bull. Am. Assn. Pet. Geol.* **41** (1957): 1748–1759.
38. G.H. Murray, Jr., *Bull. Am. Assn. Pet. Geol.* **52** (1968): 57–65.
39. J.F. Harris, MSc. Thesis, Univ. Tulsa, Tulsa, Oklahoma (1959), 56 p.
40. F.D. Gorham, Jr., L.A. Woodward, J.F. Callender, and A.R. Greer, *Bull. Am. Assn. Pet. Geol.* **63** (1979): 598–607.
41. C.A. Chapman, *Jour. Geol.* **66** (1958): 552–558.
42. N.J. Price, *Geol. Mag.* **46** (1959): 149–167.
43. M.O. Withjack and C. Scheiner, *Bull. Am. Assn. Pet. Geol.* **66** (1982): 302–316.
44. D.W. Stearns, *NSF Advanced Science Seminar in Rock Mechanics*, Special Rept. (Air Force Cambridge Res. Lab., Bedford, Mass., 1967), pp. 215–229.
45. J.D. Mollard and J.R. Janes, *Airphoto Interpretation and the Canadian Landscape* (Canadian Dept. Energy, Mines, and Resources, Hull, 1984), 415 p.
46. R.L. Reick, PhD Dissertation, Mich. State Univ., East Lansing, Michigan (1976), 216 p.
47. R.F. Flint, *Glacial and Quaternary Geology* (John Wiley and Sons, New York, 1971), 892 p.
48. B. Drake and R.K. Vincent, *Proc. 10th Intl. Symp. Rem. Sens. of Environ.* (ERIM, Ann Arbor, 1975), pp. 933–947.
49. A. Maslowski, *Northeast Oil World* (Sept. 1985): 19–21.
50. R.A. Bagnold, *The Physics of Blowing Sand and Desert Dunes* (Metheun Publishing Co., London, 1941), 265 p.
51. A.D. Howard, *Bull. Am. Assn. Pet. Geol.* **51** (1967): 2246–2259.
52. C.D. Ollier, *Tectonics and Landforms* (Longman Group Ltd., London, 1981), 324 p.
53. S.O. Al Khatieb and J.W. Norman, *Jour. Pet. Geol.* **9** (1986): 463–468.
54. C. DeBlieux and G.F. Shepherd, *Oil and Gas Jour.* **50** (1951): 86–100.
55. L.H. Lattman, *Mineral Industries* **28** (1959): 1–8.
56. W.W. Doeringsfeld, Jr., and J.B. Ivey, *Mountain Geol.* (1964): 183–195.
57. F.A. Melton, *Jour. Geol.* **67** (1959): 351–370.
58. F.A. Penney, *Rocky Mtn. Assn. Geol. Symp.* (1975): 55–61.
59. W.V. Trollinger, *Shale Shaker* (April 1968): 162–167.
60. W.P. Jenny, *Oil and Gas Jour.* (March 1982): p. 212.
61. W.P. Jenny, *Oil and Gas Jour.* (Oct. 1982): 233–234.
62. D.E. Glass, *Bull. Am. Assn. Pet. Geol. Assn. Round table* (1981): p. 930.
63. W. Lee and T.J. Payne, *Kans. Geol. Survey Bull.* **53** (1944): 79 p.
64. A.R. Sanford, *Bull. Geol. Soc. Am.* **70** (1959): 19–52.
65. J.W. Sears, MSc Thesis, University of Wyoming, Laramie, WY (1973), 100 p.
66. P.W. Huntoon and J.W. Sears, *Bull. Geol. Soc. Am.* **86** (1975): 465–472.
67. W.R. Jamison, PhD Thesis, Texas A & M University, College Station, TX (1979), 168 p.
68. O.G. Heyman, MSc Thesis, University of Wyoming, Laramie, WY (1983), 133 p.
69. R.J. Stein and J. Wickham, *Tectonophys.* **66** (1980): 225–251.
70. R.A. Hodgson, *Bull. Am. Assn. Pet. Geol.* **45** (1961): 1–38.
71. A.O. Alpay, *Soc. Pet. Eng. Ann. Fall Mtg.* SPE No. **2567** (1969): 1–11
72. S.E. Roberts, *Nat. Res. Council, Highway Res. Board, 33rd Ann. Mtg.* (1954): 341–345.
73. F.D. Stacy, *Physics of the Earth* (John Wiley and Sons, New York, 1969): 108–117.
74. P.J. Barosh, *Geol. Soc. Am. Memoir* **110** (1968): 199–217.
75. E.T. Brown and J.A. Hudson, *Earthquake Engr. and Struct. Dynam.* **2** (1974): 379–386.
76. G.L. Prost, *Int. J. Rock Mech. Min. Sci. & Geomech. Abstr.* **25** (1988): 263–272.

77. B.J. Skinner, *Geol. Soc. Am. Memoir* **97** (1966): 76–96.
78. J.S. Hallin, MSc Thesis, University of Wyoming, Laramie, WY (1973), 108 p.
79. D.L. Turcotte and E.R. Oxburgh, *Nature* **244** (1973): 337–339.
80. A. Nur, *Int. Basement Tectonics Symp. Contrib.* **22** (1978): 155–167.
81. A. Nur, *Jour. Struct. Geol.* **4** (1982): 31–40.
82. G.E. McGill and A.W. Stromquist, *Jour. Geophys. Res.* **84** (1979): 4547–4563.
83. H.J. Melosh and G.E. McGill, *EOS* **63** [abs] (1982).
84. G.L. Prost, PhD Thesis, Colo. School of Mines, Golden, CO (1986), 293 p.
85. G.H. Davis, *Structural Geology of Rocks and Regions* (John Wiley and Sons, New York, 1984), 492 p.
86. J.D. Lowell, *Structural Styles in Petroleum Exploration* (Oil and Gas Consultants International, Tulsa, 1985), 477 p.
87. F.F. Sabins, Jr., *Remote Sensing Principles and Interpretation* (W.H. Freeman and Co., New York, 1987), 2nd ed., Chap. 5, p. 140.
88. C.K. Morley, R.A. Nelson, T.L. Patton, and S.G. Munn, *Bull. Am. Assn. Pet. Geol.* **74** (1990): 1234–1253.
89. D.P. Krynine and W.R. Judd, *Principles of Engineering Geology and Geotechnics* (McGraw-Hill Book Co., New York, 1957), 730 p.
90. J.G. Moore, *U.S. Geol. Survey Professional Paper 400-B* (1960): 409–411.
91. H.R. Ritzma, *Wyo. Geol. Assn. 23rd Ann. Field Conf. Guidebook* (1971): 145–150.
92. J.D. Love, J.L. Weitz, and R.K. Hose, *Geol. Survey Wyo. Map Series* **7A** (1955): 1:500,000.
93. J.D. Love, A.C. Christiansen, and R.W. Jones, U.S. Geol. Survey Map **OF 79-1301** (1979): 1:250,000.
94. J.D. Love, A.C. Christiansen, T.M. Brown, and J.L. Earle, *U.S. Geol. Survey Map* **OF 79-962** (1979): 1:250,000.
95. B. Bryant, L.W. McGrew, and R.A. Wobus, *U.S. Geol. Survey Map* **1-1163** (1981): 1:250,000.
96. A.F. Jacob, *Rocky Mountain Foreland Basins and Uplifts* (Rocky Mtn. Assn. Geol., Denver, 1983), pp. 229–244.
97. D.L. Anderson, in *Continents Adrift*, J.T. Wilson, ed. (W.H. Freeman and Co., San Francisco, 1972), pp. 143–157.
98. R.E. Wilcox, T.P. Harding, and D.R. Seely, *Bull. Am. Assn. Pet. Geol.* **57** (1973): 74–96.
99. T.P. Harding, *Bull. Am. Assn. Pet. Geol.* **58** (1974): 1290–1304.
100. J.C. Crowell, in *Tectonics and Sedimentation: SEPM Spec. Pub.* **22**, W.R. Dickenson, ed. (1974), pp. 190–204.
101. A.G. Sylvester and R.R. Smith, *Bull. Am. Assn. Pet. Geol.* **60** (1976): 2081–2102.
102. F. Segal, and D.D. Pollard, *Jour. Geophys. Res.* **85** (1980): 4337–4350.
103. G.L. Prost, *World Oil* **211** (1990): 39–45.
104. C.K. Morely, *Jour. Struct. Geol.* **9** (1987): 331–343.
105. W.R. Jamison, *Jour. Struct. Geol.* **9** (1987): 207–219.
106. J.G. Ramsay and M.I. Huber, *Techniques of Modern Structural Geology 2: Folds and Fractures* (Academic Press, London, 1987), 548 p.
107. W.M. Dunne and D.A. Farrill, *Geology* **16** (1988): 33–36.
108. F. Bilotti, J.H. Shaw, and P.A. Brennan, *Bull. Am. Assn. Pet. Geol.* **84** (2000): 727–740.

ADDITIONAL READING

D. Brown, Will Satellites Replace Field Work? Explorer (AAPG, Feb. 1996): 10–12.
H.E. McKinstry, *Mining Geology* (Prentice-Hall, Englewood Cliffs, 1948), pp. 290–327.
C.F. Park, Jr., and R.A. MacDiarmid, *Ore Deposits* (W.H. Freeman and Co., San Francisco, 1970), pp. 64–100.

D. Powell, *Interpretation of Geological Structures Through Maps* (Longman Scientific & Technical/ John Wiley & Sons, New York, 1992), 176 p.

QUESTIONS TO CONSIDER

1. You are exploring for giant oil fields with four-way closure. What surface features would you look for in the Amazon basin? In the Andes? On the flat plains of the Tibetan Plateau?
2. How might a fault appear at the surface? A mineralized fault?
3. Where would you drill for groundwater if pre-dawn thermal imagery showed a cool zone up slope from a faulted pediment? Why?
4. What does a linear fault trace tell about the inclination of the fault plane? What does a curved fault trace tell about the fault geometry?
5. How can normal faults and strike slip faults appear similar on imagery? How can one tell them apart?
6. How can unconformities and thrust faults appear similar on imagery? How can one tell them apart?

6

Spectral Stratigraphy

CHAPTER OVERVIEW

In Chapter 4 we discussed recognizing lithology on black/white and color images. Here we look at mapping rock types and surface materials using characteristics of their reflectance curves. Multispectral instruments offer the possibility of automated mineral mapping: one flies over an area, records the reflectance spectra of every object on the ground, and matches the spectral curves to those in a reference library to determine the composition of the surface.

From the perspective of the petroleum geologist evaluating a frontier basin, it would be useful to be able to map the distribution of source rock and its maturity, and map the distribution of outcropping reservoir and seal units around the basin margins. We would also like to determine, directly or indirectly, if any seeps exist within the basin. These problems are being addressed as increasingly sophisticated sensors become available. A minerals geologist will be interested in identifying surface units (e.g., quartz monzonite porphyry vs. granite, Carlin-type black shale vs. shale, kimberlite vs. carbonatite) and will want to identify alteration types and their mineral associations. All geologists will want to understand the lithologies and geologic relationships (which unit is above, below, a lateral equivalent, or faulted out) when entering an unmapped area for the first time.

ROCK SPECTRA

In its simplest form spectral stratigraphy means using the unique combination of spectral reflectance (color), brightness, and erosional texture associated with lithologic units to make photogeologic maps. The units that are mapped are "photogeologic" units that represent volumes of rock bounded by horizons (time surfaces) recognizable on the photographs. In this sense they may differ from standard geologic units, which are bounded by horizons recognizable in the field. These units have no genetic connotations, nor is it always possible to identify the lithology involved, particularly if there is heavy vegetation cover. The definition of the units is highly dependent on the resolution of the imagery in that high resolution airphotos, for example, will allow more and better discrimination of photogeologic units than lower resolution satellite images. These units can be followed along strike until they are covered, faulted out, reach an unconformity, or change facies. Correlation of these units across areas without outcrops is tenuous, but should be based on several supporting criteria. Besides having a similar expression, the unit that one feels is correlative should fall within the same sequence of photogeologic units, i.e., it should fit in the same photostratigraphic framework. If one cannot follow an individual unit, it may still be possible to map the greater sequence of units [1].

It is a simple matter to compile reference curves of single minerals, and this has been done by Hunt and Salisbury [2,3], among others. It is more difficult to identify mixtures of minerals in varying amounts, what we call lithologies or rock types [4]. Add to this the effect of atmospheric absorption and scattering, the mixing of rock types (e.g., a limy shale), weathering products on the surface of an outcrop, soil cover, and plant cover of varying density and composition, and we have a formidable task ahead of us before we can generate geologic maps automatically.

These maps will not be "classical" geologic maps. It is possible, however, to generate a fairly accurate mineralogic or lithologic distribution map using imaging spectrometer data. These instruments quite simply can "see" more than the human eye because of the extended wavelength range available to high resolution imaging sensors. The product is an excellent first step at obtaining a reconnaissance map showing mineralogic composition and assemblages at the surface. Although not classical geologic maps, they are potentially just as meaningful, if not more so, since much of the information provided would not be available without detailed lab work such as x-ray diffraction, etc. Finally, it should be emphasized that spectral remote sensing can enhance standard photogeologic mapping by making images easier for the geologist to interpret. Among other things, it can assist in (a) the remote identification of specific lithologies (e.g., limestones vs. dolomites) that might not be readily separated on standard photos or imagery, (b) extending known lithologies into unmapped territory by means of their spectral characteristics, (c) making it easier to identify surface alteration (e.g., dolomitization, sericitization), and (d) making possible the identification of organic-rich source rock, and perhaps its maturity.

Mineral Spectra

In the early 1970s workers at the Air Force Cambridge Research Laboratory began measuring the spectral reflectance of suites of minerals grouped by anions, e.g., silicates, carbonates, oxides, etc. The curves are characterized by reflectance peaks and absorption bands caused by electronic and vibrational processes within their crystal lattice. It was determined, for example, that iron oxides have a reflectance peak in the visible red (0.73 microns) and absorption features near 0.9 microns and 0.4 microns. These minerals, which include hematite, goethite, limonite, and jarosite, have a diagnostic spectral reflectance that was used by Rowan, et al. [5] to process the earliest Landsat MSS images to identify ferric iron associated with mineralized gossans in the Goldfield mining district, Nevada. This involved a processing technique called ratioing, where brightness (reflectance) values in one channel are divided by values for the same pixel in another channel (Fig. 6.1). Ratios enhance the contrast between materials with different reflectance at a specific wavelength, and suppress the effects of shadows (topography), since low reflectance values are divided by low values. Although the range of possible values goes from zero to infinity, most ratio values fall between 0.5 and 2.0 and are then stretched, or scaled, back to the normal range of 0–255 (for 8-bit data like Landsat and SPOT). This can be performed on multiple channels: If three sets of ratios are combined, and each is assigned a different color, a

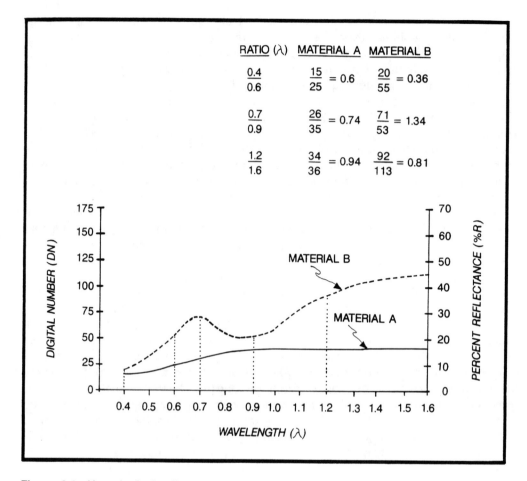

Figure 6.1 Hypothetical reflectance curves showing how ratios enhance minor reflectance variations. From Prost [6].

color composite image is generated. The ratio technique is only one of many image processing methods that use discrete, widely separated broad bandwidth multispectral data to generate images that show the distribution of specific mineral assemblages. The mining industry was an early user of this technique because there are a number of hydrothermal alteration minerals, including iron oxides and clay minerals, that are readily mapped using ratios [6–8].

Two other techniques ought to be mentioned briefly when discussing methods for enhancing the spectral information content of images. Although they do not take advantage of discrete spectral absorptions, the principal components (PC) and intensity-hue-saturation (IHS) transformations can make it easier for the interpreter to distinguish between materials with minor spectral variations. The PC transformation is used to decrease the amount of correlation between bands of data and maximize the differences between bands [9]. In most images the information content of adjacent bands is nearly identical (highly correlated). A three-dimensional plot of reflectance values in three bands (displayed as red, green, and blue, and known as "RGB space") would generate an ellipsoidal cluster of points. The tightness of this cluster is an indication of the redundancy of the data sets. The PC transformation generates a new set of orthogonal (noncorrelated) axes in space such that the principal axis is oriented through the long dimension of the ellipsoid, and the secondary and tertiary axes define correspondingly smaller distributions or densities of data points. This transformation can be performed on any number of bands and will generate axes in n-space where the n axes correspond to the number of input bands. Generally speaking, the first principal component contains the brightness information of the image, the second, third, etc., components contain progressively less information, and the last components tend to contain first the systematic then the random noise in the image. One can create a color image using any three components (Fig. 6.2). Usually the first three components are used. The second and third components tend to reveal subtle differences in surface materials. One drawback of this technique is that, unlike standard color images or ratios, it is not possible to predict what colors will correspond to specific surface materials, since each PC image is a function of the unique distribution of brightness values within the original image. This also means that the color combinations are not consistent from one image to the next, that is, the colors generated for a limestone in one image will not be the same colors as those in the same limestone in a later image, or for the same limestone in an adjacent image.

A variation of the PC image is the decorrelation stretch [10]. In the decorrelation stretch a PC image is generated, then is contrast enhanced using, for example, a histogram equalization stretch. This new data is then subjected to an inverse PC transformation that returns it to the original RGB color space. The resulting data has the same color combinations as the original color image, but the colors tend to be highly saturated and spectral differences are intensified (Fig. 6.3).

The IHS transform takes an image from the familiar red-green-blue color space to a new system of intensity, hue, and saturation [12]. Intensity is a function of brightness; hue represents the color of the object; and saturation is the purity of the color. An image is first transformed into IHS space, where the intensity, hue, and saturation components are individually stretched. They are then transformed back to the original RGB color space and result in an image with greater dynamic range in colors and tones. There are a number of variations of this transformation. In one example, the original Landsat TM image has bands 7-3-1 as red-green-blue. These are transformed to IHS space and stretched. Then the intensity band is removed and replaced by the TM band 4 and the transformation is reversed to return to RGB space. This technique has been shown to amplify subtle spectral differences in the original image (Fig. 3.15).

One can not only enhance, but can often identify many surface materials on the basis of their unique spectral curves. These curves are characterized by absorption bands (reflectance or emissivity minima) caused by the presence of OH, H_2O, CO_3, SO_4, CH, and SiO_2. Vegetation has an absorption due to chlorophyll at wavelengths shorter than 0.75 microns. Clay minerals have a characteristic absorption centered near 2.2 microns. This minima moves slightly depending on the clay type (Fig. 6.4). Calcite has an absorption feature at 2.35 microns; gypsum at 2.4 microns; silicate minerals have a reststrahlen band (emissivity minimum) between 8 and 11 microns, depending on the dominant mineral (Fig. 6.5). The minima shifts to longer wavelengths as the quartz content decreases. Organic matter (including hydrocarbons) has a characteristic minima at 1.75 microns. In the microwave region minerals can be characterized by their emissivity curves and dielectric constants [14].

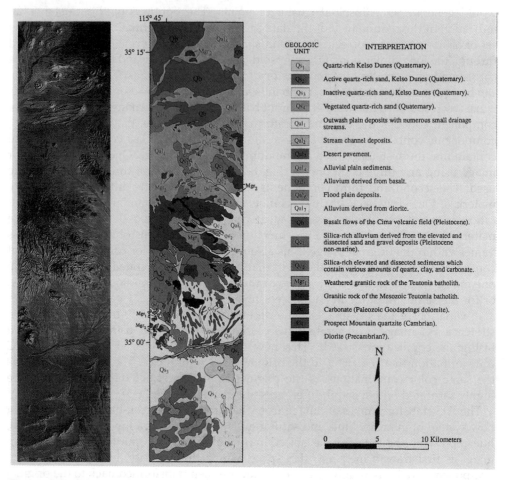

Figure 6.2 Thermal Infrared Multispectral Scanner (TIMS) image of part of the Mojave Desert near Kelso, California. This PC image was generated using bands 1 (8.2–8.6 microns), 3 (9.0–9.4 microns), and 5 (10.2–11.2 microns) assigned to blue, green, and red, respectively. The surficial map was derived from the image. In general, quartz-rich rocks are red; alluvium is reddish pink (older) to red with a blue cast (younger); vegetation is yellow; carbonates range from bright green to yellow; quartzites are orange; and basalts are dark blue-green. North is at the top of the image. Processed by P.W. Barbera, et al., Arizona State Univ. Geology Dept. in conjunction with Daedalus Enterprises Inc., Ann Arbor. (See color plate.)

Many workers have been compiling ever more complete libraries of spectral curves (see, for example, Gaffey [15], Clark, et al. [16], and Krohn [17]). They have demonstrated the ability to identify minerals, for example, the differences between calcite, dolomite, and aragonite [15]. In some cases they have shown that the fine detail in the spectra vary with elemental abundance, e.g., the 2.2 micron absorption minima in montmorillonites shifts to longer wavelengths with increasing calcium content [16].

Figure 6.3 This TIMS image of the Panamint Mountains and west side of Death Valley was processed using bands 1-3-5 as blue-green-red followed by a decorrelation stretch. Quartzites are bright red; metacarbonates are blue-green; volcanics are shades of purple and violet; alluvial materials, fan gravels, and evaporites are all clearly distinguished by hue. From Khale and Goetz [11]. (See color plate.)

HYPERSPECTRAL IMAGERY

Whereas the shape of an object can be used to recognize what it is, its spectra can be used to identify what it is made of. Recently imagery has become available with tens or even hundreds of closely spaced narrow bandwidth channels (Chapter 3). The acquisition, processing, and analysis of this data is known as imaging spectrometry or hyperspectral remote sensing [18]. Hyperspectral imagery can be thought of as a cube of information: The top of the cube consists of a map view of the earth's surface in a

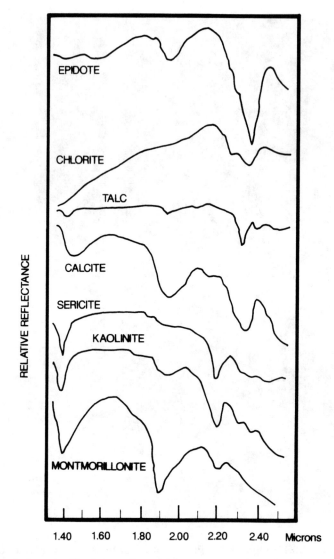

Figure 6.4 Near infrared reflectances for some common minerals. From Henderson and Rock [13].

given channel, whereas the depth of the cube consists of multiple layers, each representing surface reflectance in a different wavelength or band (Figs. 6.6, 6.7). For any given point on the surface it is possible to generate a reflectance curve by plotting the reflectance values through the depth of the cube. This technique has been shown to be useful for the identification of surface materials such as rock types, soils, and vegetation. Hyperspectral imagery is being used during environmental site characterization to map the distribution of contaminants, to identify and map the distribution of minerals and alteration during mineral exploration programs, and to map the distribution of rock types such as organic shale source rocks for petroleum basin evaluations.

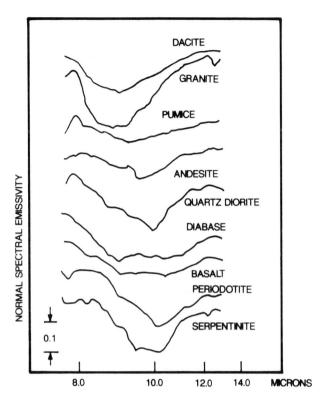

Figure 6.5 Thermal infrared emittance for some silicate minerals and rocks. From Henderson and Rock [13].

In the mid-1980s people first began working with hyperspectral data sets. Goetting and Lyon [20] developed an expert system that identified minerals by examining 16 spectral windows in the range 1.36 to 2.50 microns. This system characterized the minerals by the position of their absorption minima and the strength of the minimum relative to other absorptions. In 1990 Kruse et al. [21] used the 63 channel GERIS imaging spectrometer to identify minerals based on their field and laboratory spectra (Fig. 3.6), and then made mineral distribution maps. Since that time workers have been developing expert systems to interactively process imaging spectrometer (primarily AVIRIS) data sets to extract spectra from each surface pixel, characterize the absorptions, match them to curves in a spectral library, and generate mineral distribution maps [22–24]. These knowledge-based systems first generate continuum-removed (normalized) spectra, then characterize them using attributes of the absorption bands. A human expert then decides interactively which spectral feature attributes are characteristic of a material and assigns a weight to their importance. The system then analyzes unknown spectra and generates images showing the distribution of materials and the measure of certainty of occurrence of a given mineral at each pixel [24].

The problems of mineral mixing, vegetation cover, weathering, and soils still do not allow consistent, unambiguous mapping. This is, however, another area of active research at this time. Mustard and Pieters [25,26] proposed an unmixing technique that

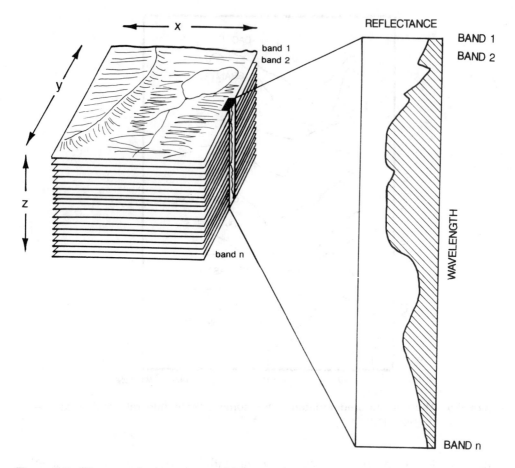

Figure 6.6 Diagram of a hyperspectral image cube consisting of "n" layers of images in "n" wavelengths. One may extract a reflectance curve for any given pixel in the image.

works well in parts of the spectrum that do not contain strong absorptions. The spatial distribution and abundance of primary surface components are calculated using a nonlinear (intimate) mixing model that requires one to input the probable unmixing end members and particle sizes. They deconvolved (unmixed) Airborne Imaging Scanner 128 band spectrometer data over Moses Rock dike, Utah, using six spectral end members (ultramafic breccia, two sandstones, gypsiferous soil, clay-rich soil, and desert varnish). Their calculations of surface composition and abundance are similar to field observations. Bierwirth [27] used a linear unmixing algorithm on eight-band NS001 aircraft imagery at a test site in Queensland, Australia. Six images were generated that show the spatial abundance of green vegetation, dry vegetation, kaolinite, hematite, goethite, and quartz. An inverse linear unmixing model was developed and applied by Boardman [28,29]. He unmixed 224 channel AVIRIS data by using a human expert to pick the most likely spectral end members (minerals, rock or soil units, or vegetation), then solves for the unknown abundances of these minerals. The end product is a set of images showing the distribution of single end members. These images can be combined in color to make lithologic distribution maps.

SPECTRAL STRATIGRAPHY 183

Figure 6.7 AVIRIS hyperspectral cubes. The front face is a color infrared image of four different areas. The sides represent the color-coded radiance of the edge pixels in 224 bands from 0.4 microns in front to 2.45 microns in the back. Reds are high reflectance values; blues are low values; and black zones are atmospheric absorption bands. From Goetz [19]. (See color plate.)

Crosta, et al. [30] examined two publicly available software programs to determine the effectiveness of hyperspectral mineral classification where no prior ground spectra or atmospheric information was available. AVIRIS data were used to examine hydrothermal alteration at the Bodie and Paramount mining districts north of Mono Lake, California. Bodie is a lode quartz vein gold deposit with stockworks developed in a zone of hydrothermal alteration along fault zones in dacitic, andesitic, and rhyolitic lavas and tuffs. Silicification near the center of the deposit is surrounded by zones of potassic, argillic, sericitic, and propylitic alteration as one moves outward. Clay mineral distribution maps were made for short wave infrared bands (2.0–2.45 microns), and iron oxide mineral maps were made for the visible-near infrared (0.42–1.34 microns). They concluded that the algorithms were capable of recognizing and mapping a number of hydrothermal minerals without independent ground information. This is important since exploration activity often does not allow for preliminary collection of ground data during the reconnaissance phase of a program.

An interesting application of hyperspectral techniques is for describing drill core and cuttings. Kruse [31] presents a technique using a portable field spectrometer (PIMA) to measure infrared reflectance from 1.3 to 2.5 microns from split core. The spectrometer has 600 channels and generates real time image cubes. This allowed extraction of individual mineral spectra, linear unmixing, and comparison to a spectral library in order to make mineral maps of the core. The technique could decrease cost and turnaround time for core descriptions during both mineral and hydrocarbon exploration programs.

LITHOLOGIC MAPPING

It is sometimes more important for the earth scientist or engineer to map suites of minerals, such as sandstones or limestones, than to map individual minerals. Again, the pioneering work was done by Hunt and Salisbury [4]. They found that similar lithologies often have similar reflectance curves (Figs. 6.8, 6.9). This was then taken as the basis for work by others, such as Kahle and Rowan [32], Conel et al. [33], Lang et al. [34], and Lang et al. [35], who set about making remote sensing geologic maps using spectral units that usually, but not always, corresponded to lithologic units (Fig. 6.10). Work in the Wind River and Bighorn basins, Wyoming, showed that a combination of photogeologic and spectral interpretation of multispectral data was useful for characterizing the attitude, thickness, and lithology of various strata. They were able to map the distribution of quartz, calcite, dolomite, smectite clays, and gypsum in the stratigraphic section using visible, near infrared, and thermal infrared multispectral data. They then co-registered digital elevation data to the imagery, traced the contacts of various units, automatically solved three-point problems, and generated strike and dip information [34]. This type of map should allow the geoscientist to rapidly locate not only structures, but also all outcrops of sandstone, organic-rich shale, and evaporites in an area of interest, and thus map the distribution of potential reservoir, source, and seal units around the margin of a basin.

Scavetti et al. [36] proposed a methodology for stratigraphic remote sensing using an example from arid northern Somalia. They integrated air photos, TM imagery, and field measurements to create a chronostratigraphic framework for mapping and correlating the Jurassic through Eocene section (Fig. 6.11). Measured sections in widely separated areas were used as control points for photohorizons mapped on 1:50,000 airphotos and regional TM images. These horizons were defined on the basis of colors, absorption features, stratal patterns (massive vs. thin beds, crossbedding), and surfaces bounded by contrasting erosional styles. The fact that the photostratigraphic surfaces were laterally extensive and could be seen to change facies (e.g., reduced number of sandstones within a unit suggesting a change from a nearshore to an offshore environment in a transgressive sequence) suggests that bounding surfaces may be equivalent to sequence stratigraphic surfaces.

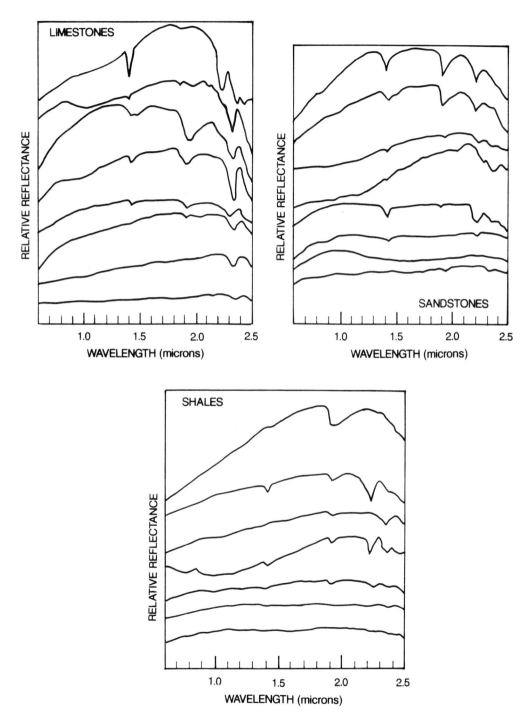

Figure 6.8 Visible and near-infrared reflectance curves for rock families, including sandstones, shales, and limestones. Spectra are displaced vertically. From Hunt and Salisbury [2].

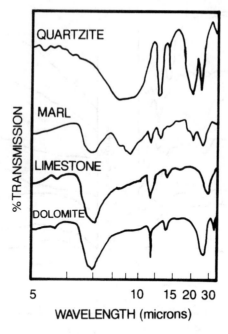

Figure 6.9 Thermal infrared transmission spectra of some sedimentary rocks from the East Tintic mountains, Utah. From Kahle and Rowan [32].

An interesting example of stratigraphic mapping in the arctic using radar imagery is given by Hanks and Guritz [37]. ERS-1 synthetic aperture radar was used to map rock types in the Porcupine Lake area of the northeastern Brooks Range, Alaska. The region contains tundra vegetation in the valleys and is essentially barren on higher slopes. The northern part of the area contains pre-Mississippian metamorphics and Mississippian to Pennsylvanian carbonates, whereas the southern part of the test site contains thrusted Mississippian to Jurassic carbonates and clastics. Carbonate slopes in this area consist of angular rubble that appears rough (bright) on the C-band radar. Shales form valley bottoms or smooth slopes that are generally covered by vegetation or water. Sandstones form tabular rubble of intermediate roughness and brightness. These units can be distinguished using radar on the basis of their surface roughness and dielectric constant, which are a function of lithology and moisture content, respectively.

The ERS-1 C-band radar (6.3 cm wavelength) was acquired during the summer and was radiometrically calibrated to convert radar brightness to a backscatter image. Terrain correction was performed using a digital elevation model to move pixels to their correct topographic position and minimize geometric distortions. The final product was a normalized incidence angle image. This assumes a surface of uniform roughness, then uses the digital elevation model to calculate the theoretical backscatter of each pixel based on the incidence angle of the radar beam on the surface at that point. The resulting artificial image is then subtracted from the corrected radar image (Fig. 6.12). Backscatter in the normalized incidence angle image then represents changes in surface properties (roughness, dielectric constant). This image, together with some knowledge of the rock types in the area, can be used to map lithologic units over large regions.

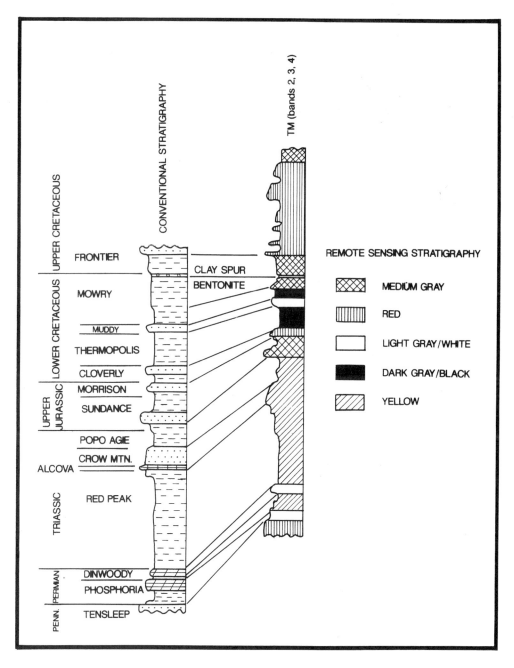

Figure 6.10 Correlation of Landsat TM spectral stratigraphy and conventional stratigraphy from the northern Casper arch, Wyoming. From Lang, et al. [34].

188 REMOTE SENSING FOR GEOLOGISTS

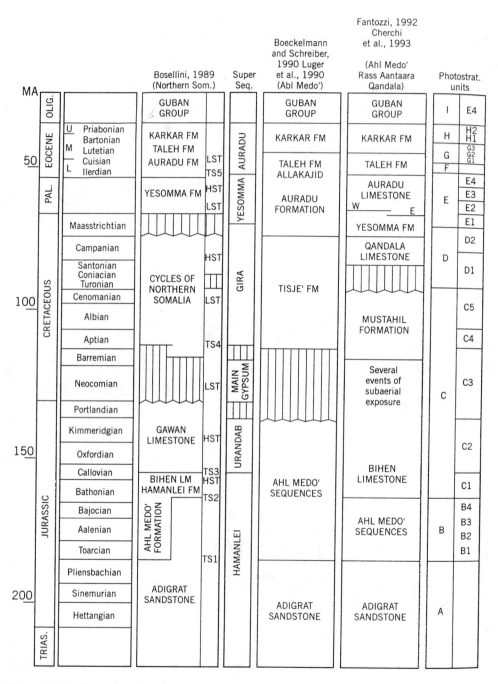

Figure 6.11 Photostratigraphy for northern Somalia, from Sgavetti, et al. [36]. (a) Stratigraphic framework; (b) correlation scheme. (Reprinted by permission of the American Association of Petroleum Geologists whose permission is required for further use.)

SPECTRAL STRATIGRAPHY 189

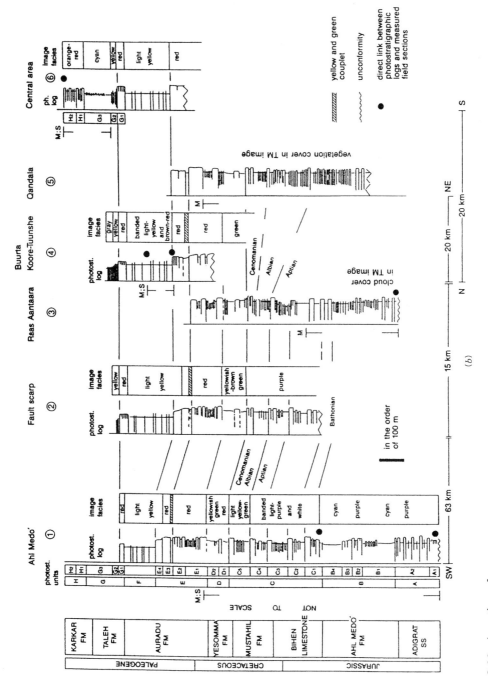

Figure 6.11 (continued)

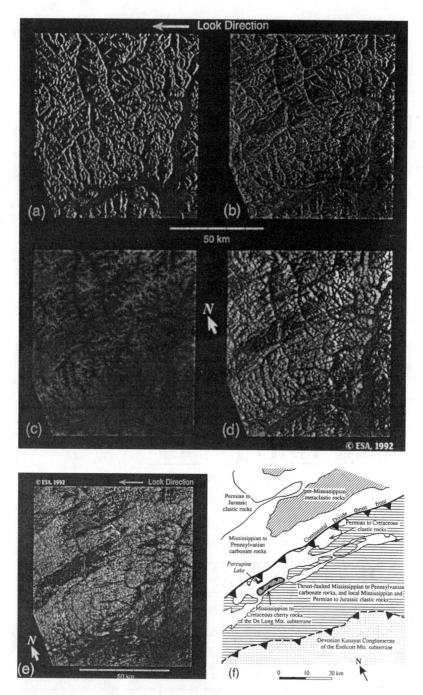

Figure 6.12 Porcupine Lake area, Alaska, from Hanks and Guritz [37]. (a) Simulated SAR image made from a DEM; (b) Radiometrically-calibrated ERS-1 image; (c) Digital elevation model; (d) Terrain-corrected SAR image; (e) Incidence angle image; (f) interpreted geologic map. (Reprinted by permission of the American Association of Petroleum Geologists whose permission is required for further use.)

APPLICATIONS – HYDROCARBONS

Source Rock Mapping

Udo and Etuk [38] have demonstrated that it is possible to evaluate the quality and thermal maturity of organic matter in rock extracts in the lab using infrared spectroscopy. Rowan, et al. [39,40] studied Landsat TM images over eastern Nevada and western Utah and related spectral reflectance of the Mississippian Chainman Shale to its maturity as measured on field samples using vitrinite reflectance and Rock-Eval pyrolysis. They found that the shape of the shale reflectance curve changes from concave-downward to nearly flat as thermal maturity increases (Fig. 3.5). They were able to map differences between mature and supermature areas using the TM ratio of band 4/band 5 (after first minimizing the effect of vegetation). Decreasing band 5 reflectance and increasing 4/5 ratio values indicate increasing thermal maturity. They were able to screen out limonitic sandstones, which had similar ratio values, by using the diagnostic reflectance curves of these iron-bearing units. Weathering of immature and mature Chainman Shale causes increasing reflectance in TM 5, but has no effect on the reflectance of supermature (vitrinite reflectance greater than 2.0) samples. An attempt to correlate reflectance in the TM bands, as well as various ratios, to the percent total organic carbon (TOC) was unsuccessful, but TOC values only went up to 1.57%. Richer source rocks might have more of a spectral response.

Seeps Mapping

It should be relatively easy to map seeps offshore, since the background (water) is more-or-less homogeneous. Whereas water generally transmits or absorbs radiation in the visible and near-visible range, oils will fluoresce in the ultraviolet and reflect in the visible and near-infrared. Thus, if a seep is large enough to be detected with a given instrument, it should be obvious after the standard contrast enhancements. Offshore gas seeps can be detected using thermal images, since the gas bubbles bring up cooler bottom water as they rise. Oil slicks can appear either cooler or warmer than background, depending on their thickness and composition [41].

This is not necessarily true for onshore seeps. Few are large enough (greater than 10 m for SPOT P, or greater than 80 m for Landsat MSS) to be resolved. If they are large enough to be visible, they could still be lost in the background noise. It has been shown that tar sands have unique spectral curves [42], and it may be possible to locate these indicators of eroded reservoirs or ancient seeps if they are large enough (Fig. 3.5). One is more likely to see the secondary effects of seeps rather than the seeps themselves. Such effects include rock and soil alteration and the vegetation changes mentioned earlier. Other means of locating hydrocarbons onshore rely on image processing techniques to characterize areas that are known to be productive, and then compare them to areas thought to contain hydrocarbons in the subsurface. A patent by Sundberg of Phillips Petroleum, for example, describes discriminant probability functions that can be used to separate "dry" from "productive" pixels on Landsat TM images by comparing reflectance values of hydrocarbon producing areas to the reflectance of adjacent areas [43]. See chapter 11, Marine Environments, for more on seeps and spills.

Stratigraphic Traps

It is obviously not possible to directly map stratigraphic traps in the subsurface, since most sensors measure reflected sunlight and cannot penetrate the earth beyond the surface veneer. Under certain circumstances one may, however, be able to guess where such features are most likely to occur. For example, northwest-elongated sandstone reservoirs in the Cretaceous Mannville Group of Alberta and in the Pennsylvanian Goosegg-Minnelusa interval and Cretaceous Teapot-Parkman interval of the Powder River basin of Wyoming were deposited as shoreline sands. In some areas, of course, the relationship of these sand bodies to surface features is hard to define or nonexistent. Yet in other areas during specific time intervals deposition was controlled by active structures, and these structures can be expressed at the present surface where recent tectonic activity or drape over the sandstone isopach thicks has revealed them [44–46]. The same can be said for northeast trending fluvial sandstone reservoirs (Cretaceous Muddy-Skull Creek interval) in the Powder River basin, where in some cases the rivers followed old fracture zones that are expressed at the surface today (Fig. 6.13).

Another type of stratigraphic play that may be evident at the surface involves reef reservoirs. Reefs can grow along the top of horst blocks, along the edge of tilted blocks, or on the high corners of rotated blocks. For example, facies and thickness changes in the Devonian Duvernay and Cooking Lake Formations at Duhamel field, Alberta, occurred along a tectonic hinge line caused by possible fault movement (Fig. 6.14). These facies changes in turn influenced the growth of the Leduc reef [48]. The

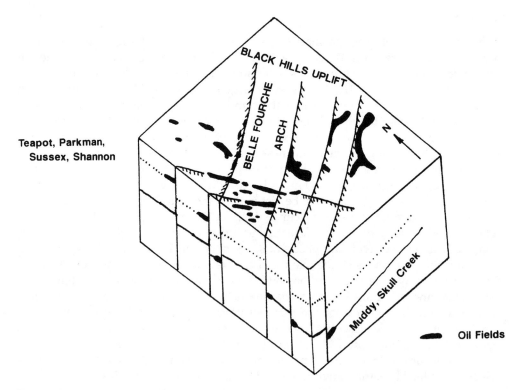

Figure 6.13 Schematic block diagram of the northern Powder River basin, Wyoming, showing paleotectonic control on the accumulation of Cretaceous deposits. Offshore bars (e.g., Teapot, Parkman, Sussex, and Shannon formations) trend northwest; river channel deposits (e.g., Muddy, Skull Creek formations) trend northeast. Modified after Slack [44].

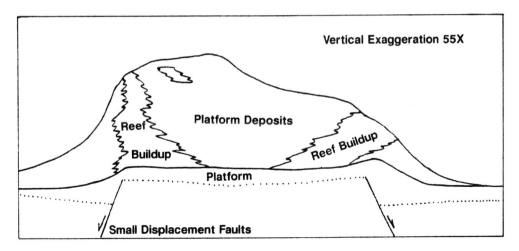

Figure 6.14 East-west cross section through the Devonian Swan Hills reef buildup, Alberta. Reefs developed above the upfaulted platform margins. The small displacement faults that border the platform have five to 13 m vertical offset, and also appear to control the distribution of dolomite zones in the reef. After Viau, et al. [47].

distribution of Devonian reefs in the Alberta basin has been related to control by northwest and northeast trending faults [49]. In the Tangent area of Alberta dolomitized, fault-bounded reservoirs in the Devonian Wabamun were found to be similar to the Ordovician Albian-Scipio play in Michigan [50]. In both of these cases the carbonates were dolomitized along fracture zones, and the dolomites provide both intercrystalline and fracture porosity for the oil accumulations.

Paleochannels, deltas, and turbidite deposits may exist in the subsurface basinward of present day drainage systems. This is because these types of deposits commonly occur as stacked vertical sequences in rapidly subsiding depositional environments such as the Gulf of Suez, the San Joaquin Valley of California, or the coast of the Gulf of Mexico [51]. In addition, many of the river systems delivering sediments are now, and have been for the past several million years, incised into the topography or controlled by the position of fault blocks. In such cases paleochannels should occur more or less vertically beneath present day channel systems.

One cannot merely locate present day fracture zones or lineaments and drill along them expecting to hit oil. However, when used in conjunction with other data such as field work or well control, and when used with a valid conceptual model of the depositional history, they can help guide potential fields and seismic programs and ultimately contribute to defining a stratigraphic play location or boundary.

APPLICATIONS – MINERALS

There are a number of economic minerals that occur as stratigraphic units or are stratiform. Among these are coal, phosphates, uranium-vanadium, banded iron formations, chemical precipitates, stratiform deposits, replacement mantos, laterites, and placers. Each has features that may be recognizable using remote sensing techniques.

Coal, Phosphate, and Uranium-Vanadium Deposits

Outcrops of coal are recognized and mapped on airphotos as dark bands, generally weathered out to a greater extent than adjacent sandstones and to the same degree as nearby shales. They may be difficult to distinguish from dark shales except where they have burned and formed a distinctive clinker or natural slag. These clinkers are brick red in color and form a relatively resistant rubble that, when eroded, begins to resemble badlands.

Phosphatic formations, such as the Permian Phosphoria in the northwestern United States, are not only recognized as excellent petroleum source rocks, but also contain valuable amounts of phosphate and other metals (e.g., vanadium). This rock, dark when freshly broken, weathers to a characteristic light color that can often be followed for great distances. Otherwise, these phosphatic rocks can be followed along strike using photogeologic methods the same as any other unit.

Uranium-vandium deposits form at roll fronts in sandstones where oxidizing and reducing groundwaters come into contact. The metals, in solution in oxygen-rich groundwater, precipitate under reducing conditions (e.g., where there is an increase in organic matter, or in areas of seeping hydrocarbons). Roll fronts may be recognized when, for example, an outcropping red-bed sandstone is bleached at the surface. This represents the contact between oxidizing and reducing zones where the iron oxide minerals have been reduced or mobilized (see Chapter 5, Color Anomalies). Such is the case, for example, with uranium occurrences in Lisbon Valley, Utah.

Base Metal Sedimentary Deposits and Chemical Precipitates

Banded iron deposits ranging from 15 to 35% iron are economically important in the Lake Superior district, in Brazil, Australia, and elsewhere. Iron minerals include magnetite, limonite, siderite, chlorite, specularite, hematite, and pyrite, among others. Typically laminated to thin-bedded, these sedimentary deposits are almost always Precambrian or Paleozoic. Oolitic iron of the Silurian Clinton Formation in the eastern United States, or Jurassic oolitic limonite between 30 and 35% iron in Luxembourg and Alsace-Lorraine, is found associated with limestone and silica. Iron minerals (limonite, siderite, hematite, chlorite) are thought to have replaced the original calcareous ooze. Iron formations can be recognized on imagery as a result of (1) their overall resistance to erosion (in many areas, as in the Superior district, the iron ores are resistant, whereas others, like Clinton ores of the Birmingham district, Alabama, are soft), (2) the characteristic iron absorption features of the iron minerals at 0.4 and 0.9 microns, and (3) the red colors (reflectance peak at 0.73 microns) of the oxidation products. Depending on the associated minerals, one can also look for absorption bands characteristic of carbonates (siderite) and silicates (chert, greenalite). Sulfide facies will weather to easily recognized iron oxides at or near the surface. Obviously, magnetic techniques would be an appropriate compliment to imagery in exploring for or extending known deposits.

Manganese deposits are sometimes associated with banded iron formations and sometimes with volcaniclastics or deep marine precipitates. Ores range from pyrolusite in an oxidizing environment to rhodochrosite and hausmannite (intermediate Eh-Ph) to alabandite or manganosite in reducing environments. The ores may be recognized by their dark color (pyrolusite), but are more likely to be found as a result of more flamboyant adjacent units such as highly altered red and green andesitic tuffs in the

Elqui River valley, Chile, the red-brown "bayate" jasperoid in Cuba or Haiti, or the more resistant iron formations of the Cuyuna range, Minnesota and Morro do Urucum, Brazil.

The Kupferschiefer of northern Europe is a metal-bearing bituminous calcareous shale of Mid-Permian age. Although it is less than a meter thick, it extends from northern England to Poland. It overlies a thin conglomerate, and is overlain in turn by limestones. Ore minerals include bornite, chalcocite, chalcopyrite, galena, sphalerite, tetrahedrite, and pyrite, with copper being the most important metal mined. The Zambian copper ores exist in Precambrian carbonaceous shales and consist of chalcopyrite, bornite, and chalcocite. The entire sequence has been subjected to greenschist facies metamorphism. These two examples of stratiform base metal deposits show that they will be expressed and can be mapped at the surface like any other sedimentary unit. The main difference is that these deposits contain pyrite that will weather to iron oxides that can be easily found using band ratios, and the characteristic ore and gangue minerals can be mapped using hyperspectral imagery.

Replacement Mantos

Gold mineralization at Carlin, Nevada, is disseminated as microscopic grains in an altered and leached Devonian limestone. Ore minerals include gold, realgar, cinnabar, and stibnite associated with extensive silicification, argillic alteration, and pyritization. Gold, derived from epithermal solutions, precipitated in permeable and porous limestone associated with illite, organic matter, pyrite, and mircrocrystalline quartz. Multispectral and hyperspectral imagery can be used to map ore and alteration minerals occurring at the surface.

Mississippi Valley-type deposits are essentially replacement of carbonates by sulfide ore minerals, chiefly sphalerite and galena with minor amounts of chalcopyrite, pyrite, and marcasite. In the Tri-State mining district of Missouri, Oklahoma, and Kansas, lead and zinc deposits occur in individual horizons in the Mississippian Keokuk and Warsaw limestones. A dolomitic core is surrounded progressively outward by the main ore zone, jasperoid zone (silicified chert breccia), a shale zone, and crinoidal limestones. Prominent lineaments exist on the surface that are thought to be basement faults that controlled mineralizing solutions. A shale seal exists above the mineralized unit, perhaps containing the fluids within the porous and reactive limestones and dolomites. In addition to multispectral/hyperspectral imagery to map ore minerals, jasperoid, and dolomitic alteration, the association of deposits with faults is an important clue to locating new deposits or extensions of existing mineral zones.

In the Gilman district, Colorado, the Mississippian Leadville limestone was dolomitized prior to mineralization. The dolomite, being somewhat more porous and permeable, reacted with mineralizing fluids and became host to native gold, gold-silver tellurides, copper-silver sulfosalts, pyrite, sphalerite, galena, chalcopyrite, bornite, and pyrrhotite. Some of the mantos are up to 100 meters wide by 50 meters thick and several hundred meters long. The Magma mine, at Superior, Arizona, saw the Devonian Martin limestone replaced by a deposit 7 meters thick and 290 meters wide, and 1500 meters along bedding. Ore and gangue minerals include specularite, pyrite, chalcopyrite, bornite, chalcocite, sphalerite, galena, and minor quartz, barite, and magnetite. While surrounding carbonates are relatively unaltered, the adjacent diabase and schist are sericitized and silicified. In both of these examples faults are thought to have played an essential role in moving fluids into the carbonates. Thus, faults, alteration zones, and diagnostic minerals are the key to mapping these types of deposits on imagery.

Fluvial and Marine Placer Deposits

Fluvial placer deposits can be important sources of gold, diamonds, and tin. Minerals include native gold and platinum, zircon, cassiterite, chromite, rutile, ilmenite, and several gemstones. Placer deposits occur because the heavy minerals fall out of suspension in areas where the velocity of water in a river or stream slows abruptly, such as below rapids or falls or where the stream gradient flattens. These are the same conditions that are conducive to braided streams, i.e., bed load exceeds the stream's capacity to carry sediment. Logical places for placers to accumulate are on point bars (inside curves of meanders), sand bars below rapids or where a stream channel is narrow and incised, and at deltas in lakes or along the coast (where longshore currents then can rework the sediments). Because most preserved placers occur at the surface and are of Cenozoic age they are most amenable to remote sensing methods. If one knows where a lode deposit exists, it is logical to look for placers downstream. In the case of paleo-placers, such as in the Sierra Nevada of California or the Victoria field of Australia, old buried stream beds exist beneath or nearly beneath present-day streams. In both examples the placers were also buried beneath lava flows. Since the lavas flowed down the stream valleys, they tend to be thicker there. After a period of erosion the adjacent, thinner lavas were worn away and the thick, valley fill deposits remain as topographic highs capping the old placer deposits (e.g., Fig. 4.3). These features are readily mapped on airphotos and imagery of an appropriate scale.

The marine placers, such as the gold deposits at Nome, Alaska, the diamond deposits of Namibia, or the titanium deposits of Florida, are usually brought to the beach by rivers, then reworked and concentrated in narrow zones parallel to the coast by both wave and wind action. At Mineral City, Florida, for example, the main concentration of titaniferous sands occurs between sand dunes and the wave-swept beach where storm waves have reworked the sand dunes, removing the fines and lighter minerals. As sea level has risen, some of these deposits are now slightly offshore in submarine bars. Since magnetite is a common heavy mineral in these deposits, aeromagnetics is a logical tool to use along with airphotos and/or hyperspectral imagery.

Sand and gravel are important ingredients in concrete and are also used for roadbeds, building site preparation, and other construction purposes. They are usually easy to recognize and map on airphotos and images because they occur in river channels, beach deposits, and sand dunes. Where covered by vegetation, they will support plant communities such as pine trees that favor well-drained soils and substrate.

Laterites

Oxidation and leaching of surface rocks can lead to gossans where sulfide deposits are at the surface, or can lead to laterites where the insoluble residue of iron-, aluminum-, manganese-, or nickel-rich parent rocks are present. Enrichment takes place where the oxidized product is stable and other constituents are selectively leached away. Iron, manganese, and aluminum form oxides and hydroxides that are relatively insoluble at the surface. Lead forms a stable sulfate; copper, lead, and zinc can form stable carbonates in some environments. Copper, zinc, nickel, and chromium can form stable silicates and oxides.

In tropical climates iron and aluminum form metal oxide-rich soils known as laterites. Iron laterites form over ferromagnesian rocks, especially in gentle topography where siliceous constituents can dissolve without eroding the soils. Many iron laterites form over serpentines. Aluminum laterites, called bauxite, tend to form over syenites and nepheline syenites. Common minerals include boehmite, gibbsite, and diaspore often associated with clay minerals. Bauxites also form over carbonate rocks associated with residual clays in a red soil called terra rosa. Manganese laterites, consisting of pellets of manganese oxides, are known as granzon. Nickel laterites form over serpentines, peridotites, and dunnites, as in the deposits of New Caledonia, which contain 6–10 percent nickel and lesser amounts of chromium and cobalt.

The common factor in all of these weathering-enriched deposits is that they occur in tropical or semi-tropical environments. The laterite generally forms a hard soil horizon that is not agreeable to many plants, so that the forest canopy will change, become stunted, or become barren over the deposits. This change should be detectable using multispectral imagery or airphotos. The association with clay minerals and hydroxides means that, where exposed, these minerals can be detected by their absorption bands between 2.2 and 2.4 microns. Terra rosa soils can be recognized by their red color, and iron laterites by the iron oxide mineral absorptions and red color. Hyperspectral imagery may be able to recognize individual diagnostic minerals.

REFERENCES

1. M. Scavetti, *Bull. Am. Assn. Pet. Geol.* **76** (1992): 708–730.
2. G.R. Hunt and J.W. Salisbury, *Modern Geology* **1** (1970): 283–300.
3. G.R. Hunt and J.W. Salisbury, *Modern Geology* **2** (1971): 23–30.
4. G.R. Hunt and J.W. Salisbury, *Modern Geology* **5** (1976): 211–217.
5. L.C. Rowan, P.H. Wetlaufer, A.F.H. Goetz, F.C. Billingsley, and J.C. Stewart, *U.S. Geol. Survey Professional Paper* **883** (1974): 35 p.
6. G.L. Prost, *Econ. Geol.* **75** (1980): 894–906.
7. M.J. Abrams, R.P. Ashley, L.C. Rowan, A.F.H. Goetz, and A.B. Kahle, *U.S. Geol. Survey Open File Rept.* **77–585** (1977), 19 p.
8. M.D. Krohn, M.J. Abrams, and L.C. Rowan, *U.S. Geol. Survey Open File Rept.* **78–585** (1978), 66 p.
9. M. Loe've, *Probability Theory* (Van Nostrand, Princeton, 1955).
10. A.B. Kahle, D.P. Madura, and J.M. Soha, *Appl. Optics* **19** (1980): 2279–2290.
11. A.B. Kahle and A.F.H. Goetz, *Science* **222** (1983): 24–27.
12. N.M. Short and L.M. Stuart, *The Heat Capacity Mapping Mission*, **NASA SP 465** (U.S. Gov' Printing Office, Washington, D.C., 1982).
13. F.B. Henderson III and B.N. Rock, *Fourth Geosat Workshop* (Am. Soc. Photogram, Falls Church, 1983).
14. X. Jinhai, *Chinese Jour. Geochem.* **9** (1990): 169–177.
15. S.J. Gaffey, *Geology* **13** (1985): 270–273.
16. R.N. Clark, T.V.V. King, M. Klejwa, G.A. Swayze, and N. Vergo, *Jour. Geophys. Res.* **95 B8** (1990): 12,653–12,680.
17. M.D. Krohn, *Jour. Geophys. Res.* **91 B1** (1986): 767–783.
18. A.F.H. Goetz, G. Vane, J.E. Solomon, and B.N. Rock, *Science* **228** (1985): 1147–1153.
19. A.F.H. Goetz, *Episodes* **15** (1992): 7–14.

20. H.R. Goetting and R.J.P. Lyon, *Proc. of the 5th Thematic Conf. on Rem. Sens. for Exploration Geology* (ERIM, Ann Arbor, 1987), pp. 513–520.
21. F.A. Kruse, K.S. Kierein-Young, and J.W. Boardman, *Photogram. Engr. and Rem. Sens.* **56** (1990): 83–92.
22. F.A. Kruse, A.B. Lefkoff, J.W. Boardman, K.B. Heidebrecht, A.T. Shapiro, P.J. Barloon, and A.F.H. Goetz, *Remote Sensing Environ.* **44** (1993): 145–163.
23. F.A. Kruse, A.B. Lefkoff, and J.B. Dietz, *Remote Sensing Environ.* **44** (1993): 309–336.
24. F.A. Kruse and A.B. Lefkoff, *Remote Sensing Reviews* **8** (1993): 3–28.
25. J.F. Mustard and C.M. Pieters, *Jour. Geophys. Res.* **92 B4** (1987): E617–E626.
26. J.F. Mustard and C.M. Pieters, *Jour. Geophys. Res.* **92 B10** (1987): 10,376–10,390.
27. P.N. Bierwirth, *Int. Jour. Remote Sensing* **11** (1990): 1999–2017.
28. J.W. Boardman, PhD Dissertation, University of Colorado, Boulder (1991), 212 p.
29. J.W. Boardman and F.A. Kruse, *Proc. of the 10th Thematic Conf. on Geologic Rem. Sens.* **1** (ERIM, Ann Arbor, 1994), pp. I-407 to I-418.
30. A.P. Crosta, C. Sabine, and J.V. Taranik, *Remote Sensing Environ.* **65** (1998): 309–319.
31. F.A. Kruse, *Int. Jour. Remote Sensing* **17** (1996): 1623–1632.
32. A.B. Kahle and L.C. Rowan, *Geology* **8** (1980): 234–239.
33. J.E. Conel, H.R. Lang, E.D. Paylor, and R.E. Alley, *IEEE Trans. on Geoscience and Rem. Sens.* **GE-23**, (1985): 562–573.
34. H.R. Lang, S.L. Adams, J.E. Conel, B.A. McGuffie, E.D. Paylor, and R.E. Walker, *Bull. Am. Assn. Pet. Geol.* **71** (1987): 389–402.
35. H.R. Lang, M.J. Bartholomew, C.I. Grove, and E.D. Paylor, *Jour. Sed. Pet.* **60** (1990): 504–524.
36. M. Scavetti, M.C. Ferrari, R. Chiari, P.L. Fantozzi, and I. Longhi, *Bull. Am. Assn. Pet. Geol.* **79** (1995): 1571–1589.
37. C.L. Hanks and R.M. Guritz, *Bull. Am. Assn. Pet. Geol.* **81** (1997): 121–134.
38. O.T. Udo and E.E. Utuk, *Jour. Geochem. Explor.* **37** (1990): 285–300.
39. L.C. Rowan, M.J. Pawlewicz, and O.D. Jones, *Bull. Am. Assn. Pet. Geol.* **76** (1992): 1008–1023.
40. L.C. Rowan, F.G. Poole, and M.J. Pawlewicz, *Bull. Am. Assn. Pet. Geol.* **79** (1995): 1464–1480.
41. R.H. Goodman in A.E. Lodge, ed., *The Remote Sensing of Oil Slicks* (John Wiley and Sons, Chichester, 1989): 39–65.
42. E.A. Cloutis, *Science* **245** (1989): 165–168.
43. K.R. Sundberg, U.S. Patent 4,908,763 (1990), 59 p.
44. P.B. Slack, *Bull. Am. Assn. Pet. Geol.* **65** (1981): 730–743.
45. R.W. Marrs and G.L. Raines, *Bull. Am. Assn. Pet. Geol.* **68** (1984): 1718–1731.
46. R.C. Michael and I.S. Merin, *Bull. Am. Assn. Pet. Geol.* **70** (1986): 453–455.
47. C.A. Viau and A.E. Oldershaw, in Carbonates in Subsurface and Outcrop, Canadian Soc. of Pet. Geol. Core Conf. (1984), pp. 103–131.
48. J.M. Andrichuk, *Bull. Am. Assn. Pet. Geol.* **45** (1961): 612–632.
49. R.G. Greggs and D.H. Greggs, *Jour. Pet. Geol.* **12** (1989): 377–404.
50. P.L. Churcher and A.H. Majid, *Bull. Canadian Pet. Geol.* **37** (1989): 241–245.
51. W.L. Fisher and J.H. McGowan, *Bull. Am. Assn. Pet. Geol.* **53** (1969): 30–54.

ADDITIONAL READING

W.G. Ernst and E.D. Paylor II, *Bull. Am. Assn. Pet. Geol.* **80** (1996): 1008–1026.
W.T. Jansen, *Geobyte* (October 1992): 46–49.
H.R. Lang, Stratigraphy, *Manual of Remote Sensing* **3** (John Wiley and Sons, 1999), 3rd ed., Chap. 6.
J.V. Taranik and A.P. Crosta, *Intl. Archives of Photogram. and Rem. Sens.* **XXXI B7** (1996): 689–698.
D.M. Uhlir, *Earth Observation Mag.* (February 1995), 3 p.

QUESTIONS TO CONSIDER

1. Why should you be able to distinguish shales from carbonates in an arctic environment using radar?
2. On what basis and criteria would you establish a stratigraphic framework when working in a new and unknown area?
3. You are in charge of a program to find new diamond deposits. What sensors would you use, and what would you look for? What if you were looking for coal? For Mississippi Valley-type lead-zinc deposits?
4. Why could deeply buried channel or beach bars have some expression at the surface? How might they be expressed?
5. What is the advantage, if any, of hyperspectral imagery over multispectral imagery? Are there any drawbacks?
6. What, if anything, prevents us from flying an instrument or suite of instruments and generating geologic maps automatically, i.e., without an interpreter?

7

Exploration Case Histories

CHAPTER OVERVIEW

The purpose of this section is to review some examples where remote sensing has contributed to the discovery of minerals and hydrocarbons. Petroleum case histories are given for basins in different stages of exploration maturity, for structural and stratigraphic traps, and using different sensor systems. Examples are provided for hydrothermal gold exploration, porphyry copper systems, and diamond exploration. It is necessary to limit this list here, and readers are referred to the literature for more extensive readings. It is also important to repeat that remote sensing in and of itself is not a stand-alone exploration tool. None of the discoveries reviewed here are exclusively the result of a remote sensing/photogeologic study. Indeed, in most exploration programs the contribution of remote sensing comes at the start of the program, that is, it is used to define an area wherein to stake claims, acquire leases, or to design follow-up surveys. Therefore, although there are some cases of discoveries made using only remote sensing, in most cases remote sensing is but one early step in a long process. The following case histories are culled from the remote sensing and industry literature and are meant to serve as examples of different approaches to exploration programs.

FRONTIER PETROLEUM EXPLORATION: STRUCTURAL TRAPS AT TRAP SPRINGS, NEVADA

The following example is taken from a review of this discovery by Foster [1] and Duey [2].

The Trap Springs field in Railroad Valley, Nevada, was discovered by Northwest Exploration Company on the basis of a detailed study of a known field and possible analogs followed by photogeology, field work, and a seismic program (Fig. 7.1). This is a classic example of photogeology playing a major role in the discovery of an oil field.

The geologists involved in this discovery began by analyzing the Eagle Springs field, discovered in Railroad Valley in 1954. Through analysis of wells and seismic data over the Eagle Springs field they learned that the major factor controlling entrapment was the erosional and fault truncation of the reservoir units, the Oligocene Garrett Ranch volcanics, and the Eocene Sheep Pass Formation (Fig. 7.2). This truncation occurred against a fault-bounded uplift at the valley margin.

EXPLORATION CASE HISTORIES 201

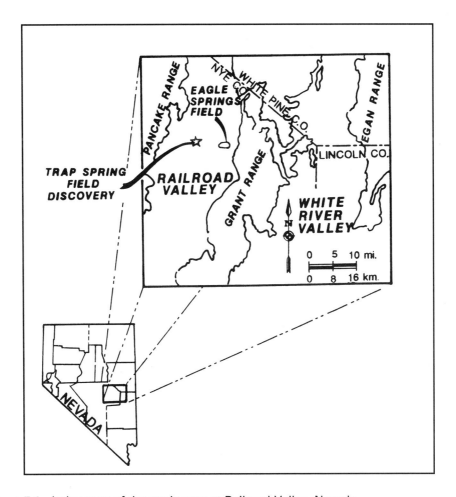

Figure 7.1 Index map of the study area at Railroad Valley, Nevada.

Their work showed that the reservoir beds were present in a structural low, and that the structural highs along the basin margin were "bald" Paleozoic knobs that had been stripped of their Tertiary reservoir units. This exploration model differed from previous (unsuccessful) exploration efforts in that all previous programs had drilled the structural highs in the mistaken belief that the Eagle Springs field was located on a structural high rather than in the low adjacent to it. These other programs invariably drilled from basin fill immediately into Paleozoics.

Once they understood the trapping mechanism they set about mapping structures on airphotos and in the field. It turned out that the mountain front was not the location of the main basin bounding fault, because the front had been eroded back and a pediment had formed. The faults could be seen on airphotos as linear tonal and vegetation anomalies caused by moist soil and springs localized along the faults (Fig. 7.3). Furthermore, the highs were bounded by faults that extended from the valley into the adjacent ranges (Fig. 7.4). This system of faults, then, delineated a series of high and low blocks that were interpreted as analogs to the Eagle Springs field.

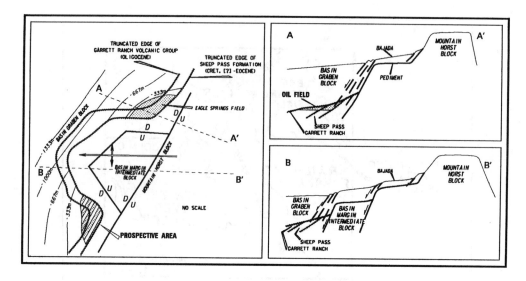

Figure 7.2 The Eagle Springs field is located on an intermediate block adjacent to a "bald" high. Cross section AA¹ cuts across the Eagle Springs field; section BB¹ cuts across the high block and shows how the reservoir has been removed there.

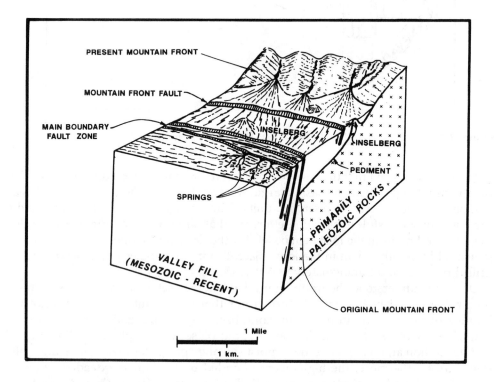

Figure 7.3 Block diagram showing that the mountain front does not coincide with the main mountain front fault or main graben-bounding fault.

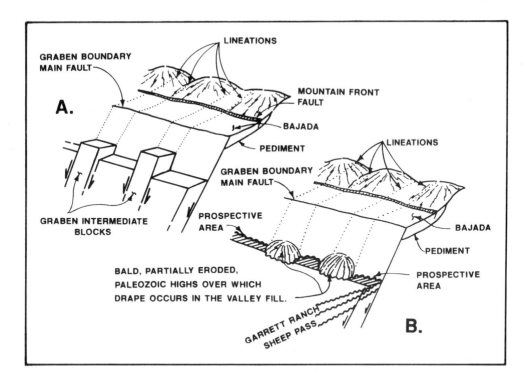

Figure 7.4 The downdropped fault blocks are broken along crosscutting faults that extend from the basin into the mountains. The reservoir units were eroded off the tops of the intermediate blocks, but were preserved adjacent to them.

The photointerpretation was used during a field program that identified the mountain front and master graben-bounding faults. One problem the geologists encountered was old lake shorelines that in some areas were confused with faults during the photointerpretation. They differentiated between these by noting that faults cut across topographic contours, whereas paleoshorelines followed the contours.

Acreage was acquired after field work and a preliminary seismic program, and further detailed seismic data was acquired. The discovery well was drilled in November, 1976. Over 550,000 barrels of oil were produced by the end of 1978.

MATURE BASIN PETROLEUM EXPLORATION: PARADOX BASIN STRATIGRAPHIC TRAPS, UTAH

The following review is taken entirely from Merin and Michael [3].

This project began with a review of the literature regarding controls on accumulation and production in the Paradox basin of southeastern Utah (Fig. 7.5). These workers learned from previous studies that the existing fields are stratigraphic traps which appear to be structurally controlled in the sense that the desired reservoir facies, Pennsylvanian algal mounds, are located along fault-controlled paleobathymetric hingelines or along the margins of faulted blocks. This provided the incentive for an

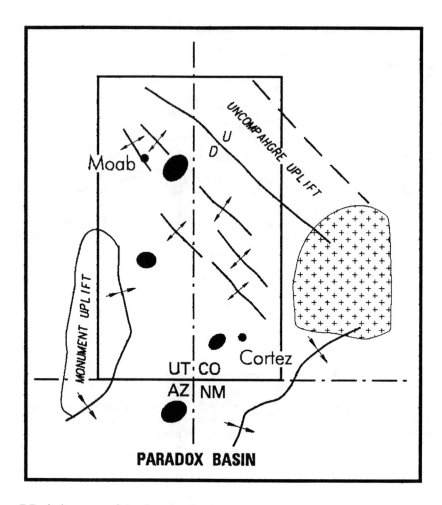

Figure 7.5 Index map of the Paradox basin stratigraphic trap play area. Major fields are shown in black.

integrated Landsat-potential fields-field mapping study of major structures in the basin.

Landsat imagery was acquired and interpreted to help locate these reservoirs by mapping the regional structural framework and local structural details associated with producing fields and trends. The satellite structure map was then compared to gravity and magnetics maps and existing isopach data, and then was checked using field mapping to confirm and/or deny the interpretation. The goal was to test this technique and provide a strategy and procedure for future exploration efforts.

The Pennsylvanian algal mounds in the Paradox Formation are lens-shaped, partly dolomitized carbonate muds with porosity enhanced both by subareal leaching during deposition and later diagenetic dolomitization. These pods grade laterally and vertically into tight micrites.

The Landsat images were interpreted at a scale of 1:250,000. All evidence of structure, including folds, faults, and fracture zones were mapped. The authors found that northeast and northwest lineaments were dominant throughout the area. The interpretation was checked by mapping structures in the field. Those lineaments that crossed outcrops were found in most cases to have either measurable offset or zones of lineament-parallel joints. Many of these zones were also found to coincide with gravity and magnetics lineaments, suggesting that they originated in offsets of basement.

The results of this investigation show that individual carbonate mounds trend northwest, yet many are abruptly terminated or segmented by northeast oriented structures (Figs. 7.6, 7.7). Isopach maps reveal that the algal buildups occur on northwest-oriented paleostructural highs, and that within the mounds individual pods may be oriented northeast. Several fields, including the Ismay, Bug, Papoose Canyon, Tricentral, Gothic Mesa, and Cache fall into this pattern.

The methodology proposed as a result of this work is to first learn what controls the distribution of known production. In this case stratigraphic traps were indirectly controlled by structure since the bioherm reservoirs were located on paleobathymetric highs that were structurally controlled. The next step was to determine the structural framework of the area by interpreting Landsat images and then comparing the structure interpretation to isopach maps to learn where the bathymetric highs had been. This step is used to eliminate those structures that are less likely to have exerted

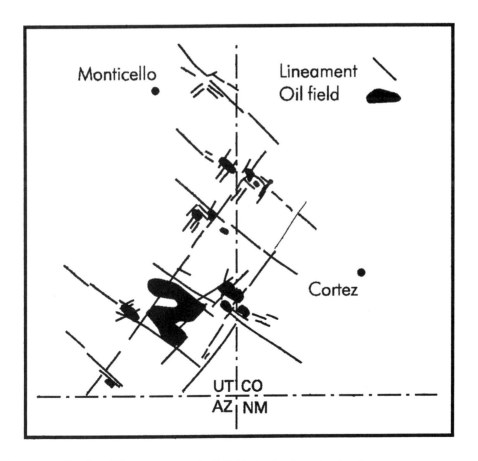

Figure 7.6 Regional lineaments and oil fields in the Paradox basin area.

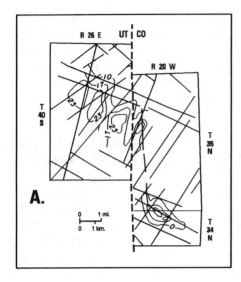
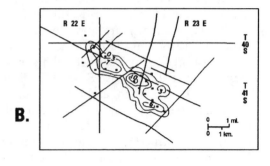

Figure 7.7 The correlation of isopachs and Landsat lineaments shows how the algal mounds appear structurally controlled. A. Ismay isopach (in meters), Ismay field and Cache field. B. Lower Desert Creek isopach, Gothic Mesa field.

control (facies or thickness changes) on the desired stratigraphic intervals. Field mapping in critical areas provided some confidence that the features being mapped were indeed faults and fracture zones. A map showing the distribution of existing fields was used to identify those producing trends that could be extended into new areas, and to suggest areas without production that appear analogous to producing fields.

MATURE BASIN PETROLEUM EXPLORATION; DENVER BASIN FRACTURED RESERVOIRS, COLORADO-WYOMING

The following discussion is taken from work by Merin and Moore [4].

The Cretaceous Niobrara Formation in the Denver Basin is a fractured reservoir unit that consists of limestones and interbedded calcareous shales (Fig. 7.8). It is also an oil-prone, mature source rock that has been in the oil generation window since the Eocene. Within the study area, the Niobrara lies at depths between 1800 and 2100 m and has matrix permeabilities generally less than 0.01 md and porosities less than 10%. This is a poor reservoir unless enhanced by fracturing.

Two types of fracture-controlled oil production have been described for the Niobrara Formation: production from fractures associated with flexures and folding along the west side of the basin (e.g., Loveland and Berthoud fields), and production from fractures in relatively undeformed parts of the basin (e.g., Silo field). A program to define fractured production fairways within the basin utilized Landsat multispectral scanner images at a scale of 1:125,000; compared the image interpretation to generalized structure contour, isopach, and resistivity maps; and then analyzed changing crustal compressive stresses to determine the fracture history.

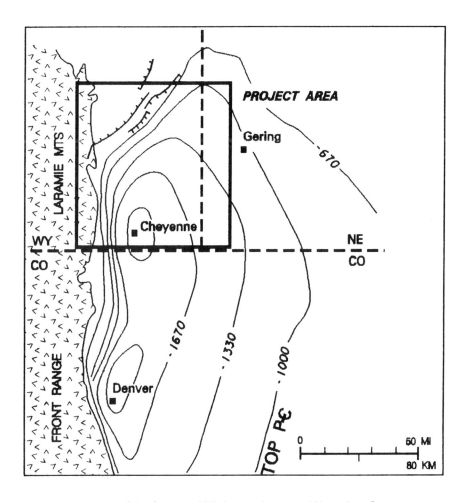

Figure 7.8 Index map of the fractured Niobrara play area, Wyoming. Structure contour on the Precambrian. Contour interval = 330 m.

Landsat imagery was examined and structure maps made using standard photointerpretation techniques (Fig. 7.9). Lineaments were mostly aligned stream segments and tonal patterns and geomorphic features that might indicate faults or joint zones. The interpretation was later digitized and lineaments were plotted as length-weighted and frequency-weighted rose diagrams to help identify significant trends. The authors feel confident that the lineaments represent extensions of previously mapped structures or previously unrecognized structures because the lineaments are parallel or subparallel to known faults and tectonic zones. They see the basin as a mosaic of tectonic blocks bounded by northeast and northwest fault zones that can be mapped at the surface as lineaments.

Well logs were examined to extract information for a basin-wide resistivity map that shows where the Niobrara is oil saturated (7.10). The logs were also used to make isopachs, and Niobrara thins were correlated with paleohighs and areas of increased organic carbon content (richer source rock). Neither of these data sets correlated well with the surface fracture zones except along the northwest margin of the basin.

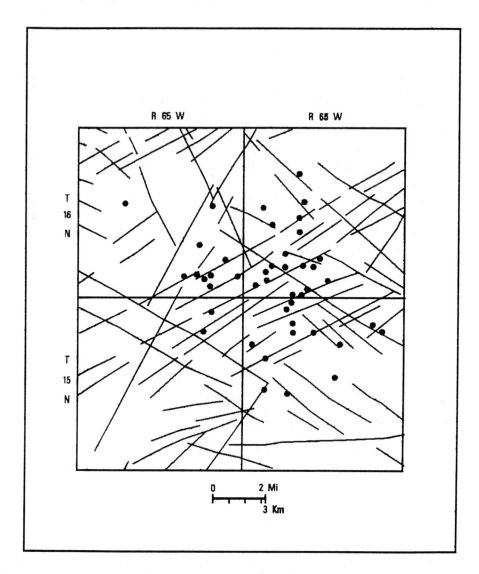

Figure 7.9 Distribution of Landsat lineaments and oil wells, Silo field area, Laramie County, Wyoming.

The authors then looked at the stress history of the basin since Laramide and found that early Laramide (Late Cretaceous through Paleocene) maximum horizontal compressive stresses were oriented east-northeast, whereas later Laramide (Eocene) compressive stresses were aligned northeast (Fig. 7.11). They make the assumption that the fractures in the basin are extensional and that they formed parallel to the maximum compressive stress. As these stresses rotated through time successive new fractures formed due to the generation of Niobrara oil. Their model predicts that northeast fractures acted as conduits and formed storage capacity (fracture porosity) for this otherwise tight formation. The northeast fractures would have been preferentially open and filled with hydrocarbons during late Laramide compression.

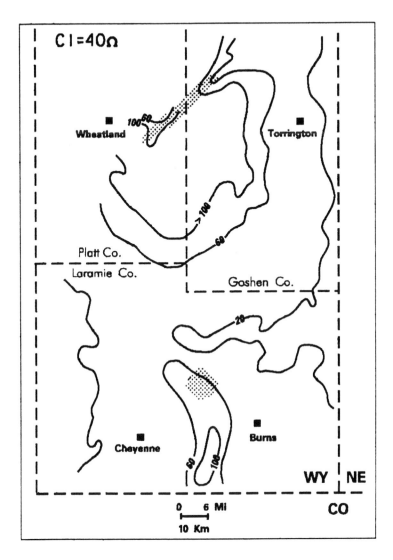

Figure 7.10 Resistivity map of the Niobrara. Contour interval = 40 ohms. Shading indicates wells with oil or shows.

The best reservoirs, then, should be at intersections of surface lineaments or along braided and intersecting northeast fracture zones.

A statistical study of producing wells versus distance to surface lineaments showed that 79% of the oil-producing wells are located within 0.16 km of a lineament, and that about two-thirds are located along northeast lineaments. In four independent attempts at randomly plotting wells on the lineaments the success rate was between 32 and 45%. In addition, core from three of the wells shows oil is produced from zones with open vertical fractures.

The results of this work suggest that productive fairways can be defined for fractured reservoir plays using careful remote sensing structural interpretations in conjunction with subsurface data.

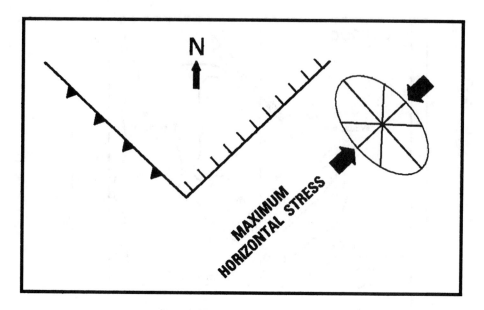

Figure 7.11 Orientation of late Laramide (Eocene) stresses proposed in this paper, and probable fracture sets associated with these stresses. Northeast joints are most likely to be open and contain oil.

MINERAL EXPLORATION: ALTERATION ASSOCIATED WITH GOLD AT GOLDFIELD, NEVADA

The following example is taken from a poster session by Hauff, et al. [5].

The Goldfield mining district, located in southern Nevada (Fig. 7.12), has an arid climate with limited vegetation cover, mainly sagebrush, rabbit brush, cactus, Joshua trees, and grasses. It is a quartz-alunite type epithermal gold deposit in silicified early Miocene intermediate volcanics, and has produced over 4,000,000 ounces of gold since coming on production in 1910. Mineralized areas, which also contain significant amounts of silver and copper, are known as ledges, and these form along an arcuate zone of fractures with extensive hydrothermal alteration containing quartz, alunite, kaolinite, dickite, diaspore, pyrophyllite and illite. Some ledges are barren, whereas others are ore-bearing. No one has been able to differentiate the barren areas from productive zones until now.

Several open pits exist within the mining district and expose the alteration zoning. Minerals in the central ore zone include dickite with silica ± diaspore; alunite occurs along the edges, and kaolinite in the argillic envelope. This study shows that gold is likely to be present whenever dickite occurs in a quartz-alunite ledge. Apparently alunite-kaolinite-quartz alone is not gold-bearing. Kaolinite is considered indicative of unmineralized rock. An attempt was made to confirm these observations through the use of remote sensing data.

Three types of data were used in this project. First, a Landsat TM image was created to provide a regional overview. This image was processed to a color ratio composite consisting of the band ratios 3/1 (red), 5/4 (green), and 5/7 (blue). Each band ratio was independently stretched to obtain the best contrast and color balance. The ratio

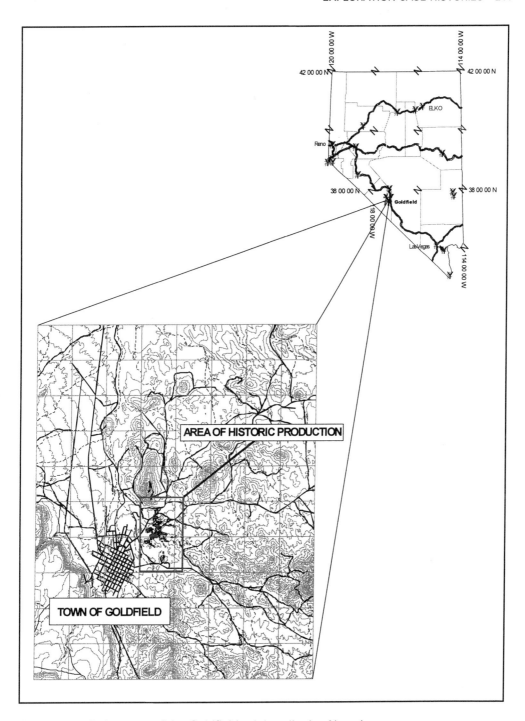

Figure 7.12 Index map of the Goldfield mining district, Nevada.

composite was used to predict areas of alteration and the general composition of alteration in the mining district. Vegetation and water were masked to enhance the

212 REMOTE SENSING FOR GEOLOGISTS

mineralization. Iron-rich minerals are red and orange; silica-rich rocks are green; clays are shades of blue; and the predicted alteration is shown as light shades of pink, yellow, and blue to white.

Ground spectral data (Fig. 7.13) were collected using the PIMA-II portable field spectrometer. This instrument provides a spectral reflectance curve for small areas that

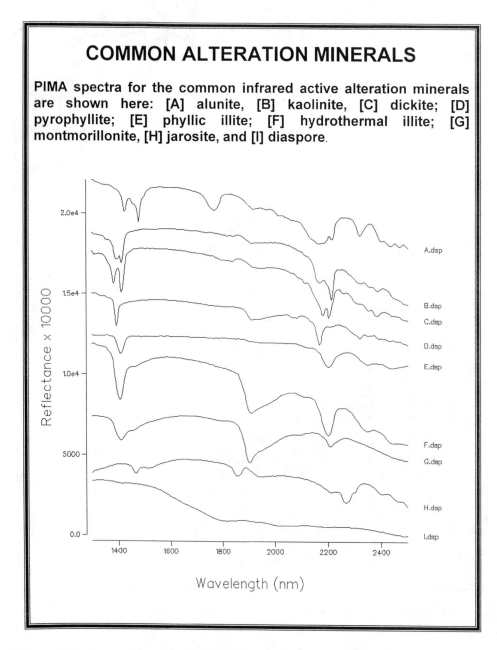

Figure 7.13 Ground spectra of dickite and kaolinite, among others, collected in the Goldfield district by Spectral International Inc.

have more or less homogeneous mineralogy and/or surface cover. It is used to calibrate the hyperspectral imagery by providing spectral control points where mineralogy, alteration, soil, and ground cover are known. This calibration database can then be used to classify hyperspectral image maps.

Classification maps were made from AVIRIS and SFSI images (Chap. 3, Airborne Scanners). The classification was based on the ground control points mentioned above. Recognizing the mineral dickite was particularly difficult using AVIRIS, since the spectral curve is similar to that of kaolinite at the resolution of the AVIRIS instrument. The SFSI instrument was better able to resolve these differences and predict the location of dickite alteration. The resulting images show blue and red where gold is most likely to occur (Fig. 7.14). These images demonstrate that dickite is a dominant

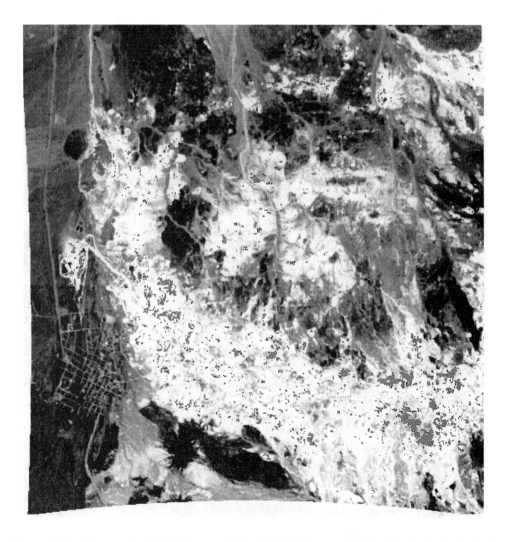

Figure 7.14 Gray-scale AVIRIS image of the Goldfield area with Alunite-dominated pixels displayed in blue and Dickite-dominated pixels displayed in red. Processing by Spectral International Inc. (See color plate.)

alteration mineral in the gold producing zones. The authors strongly feel, based on their field and hyperspectral image classification, that dickite, rather than kaolinite or alunite, is the primary indicator of gold mineralization.

STRUCTURAL MAPPING AS A GUIDE TO PORPHYRY COPPER DEPOSITS, NORTHEAST CHINA

The following discussion is taken from Jiang, et al. [6].

The purpose of this study was to develop criteria for mapping porphyry copper deposits using Landsat Thematic Mapper imagery in a heavily forested area in northeastern China. Thick soil and vegetation cover make conventional mapping difficult. The project area lies along the continental margin at the Daxinganling basin and consists of Early Ordovician to Permian interbedded volcanics and terrigenous strata. Two large porphyry-associated copper deposits were discovered here during the 1970s.

Intrusives are Variscan (Late Devonian-Permian) granodiorite and granodiorite porphyry, although only the granodiorite is known to contain ore deposits. The intrusives and country rocks are extensively hydrothermally altered, including potassium feldspar/silicification (central), sericitic (intermediate), and propylitic (outermost) zones. The intrusives appear to be located along or at the intersection of regional faults.

Landsat Thematic Mapper images were processed to false color images using bands 4, 5, and 7 printed as red, green, and blue, respectively. Color ratio composites, principal component images, and vegetation index images were also generated. Known copper deposits show up clearly on images because the intrusives are low relief features with obvious textural contrast to adjacent rocks. Thick vegetation cover makes it difficult to distinguish stratigraphic information, including color and alteration.

Structural features, in particular regional faults and fault intersections, appear to be the key to localizing the intrusives and mineralization. Faults provided the paths for migration of hydrothermal fluids and served as the locus for precipitation of mineralizing solutions. A series of nested circular structures (ring faults and/or ring dikes) associated with the intrusives are described for the first time in this paper (Fig. 7.15). The circular structures are expressed as valleys and have a maximum radius of about 12 km. The circular features coincide with radial faults, and it is felt that, whereas the regional fault intersections localized the intrusives, multiple intrusions over time formed the circular and radial fault patterns. Radial faults are also expressed by drainage patterns.

The circular structure is developed in Ordovician submarine volcanics and Silurian marine sediments. The granodiorite intrusions are thought to have developed at a weak point focused at the intersection of regional faults. Intrusions penetrated the existing sediments and formed extrusive volcanic flows and tuff deposits on the paleosurface. Two large copper deposits exist at the southern and southwestern margins of the circular structure. Some mineralization also occurs along the southwestern margin at the intersection of northwest and northeast-trending faults. Observations during this study suggest that northeast faults control mineral occurrence, whereas northwest faults control the shape of ore bodies. Both alteration and mineralization are controlled by the location and intensity of faulting.

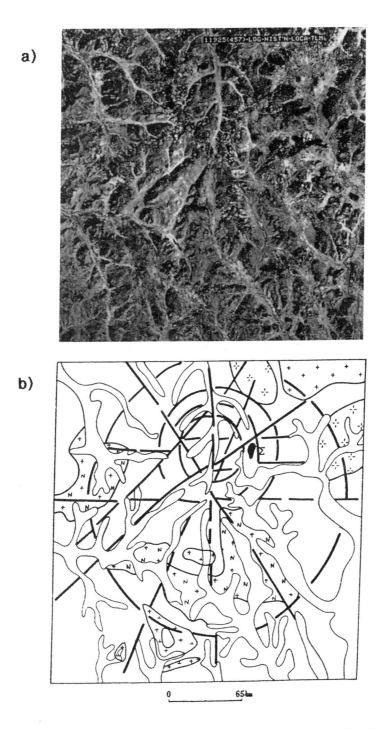

Figure 7.15 (a) Landsat TM image; (b) Structural interpretation showing ring fractures/ dikes and radial fractures that appear to control porphyry copper mineralization in Devonian-Permian granodiorites of the Daxinganling basin, northeast China.

This work demonstrates that intrusions and associated porphyry copper-type mineralization can be found by mapping major regional fault intersections that coincide with topographic and textural indicators of intrusives even in areas with heavy vegetation and soil cover.

EXPLORATION FOR KIMBERLITES AND DIATREMES, UTAH, COLORADO, AND WYOMING

A project to determine whether the Airborne Visible/Infrared Imaging Spectrometer (AVIRIS) could characterize the mineral assemblages associated with kimberlites was undertaken by Kruse and Boardman [7,8].

Two sites were examined, one along the Comb Ridge monocline in southeast Utah, the other in the State-Line district of Colorado/Wyoming. Comb Ridge is an east-dipping roughly north-south trending monocline consisting of Pennsylvanian to Jurassic age sandstones, shales, and limestones. This part of Utah is arid, and vegetation away from rivers is restricted to some grass, sagebrush, and cactus. The State-Line district is primarily Precambrian igneous and metamorphic rocks (Fig. 7.16). The area is temperate, with conifer forests and grass in clearings.

Kimberlites are diamond-bearing mantle rocks (mica peridotite) that are injected into the crust and contain predominantly ultramafic minerals including olivine, phlogopite, perovskite, spinel, chromite, diopside, monticellite, apatite, calcite, and iron-rich serpentine. Alteration to serpentine and calcite is common. Xenoliths from the mantle, crust, and overlying sediments are also found in the diatremes. A diatreme is the cone-shaped near-vertical pipe that contains the kimberlite. Their expression at the surface is typically circular to elliptical, with sharp contacts with the surrounding country rock, and the kimberlites often weather to topographic depressions.

The AVIRIS instrument contains 224 channels with a 0.01 micron spectral resolution (see Chapter 3, Airborne Multispectral/Hyperspectral Scanners). Processing consisted of an atmospheric correction, selection and identification of spectral end member minerals, and mapping of end member occurrence and abundance (see Chapter 6, Hyperspectral Imagery). The results are presented as black/white images where brighter pixels represent higher abundances of a mineral, or as black background images showing only the distribution of targeted minerals.

The Utah site contains three diatremes: the Mule Ear diatreme in the north, the Moses Rock diatreme 6 km south, and the Cane Valley diatreme 3 km southwest of Moses Rock. These diatremes contain varying amounts of country rock and "serpentine tuff." Ten end member minerals were defined using the AVIRIS bands from 0.4 to 1.3 microns, including hematite, goethite, five soil classes, green vegetation, water, and shadow. Thirteen end members were defined for the AVIRIS channels between 2.0 and 2.4 microns, including dolomite, calcite, kaolinite, illite, iron-rich illite (?), silica in sandstone, and probable weathered serpentine. These end members were used to map the diatremes. AVIRIS data show the Mule Ear diatreme is characterized by two dolomites, minor calcite, illite, kaolinite, and goethite. The Cane Valley diatreme is similar. Dolomite and a mineral that may be weathered serpentine (absorption feature at 2.35 microns) are the primary minerals at the Moses Rock dike.

Diatremes in the State Line district are localized along prominent joint and fault zones associated with the Virginia Dale ring dike complex (Fig. 7.17). The kimberlite assemblage consists of breccia containing kimberlite, serpentine, calcite, dolomite,

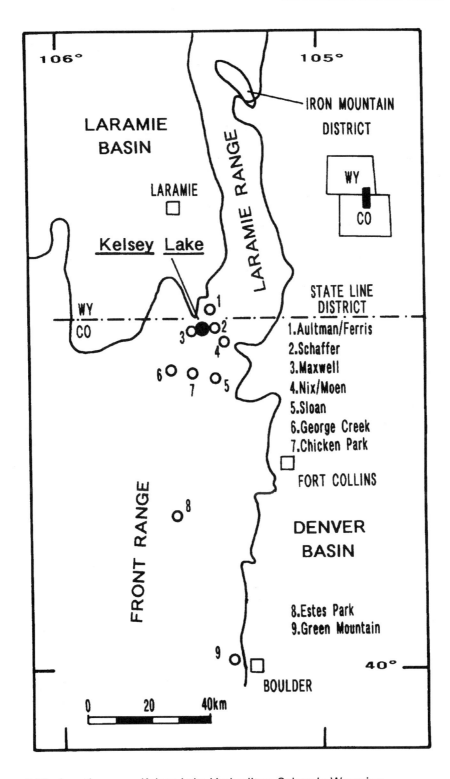

Figure 7.16 Location map, Kelsey Lake kimberlites, Colorado-Wyoming.

Figure 7.17 AVIRIS saturation-enhanced true color image of the State Line kimberlite district, Colorado-Wyoming. Processing by Analytic Imaging and Geophysics LLC. (See color plate.)

phlogopite, magnetite, perovskite, chlorite, talc, and hematite. Paleozoic xenoliths of dolomite, conglomerate, and sandstone also occur in the diatremes. The kimberlites are deeply weathered and altered to serpentine and calcite. They are also poorly exposed, so that most spectral signatures are dominated by green and dry vegetation. In some cases dolomite, calcite, serpentine, phlogopite, and kaolinite can be mapped associated with the diatremes.

Five end member minerals were mapped using AVIRIS (Fig. 7.18). The end member distribution was then extracted from the image of the Kelsey Lake mine area (Fig. 7.19). These end member minerals were compared to the spectral library and found to conform closely to the spectra for serpentine and phlogopite (Figs. 7.20 and 7.21).

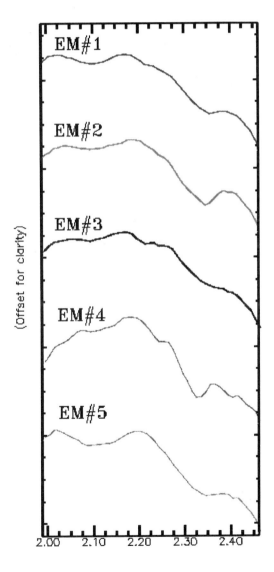

Figure 7.18 Five end-member mineral spectra mapped in the Kelsey Lake mine area, State Line district. By Analytic Imaging and Geophysics LLC. (See color plate.)

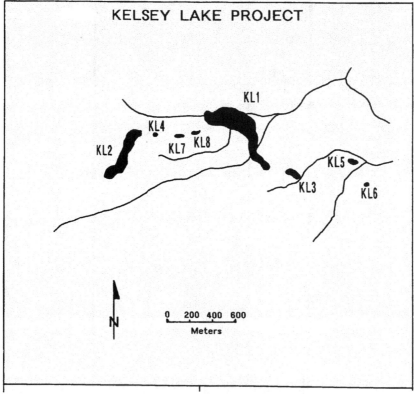

Figure 7.19 (a) AVIRIS prospects, Kelsey Lake mine area. Colors match end members in Fig. 7.18. (See color plate.) (b) Kelsey Lake mine known kimberlite locations. Processing by Analytic Imaging and Geophysics LLC.

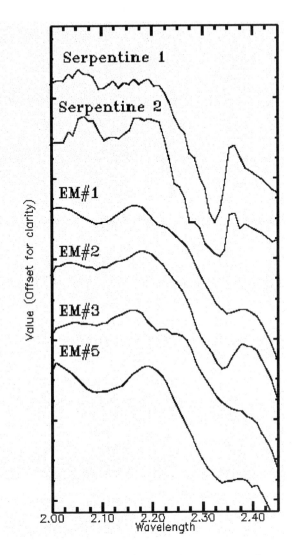

Figure 7.20 End-member spectra and reference library spectra for serpentine. By Analytic Imaging and Geophysics LLC.

This work demonstrates that minerals associated with kimberlites, particularly serpentine, calcite, dolomite, clays and iron oxides can be mapped using hyperspectral sensors. In areas with soil and/or vegetation cover it may be necessary to supplement the mineral maps with structure mapping to locate diatremes at fault intersections or along ring dikes.

MINERAL EXPLORATION IN MONGOLIA

Remote sensing and a geographic information system were used by Harrods Natural Resources and The Map Factory to document, evaluate, and plan a mineral exploration program in Mongolia [9].

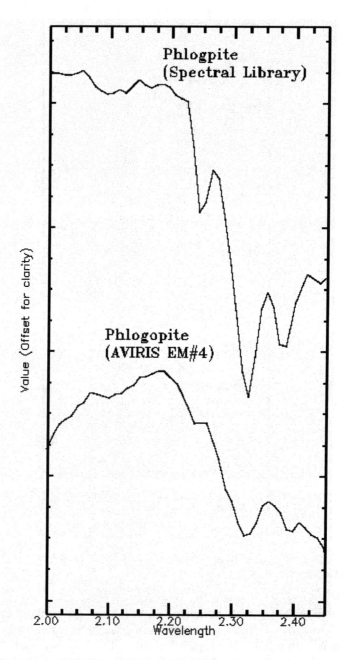

Figure 7.21 End-member spectra and reference library spectra for phlogopite. By Analytic Imaging and Geophysics LLC.

The project began with the incorporation into a GIS of existing Russian and Mongolian maps; the Digital Chart of the World (DCW); scanned geologic, tectonic, and structure maps; field mapping; maps from consultants; and alteration mapping from Landsat TM and hyperspectral imagery. These had to be converted to a common format, projection, and coordinate system. As part of the field mapping, a hand-held GPS unit was used to correct the TM imagery to accuracies better than 100 meters, and in some cases to 30–50 meters. The arid conditions made TM imagery usable at scales of 1:50,000. Digital elevation models were generated from the DCW (1:1,000,000) and 1:200,000 topographic maps. Maps and imagery were then draped over the digital topography and presented as perspective views and stereo pairs to enhance structural observations.

Image processing involved evaluation of individual pixels for their mineral characteristics, band ratios of entire TM scenes to highlight areas of hydrothermal clay and iron oxides, and principal component analysis to enhance map units. The final GIS included over 60 layers of maps, imagery, and interpretations. The project was able to proceed from the entire country to leases covering some two million acres within a one year period.

The leases were then evaluated using an airborne hyperspectral survey. This survey used a 128 channel system (Integrated Spectronics' HyMap) that ranged from 0.45 to 2.48 microns with a 16 nanometer band width. The aircraft was flown at an altitude to achieve five meter ground resolution and a 2.5 km swath. The imagery was calibrated to apparent reflectance and then processed to highlight hydrothermal alteration minerals using standard AVIRIS methods (Fig. 7.22) [10]. The results of this work will further refine the exploration effort.

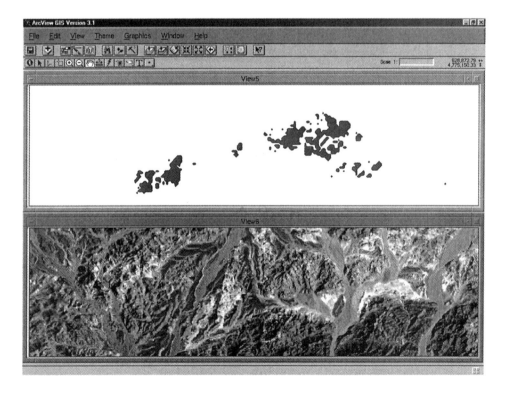

Figure 7.22 GIS smectite map extracted from HyMap hyperspectral imagery over mineral exploration acreage, Mongolia. Processing by The MapFactory, Inc. for Harrods Natural Resources, Inc.

REFERENCES

1. N.H. Foster, in G.W. Newman and H.D. Goode, eds., *Basin and Range Symposium and Great Basin Field Conf.* (Rocky Mtn. Assn. Geol., Denver, 1979), pp. 477–486.
2. H.D. Duey, in G.W. Newman and H.D. Goode, eds., *Basin and Range Symposium and Great Basin Field Conf.* (Rocky Mtn. Assn. Geol., Denver, 1979), pp. 469–476.
3. I.S. Merin and R.C. Michael, *Proc. of the 4th Thematic Conf. for Rem. Sens. in Explor. Geol.* (ERIM, Ann Arbor, 1985), pp. 183–192.
4. I.S. Merin and W.R. Moore, *Bull. Am. Assn. Pet. Geol.* **70** (1986): 351–359.
5. P. Hauff, R. Bennett, P. Chapman, G. Edmondo, G. Borstad, R. Neville, W. Peppin, and S. Perry, Goldfield, Nevada: An old problem revisited with Hyperspectral Technology, *Proc. of the 13th Intl. Conf. Applied Geologic Rem. Sens.* (ERIM, Ann Arbor, 1999).
6. D. Jiang, P. Wang, and F. Meng, *Proc. of the 10th Intl. Conf. Geologic Rem. Sens.* **2** (ERIM, Ann Arbor, 1994), pp. II-611–II-618.
7. F.A. Kruse and J.W. Boardman, *Proc. of the 12th Intl. Conf. Applied Geologic Rem. Sens.* **1** (ERIM, Ann Arbor, 1997), pp. I-21–28.
8. F.A. Kruse and J.W. Boardman, *Proc. of the 7th JPL Airborne Earth Science Workshop Pub. 97–21*, **1** (Jet Propulsion Laboratory, Pasadena, 1998), p. 259.
9. P.B. Goodwin, J.M. Ellis, M.R. Choiniere, M.B. Quinn, and H.H. Davis, *Proc. of the 13th Intl. Conf. Applied Geologic Rem. Sens.* **I** (ERIM, Ann Arbor, 1999), pp. 271–278.
10. F.A. Kruse, J.F. Huntington, and R.O. Green, *Proc. Second Intl. Airborne Rem. Sens. Conf.* **I** (1996), pp. 211–220.

ADDITIONAL READING

Petroleum

J.L. Berry and G.L. Prost, *Hydrocarbon Exploration, Manual of Remote Sensing* **3** (John Wiley and Sons, New York, 1999), 3rd ed., Chap. 9.

Z. Berger, T.H. Lee Williams, and D.W. Anderson, *Bull. Am. Assn. Pet. Geol.* **76** (1992): 101–120.

M.dF. Botella and G. Lawrence, Satellite radar guides exploration in deepwater offshore Brazil, Offshore (Sept. 2000): 42–43, 178.

W.W. Doeringsfeld Jr., and J.B. Ivey, *Mountain Geol.* **1** (Rocky Mtn. Assn. Geol., Denver, 1964), pp. 183–195.

J.M. Ellis, *Papua New Guinea Dept. Mining and Petroleum Open File* **F1/R86-126** (1986).

P.R. Lamerson, *Papua New Guinea Dept. Mining and Petroleum* Open File **F1/R/88-52** (1988).

T. Nishidai and J.L. Berry, *Proc. of 8th Thematic Conf. on Rem. Sens.* **1** (ERIM, Ann Arbor, 1981), pp. 373–389.

F.A. Penny, *Rocky Mtn. Assn. Geol. Symp.* (1975): 55–61.

W.M. Reid, *Bull. Am. Assn. Pet. Geol.* **72** (1988): 239.

L.C. Rowan, M.J. Pawlewicz, and O.D. Jones, *Bull. Am. Assn. Pet. Geol.* **76** (1992), pp. 1008–1023.

G.L. Valenti, J.C. Phelps, and L.I. Eisenberg, *Proc. of 11th Thematic Conf. on Geol. Rem. Sens.* **1** (ERIM, Ann Arbor, 1996), pp. 97–108.

Minerals

V.J. Barniak, R.K. Vincent, J.J. Mancuso, and T.J. Ashbaugh, *Proc. of the 11th Thematic Conf. on Geologic Rem. Sens.* **2** (ERIM, Ann Arbor, 1996), pp. 188–197.

S.A. Bennett, *Int. Geol. Rev.* **35** (1993), pp. 1009–1029.

T.L. Bowers and L.C. Rowan, *Photogramm. Engr. and Rem. Sens.* **62** (1996): 1379–1385.

A.P. Crósta, I.D.M. Prado, and M. Obara, *Proc. of the 11th Thematic Conf. on Geologic Rem. Sens.* **2** (ERIM, Ann Arbor, 1996), pp. 205–214.

A.P. Crósta, C. Sabine, and J.V. Taranik, *Rem. Sens. of Environment* **65** (1998): 309–319.

D. Davidson, B. Bruce, and D. Jones, *Proc. of the 9th Thematic Conf. on Geologic Rem. Sens.* **2** (ERIM, Ann Arbor, 1993), pp. 845–859.

L.A. Dick, G. Ossandon, R.G. Fitch, C.M. Swift, and A. Watts, *Prog. & Abs. Integrated Methods in Exploration and Discovery* (SEG, 1993): AB21–23.

J.R. Harris, A.N. Rencz, B. Ballantyne, and C. Sheridan, *Photogram. Engr. and Rem. Sens.* **64** (1998): 309–322.

J. Ma, V.R. Slaney, J. R. Harriss, D. F. Graham, S. B. Ballantyne, and D. C. Harris, *Proc. of the 14th Canadian Symp. Rem. Sens.* (1991), pp. 419–422.

F.P. Miranda, A.E. McCafferty, and J.V. Taranik, *Geophysics* **59** (1994): 733–743.

A.N. Rencz, C. Bowie, and B. Ward, in A.N. LeChaimant, D.G. Richardson, R.N.W. DiLabio, and K.A. Richardson, eds., *Searching for Diamonds in Canada*, Geol. Survey Can. Open File **3228** (1996): 255–257.

A.N. Rencz, J.R. Harriss, and S. B. Ballantyne, *Current Research Bull. Geol. Survey Can.* (1994): 277–282.

L.C. Rowan, K. Watson, J.K. Crowley, C. Anton-Pacheco, P. Gumiel, M.J. Kingston, S.H. Miller, and T.L. Bowers, *Proc. 9th Thematic Conf. on Geologic Rem. Sens.* **1** (ERIM, Ann Arbor, 1993), pp. 195–197.

L.C. Rowan, R.N. Clark, and R.O. Green, *Proc. 11th Thematic Conf. on Geologic Rem. Sens.* **1** (ERIM, Ann Arbor, 1993), pp. 175–176.

C. Sabine, *Remote Sensing Strategies for Mineral Exploration, Manual of Remote Sensing* **3** (John Wiley and Sons, New York, 1999), 3rd ed., Chap. 8.

V.H. Singhroy and F.A. Kruse, *Proc. of 8th Thematic Conf. on Geologic Rem. Sens.* **1** (ERIM, Ann Arbor, 1991), pp. 361–372.

D.M. Spatz and R.T. Wilson, *Proc. 10th Thematic Conf. on Geologic Rem. Sens.* **1** (ERIM, Ann Arbor, 1994), pp. 227–240.

D.M. Spatz and R.T. Wilson, *Proc. 12th Thematic Conf. on Geologic Rem. Sens.* **1** (ERIM, Ann Arbor, 1997), pp. 1–12.

K. Watson, F.A. Kruse, and S. Hummer-Miller, *Geophysics* **55** (1990): 70–79.

D.S. Windeler and R.J.P. Lyon, *Photogramm. Engr. and Rem. Sens.* **57** (1991): 1171–1178.

QUESTIONS TO CONSIDER

1. A salesman calls and tells you he has a remote sensing study that can help you locate gold deposits in central Asia. What are some of the questions you should ask about the study (other than "how much does it cost?")?
2. You are planning a remote sensing study of a frontier oil basin. What other kinds of data would you want to have to support your work?
3. Before deciding what to look for on imagery, what kind of "up-front" work can you do?
4. What are some ways that geologic structures can contribute to stratigraphic hydrocarbon traps?
5. What alteration minerals are easiest to map using multispectral/hyperspectral scanners in the visible-near infrared (to 1.7 microns) range? In the region from 2.0–2.5 microns?

Section 3
Exploitation and Engineering Remote Sensing

INTRODUCTION

At first glance it might not appear obvious that remote sensing could have a role to play in oil field exploitation or mining and engineering projects. After all, can one obtain the detail needed to site wells or map veins using satellite imagery? Yet there are many potential applications for both detailed airphoto and high-resolution satellite interpretations. The simplest use for imagery is to serve as a map base when topographic maps are not available, when more detail is needed than is available on existing topographic or planimetric maps, or when maps are so out of date that they are useless for engineering purposes. Most commonly roads and streams have changed since the original maps were made, or development has occurred, or landslides, etc., have changed the landscape. These image maps can be uncorrected (i.e., no standard projection), or they can be processed to eliminate most of the distortions inherent in images and photos and converted to a useful map projection.

EXPLOITATION APPLICATIONS

Exploitation applications include:

(1) helping choose infill and stepout well locations based on the extension of known surface trends
(2) assisting with enhanced oil recovery (e.g., waterflood) programs
(3) mapping the location and extent of fractured reservoirs
(4) predicting horizontal well drilling directions
(5) extending mineralized zones from known deposits into surrounding areas
(6) solving mine safety problems.

On rare occasions remote sensing data have been successfully used to select a well site. In these cases, airphotos have most often provided the detail that is generally required. This application of remote sensing tends to be more common in small and medium-sized companies that cannot afford costly seismic programs, or in areas that have not been explored recently. In some areas it may be less costly to drill wells than to run a seismic program, making some cost-effective method of site selection necessary. In certain plays, specifically fractured reservoirs, or when exploration is in areas with surface carbonates, volcanics, or permafrost, indications of structure on imagery may be more reliable than interpretations of conventional seismic programs. In areas with deep weathering or near vertical units, seismic becomes costly to obtain and process properly, and remote sensing becomes a valuable adjunct to assist with the interpretation of geophysical data.

Much mineral exploration is just the extension of known deposits or trends into the surrounding region. Many of these areas contain rugged topography and are not easily traversed. Field mapping is both time-consuming and expensive. It helps to have a tool that will focus the mapping and sampling programs on the most favorable target areas. Multispectral/hyperspectral imagery can help do this.

Mining is inherently dangerous. In addition to roof falls in underground mines, slope stability is a critical issue in open pit mining. Finally, after mining has been completed, there is the issue of subsidence of old, abandoned mines.

HYDROLOGIC APPLICATIONS

Surface and groundwater are vital for agriculture, industry, and urban development, and are also integral parts of wilderness ecosystems. Typical remote sensing projects involving surface and groundwater include:

(1) locating sources of water,
(2) monitoring surface water,
(3) locating geothermal resources,
(4) monitoring erosion, including coastal erosion, and
(5) flood control.

Again, remote sensing in itself cannot locate a source of groundwater, determine water quality, or identify geothermal hot spots. Yet, together with ground measurements and sampling, surface mapping, and geophysical surveys, imagery can play a major role in accomplishing all of these.

ENGINEERING APPLICATIONS

Engineering applications include:

(1) planning access into and out of an area (road building, barging),
(2) choosing facilities sites,
(3) planning pipeline, slurry, conveyor belt, and powerline routes,
(4) locating sources of aggregate for drill pads or artificial islands and construction,
(5) locating old well sites, old seismic lines, old mine shafts, and adits,
(6) mapping the distribution of permafrost, bad ground, or outcrops,
(7) mapping ice movements in the Arctic,
(8) locating drilling platforms offshore,
(9) accurately locating shorelines (at both high and low tide), and
(10) mapping water depths for port facilities or to avoid dangerous shoals.

The following chapters will deal with these issues by suggesting ways that remote sensing can be used to make these operations more cost-effective, time-efficient, and safe.

8

Exploitation Remote Sensing

CHAPTER OVERVIEW

The problems that one faces during exploitation of a field or mineral deposit include making production more efficient and cost effective, adding reserves, and finding ways to extend the life of a mine or field by finding bypassed mineralization or pay. Airphotos and high resolution satellite imagery can generally be used to plan and carry out these objectives. Several examples are provided to demonstrate remote sensing-based approaches to locating and exploiting new reserves. Cases from the petroleum industry illustrate using imagery to help locate infill and stepout wells (Bravo Dome, New Mexico), producing from fractured reservoirs (Denver and Williston basins), coalbed methane (Piceance basin), and waterfloods (Cottonwood Creek field, Wyoming). An example from the Huancavelica mining district, Peru, demonstrates the same process applied to the hard rock mining industry. Finally, remote sensing contributions to mine safety are discussed.

CHOOSING INFILL AND STEPOUT WELL LOCATIONS

Infill well locations are usually determined by a detailed examination of existing well data and extrapolating trends such as thickness changes, facies changes, and structural position. If detailed seismic data exists, it can be used to assist in the evaluation of the field, for instance, suggesting where the high point on a structure is located, showing lateral facies changes, the depth of the gas-water contact, or where there could be fault closure. But detailed seismic data often does not exist within a field. In such cases it may be possible to use surface data provided by imagery to (1) map structural closure, (2) map faults and joint zones, and (3) provide some idea as to the dominant trend of fracturing within the field and how it changes across the area. To the extent that unit thickness and facies were controlled by paleostructure, it is possible that these structures are expressed at the surface and could suggest depositional patterns (e.g., Chapter 7, Paradox Basin stratigraphic traps).

In stratigraphically trapped fields the distribution of the reservoir is a function of paleotopography. Reservoir sands, for example, may be deposited as an apron or sheet around basement highs and are missing over the uplifts. Under such circumstances it would be a waste of money to drill on these highs, even though they may appear interesting on seismic data. Because the overlying section settles through time, the basement highs may now appear as surface highs (Fig. 8.1). These subtle rises can often be mapped on imagery using techniques described in Chapter 5.

Bravo Dome Case History

An example of paleotopography expressed on the present day surface can be seen at the Bravo Dome carbon dioxide field, New Mexico [1]. The principal reservoir is the Permian Tubb sandstone, which lies at depths from 580 to 900 m. The dome itself is a

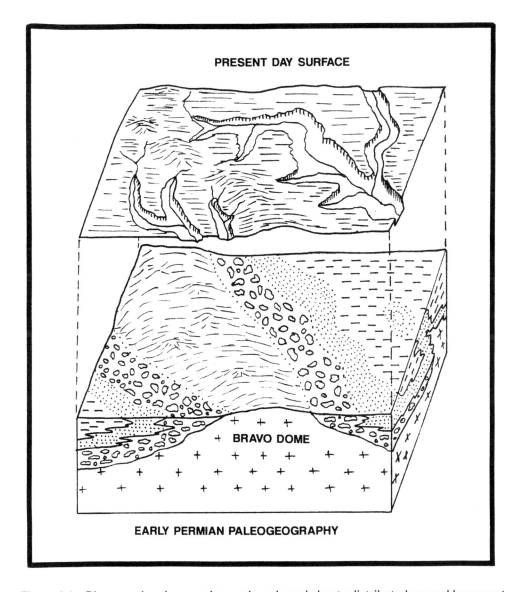

Figure 8.1 Diagram showing granite wash and sand sheets distributed around basement highs. Drape of younger units as a result of compaction causes present day topographic and drainage indications of structure.

southeast-plunging projection of the Pennsylvanian Sierra Grande uplift (Fig. 8.2). By early Permian time arkosic sands, derived from the granitic uplift, were being deposited as clastic wedges around the flanks of the uplift. The carbon dioxide is produced from the finer outlying sands rather than the granite wash closest to the inselbergs. The Cretaceous cover was gently folded during Laramide deformation in this area, and some of the small folds that are now at the surface are probably a result of Laramide movements. Although Bravo dome itself is a subsurface feature that is not revealed by outcrop patterns at the surface, it is expressed by radial drainage. Some of

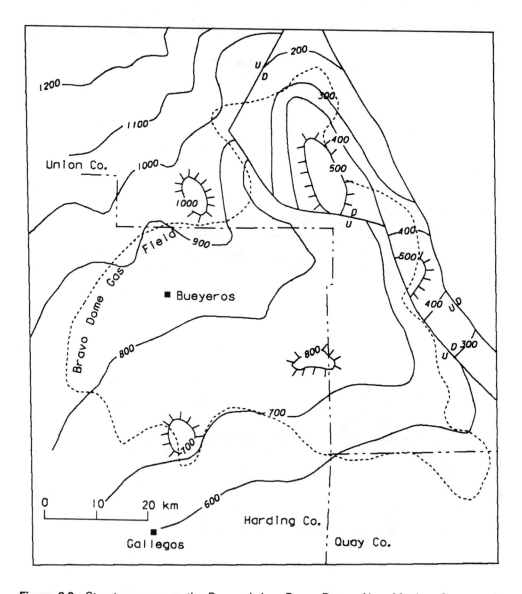

Figure 8.2 Structure map on the Precambrian, Bravo Dome, New Mexico. Contours in meters. Note that the dome is a southeast-plunging nose with scattered knobs. From Broadhead [1].

the small buried Precambrian knobs (1 to 5 km in diameter) that lack Tubb sandstone have been detected at the surface using airphotos to map drainage and tonal patterns.

Oil and gas companies have a vested interest in keeping their operating costs down, and so would rather not drill a well on or near one of the granite knobs where the reservoir is missing. By looking at structure contour maps derived from existing wells and a 5 km (three-mile) seismic grid, and comparing this map to gravity and a remote sensing surface structure map, project geologists at Amoco have determined that 50% of the circular anomalies identified by remote sensing geomorphic analysis at Bravo Dome are related to true basement structures. The surface work is used to add confidence to the other data sets in that it helps define and fill in detail on the size and location of knobs where other data sets are weak.

In other cases a reservoir unit may have been deposited along a fault or may have been offset vertically or laterally along a fault. These faults can not only bound reservoir units, but often can also provide reservoir closure on one or both sides of the fault (Fig. 8.3). In much the same way, mineral deposits are frequently cut by faults and parts of the orebody are moved some distance vertically and/or laterally. In such cases it is desirable to be able to map the location of the fault, the sense of displacement, and where the mineralization or hydrocarbons should be located.

Evidence for such faults is not always visible at the surface, yet one can often see some indication of where the section was faulted, and perhaps even get a sense of the displacement direction from the present surface (Fig. 8.4). This information will assist with planning a core drilling program to find extensions of veins, reefs, or other mineralized body. The same holds for drilling infill and stepout wells, particularly when the reservoir is a channel sand, barrier bar, or other hard to follow target. If a channel, for example, was controlled by the fault location, the present indications of the fault location and extent should indicate the distribution of the channel fill. If the channel has been offset in a normal sense and all the units dip uniformly, it should be possible to predict where the sand might be encountered and where stepout wells should have the best chance of hitting the objective (Fig. 8.5). Likewise, if there are indications of lateral offset along a zone based on offset surface features, fracture patterns, or en echelon structures, for example, one should be able to better predict the location of the offset mineral deposit or reservoir sand at depth.

FRACTURED RESERVOIRS

Exploitation of fractured reservoirs requires a knowledge of the location, density, and trend of subsurface fractures. In many cases these units would not be reservoirs at all were it not for the capacity of the fractures to hold and deliver hydrocarbons. The Cretaceous Austin Chalk in Texas has porosities that range from 3 to 9%, but unless the rock is fractured the matrix permeability averages less than 0.1 millidarcy [2]. Yet ultimate recovery has been estimated between four and eight billion barrels of oil, making this a target worth pursuing [3]. The fractured and overpressured Mississippian Mission Canyon limestone in the Williston basin is estimated to contain 100 to 500 million barrels of recoverable oil in the Mondak field alone [4]. This accumulation is not associated with any structural closure. Normal matrix porosity is between 2 and 4%, and permeability is generally less than 0.01 md. The Cretaceous Niobrara in the Denver basin has matrix porosities generally less than 10% and permeabilities of 0.01 md or less at depths greater than 1800 m. Yet initial production

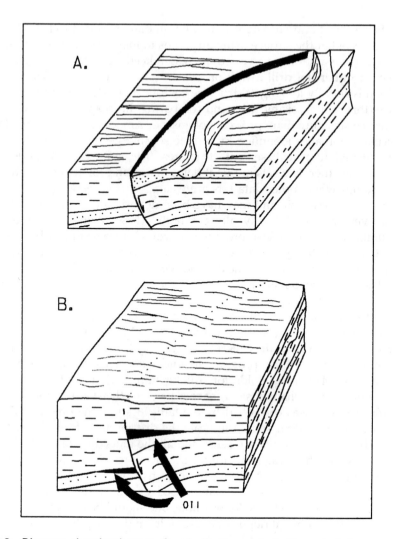

Figure 8.3 Diagram showing how surface indications of a fault can help locate a reservoir unit, and can provide the updip trapping mechanism as well. A. Paleogeography showing structural control of channel sands. B. Present day surface. The fault trace provides the clue to reservoir trend.

(IP) rates as high as 865 barrels of oil per day have been reported out of fractured zones [5]. Fractures provide primary porosity and deliverability in many world-class fields, such as those around Lake Maricaibo, for example, and are absolutely essential for production of coalbed methane since coals have effectively no primary porosity.

The procedure for evaluating fracturing within an area consists of mapping all indications of faults and joint zones using the techniques discussed in Chapter 5. Since most existing maps do not indicate joints, and may not show faults unless significant offset is measured, the image interpretation will usually provide much new information. Interpreted fractures cannot, in many cases, be identified as faults or joints, nor can the dip of the fracture surface be determined from imagery alone.

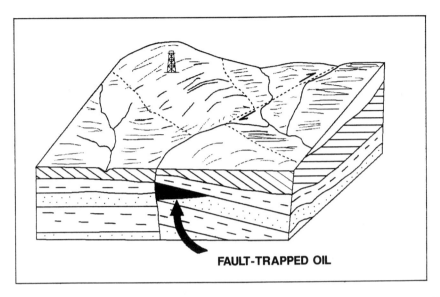

Figure 8.4 Surface indicators of offset can help determine where to drill an offset to existing production.

Fractures dipping less than 45° tend to have irregular traces and thus are seldom mapped. Outcrops have more obvious fractures than unconsolidated materials, and brittle units have more and better expressed joints than ductile units. Thus, it is obvious that image interpretations can locate many but not all fractures, can estimate their lateral lengths, and measure their azimuths, but at least some ground measurements are required as follow-up to determine the inclination and continuity of joints and joint trends from layer to layer.

Fracture maps are the best way to display the location of faults and major joint zones (Fig. 8.6). The fractures can be digitized to make the manipulation of fracture data easier and faster. Rose diagrams are a convenient way to display the orientation of major fracture sets. The usual technique is to superimpose a grid of predetermined size on the fracture map and count the number of interpreted fractures in a given azimuth range, say 15° increments from west to north to east (Fig. 8.7). This gives a frequency-based rose diagram for each grid cell. It is also possible to count the cumulative lengths of fractures in every 15° increment per grid cell. This provides a length-weighted rose diagram. The frequency-weighted roses tend to overemphasize the importance of short fractures, since all fractures have the same importance. Length-weighted roses give more weight to longer fractures. In both cases the roses are a statistical device (circular histogram) that allows one to lump many fractures together and see what the principal trends are and how they change across an area. It is also possible to run measures of statistical significance on large numbers of fractures to determine which fracture trends (or trends that lack fractures) are significant at a given confidence level (Fig. 8.8).

One may be interested in finding the areas of most intensely fractured rock, zones that often enhance the migration of fluids (e.g., hydrothermal solutions, oil, or groundwater). In this case it would be useful to generate a map of fracture intensity for all fractures, either frequency- or length-weighted (Fig. 8.9) [6]. One can also generate a

236 REMOTE SENSING FOR GEOLOGISTS

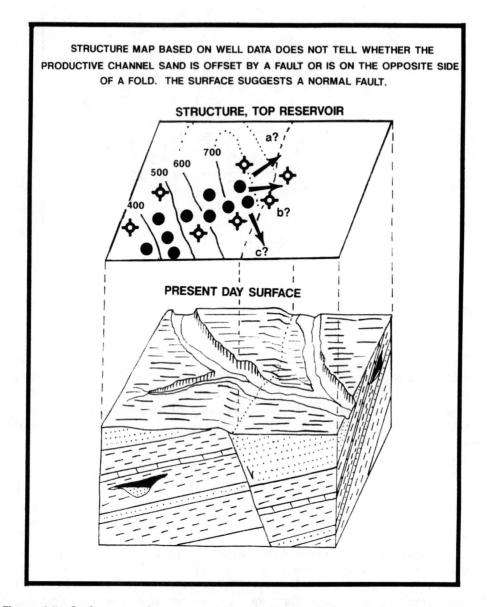

Figure 8.5 Surface mapping can suggest where to look for offset reservoirs. In the example shown here the well data may suggest a fold, but surface indicators suggest a normal fault. The channel sand reservoir was displaced in the "a" direction.

map that contours the intensity of fracture intersections based on the assumption that areas with the most intersections will be most intensely fractured (Fig. 8.10). On the other hand we may know that one fracture set is open and has good permeability, or has economic metals or minerals, whereas the other sets tend to be filled with barren veins or closed. If this is the case, we can make a fracture intensity contour map of just the trend of interest (Fig. 8.11).

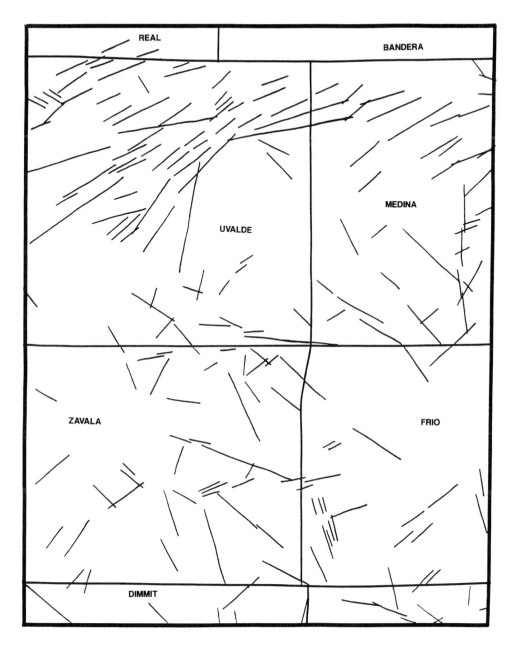

Figure 8.6 Fracture map interpreted from Landsat TM imagery over part of south Texas. Fractures have been broken into straight line segments for input into a fracture digitizing program.

Examination of cumulative production maps in oil fields that produce from fractured reservoirs indicate that adjacent wells can have highly variable values. The same would hold true for gas and groundwater. This suggests that drilling in a fracture zone can improve the chances of good production, but it is essential to hit an open

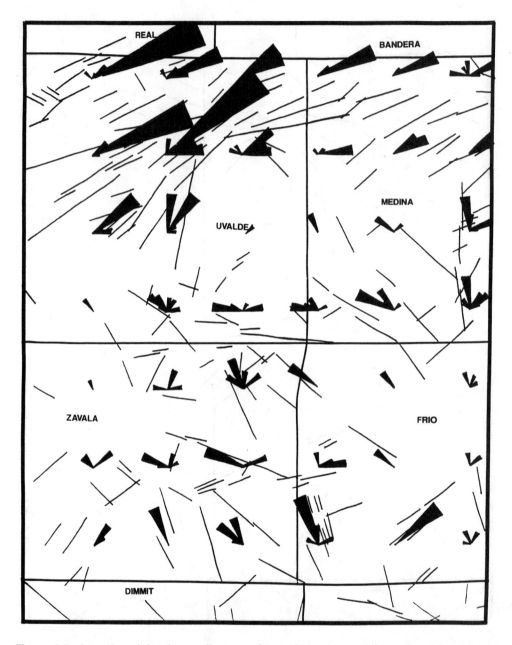

Figure 8.7 Length-weighted rose diagrams for each ten-by-ten kilometer grid cell in the area of figure 8.6. Rose petals are for 15° increments of azimuth.

fracture and tap into the reservoir. Since most fractures in gently dipping units are near vertical, horizontal drilling is a very important way to improve the odds of tapping into the fracture system.

If we can assume that fracturing is fractal, that is, that patterns of fracturing are consistent at various scales, then the relative spacing between sets (the spacing ratio)

EXPLOITATION REMOTE SENSING 239

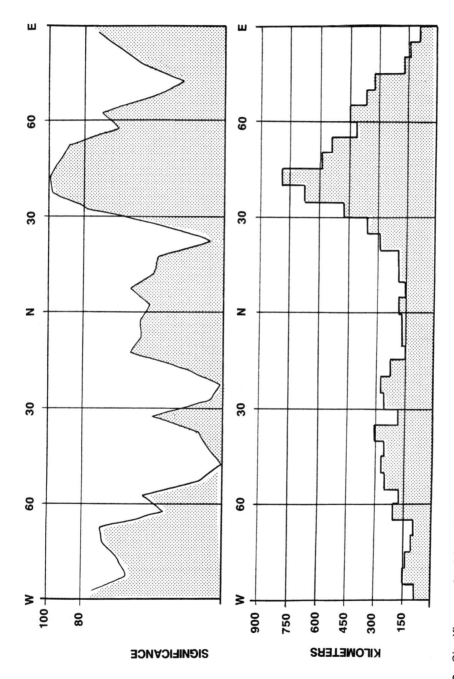

Figure 8.8 Significance plot (above) shows statistically significant trends at the 80% confidence level. (35–55E). Note that an absence of fracturing in a given direction can also be significant. Histogram (below) of total line-kilometers of fractures in each 5° azimuth range for the area of south Texas shown in Figure 8.6.

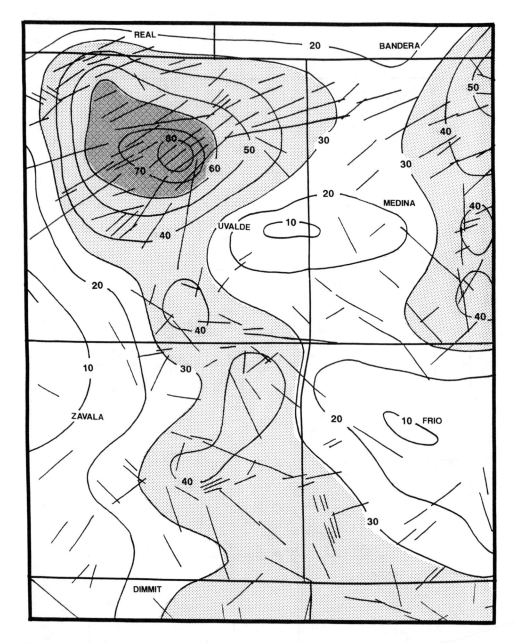

Figure 8.9 Map of fracture intensity for the area shown in Figure 8.6. Contours are kilometers of fractures per ten-by-ten kilometer grid cell.

and orientations of fracture sets should be the same at all scales from satellite interpretations to outcrop maps to core. This assumption, while imprecise, allows us to use a nomograph developed by Nolen-Hoeksema and Howard [7] to estimate the optimum direction for drilling horizontal wells (Fig. 8.12). From the imagery it is necessary to determine the primary and secondary fracture sets, the closest spacing

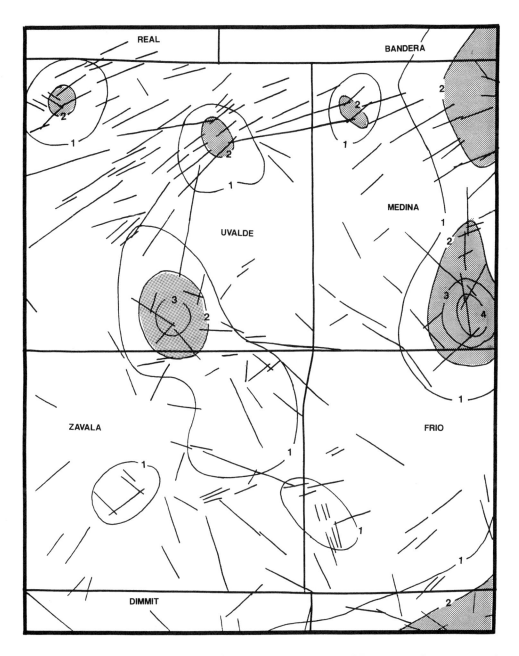

Figure 8.10 Fracture intersection density map for the area of Figure 8.6. Contours are the number of intersections per ten-by-ten kilometer grid cell.

between fractures of the primary set (S_1^*), the closest spacing between fractures of the secondary set (S_2^*), the azimuth of the primary set (theta 1), and the azimuth of the secondary set (theta 2). The optimum drilling direction derived from Fig. 8.12 is the orientation of the horizontal well that will intersect the maximum number of fractures perpendicular or near-perpendicular to the well bore.

242   REMOTE SENSING FOR GEOLOGISTS

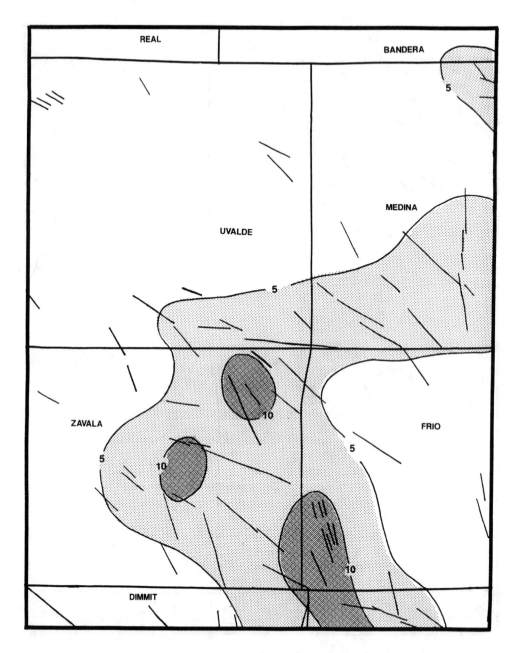

Figure 8.11 Fracture intensity map of all northwest-trending fractures in the area of Figure 8.6. Contours are the line-kilometers of fractures per ten-by-ten 10 × 10 kilometer grid cell.

The most reliable information obtained from fracture maps is orientation data, since it is a statistical composite. In most cases one can map surface fracture patterns and expect that these fracture orientations will exist at depth, although any one of the surface sets may be dominant in a particular horizon.

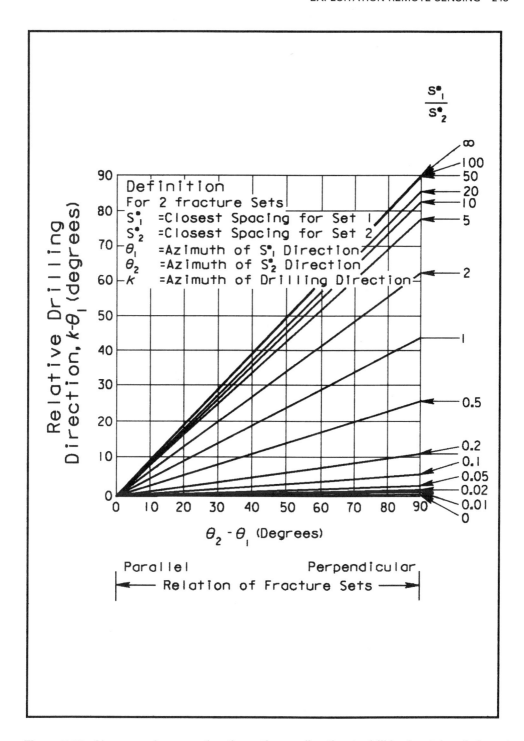

Figure 8.12 Nomograph suggesting the optimum direction to drill horizontal wells based on the orientation of the primary and secondary fracture sets. From Nolen-Hoeksema and Howard [7].

The location of a surface fracture usually will not indicate its exact position at depth, since the fracture may become diffuse or inclined, and in any case most individual fractures do not extend from the surface to great depths (if they did, the hydrocarbons would all escape!). Thus, locating a well on a surface fracture will not guarantee hitting the fracture at the reservoir level.

Fracture intensity maps are valid only under ideal conditions of uniform surface cover (the same lithology) and relatively gently inclined strata. Changing lithologies across an area will cause changing fracture intensities and trends, since these features are strongly dependent on the mechanical properties of the rock. Moderate to steeply dipping beds will cause a succession of narrow outcrop bands of different lithologies across an area, resulting in bands of varying fracture intensity. Under such conditions the fracture intensities should be compared within a unit along strike, but not from one unit to another.

COALBED METHANE

Coalbed methane is increasingly recognized as a significant source of natural gas. The Colorado Geological Survey, for example, estimated coalbed methane resources of 878 billion cubic meters (31 tcf) in the south half of the Piceance basin [8]; 700 billion cubic meters (25 tcf) of gas in-place has been estimated for coals in the San Juan basin [9]. Additional important coalbed gas resources in North America exist in the Black Warrior basin of Alabama and in the Powder River basin of Wyoming and Montana, among others. Detecting fracture systems in coal horizons may be essential to producing this gas most efficiently.

Cleat in coal is essential to producibility of gas. Coal fractures consist of nearly parallel, thoroughgoing joints, known as the face cleat, and a shorter, less continuous joint set known as the butt cleat, generally perpendicular to the primary set. Fractures provide permeability and surfaces that allow gas to desorb from the coal particles. Although coal can form fractures early during lithification in a prevailing stress field [10], many workers describe coal cleat as generally parallel to, but better developed than joints in adjacent rock [e.g., 11]. The following case history describes a project involving the evaluation of a coalbed methane prospect using remote sensing techniques.

Piceance Basin Coal Bed Methane Case History

A detailed photointerpretation was performed over three exploratory coal bed methane well sites in the Piceance basin, Colorado. Length-weighted rose diagrams of fracture trends were generated for each 10 square kilometer area. This work was done prior to drilling and evaluating the area with test wells.

In the first area, airphoto lineaments interpreted from 1:58,000 color-infrared stereo airphotos showed a dominant fracture trend of N60W. Surface fractures were measured at the future site of the Powers Federal #1. The Ohio Creek sandstone is at the surface, 710 meters stratigraphically above the top of the Cretaceous Williams Fork Formation coal. Fracture measurements at two surface locations provided two dominant trends: north to N15E, and N60W to N75W (Fig. 8.13). Oriented cores from the Williams Fork Formation at depths from 725 to 727 meters identified face cleat at five places on the core. This cleat varied from N36W to S43W, probably as a result of

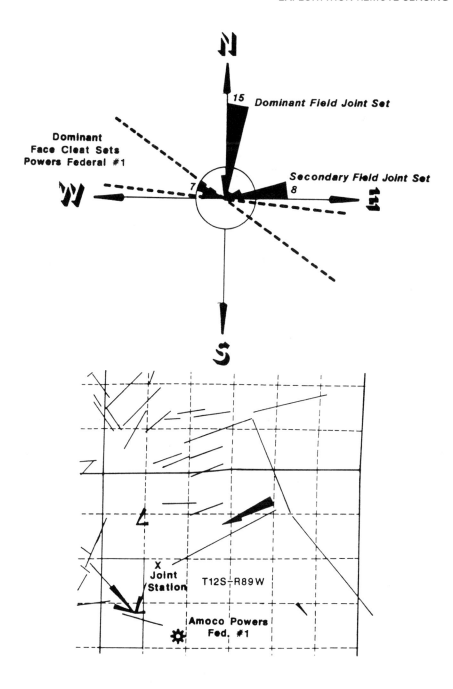

Figure 8.13 Airphoto interpretation with rose diagrams for a ten square kilometer grid, and field and oriented core measurements of fractures at the Amoco Powers Federal #1, Gunnison County, Colorado. The number of field-measured joints in the dominant sets are noted on the rose. In this area photogeologic joint sets near the well give a better indication of the face cleat in coals (725–727 m depth) than surface measurements taken at an outcrop 1.6 km north of the well.

the inaccuracies inherent to core orientation technology (W.B. Hanson, oral comm., 1988). Cleat frequency (inverse of spacing), also measured at each of the five places on the core, was used to "weight" the orientations. This yielded a dominant face cleat direction of N45–85W. In this area the joint sets interpreted from the airphotos were closer to the cleat trends than measurements from outcrops about 1.6 km from the well site.

Airphoto fractures in the area of the Electric Mountain #1 trend N60–75E, N20–30W, and N60–80W (Fig. 8.14). Surface fracture measurements were made near the well site in sandstones and mudstones of the Eocene Wasatch Formation. These indicate that a primary fracture set trends N60–75E, with a secondary set trending N30–45W. Oriented core was cut from 1477 to 1485 meters depth in the Williams Fork Formation. On the basis of one coal sample the face cleat was determined to be N74E, and butt cleat approximately N16W. Both types of surface measurements predicted the subsurface cleat trends with a high degree of accuracy.

Analysis of airphoto fracture trends at the Ruth Mountain #1 well suggests the dominant surface set trends N30–40E, and that the secondary set trends N20–30W (Fig. 8.15). No outcrops were available at this well site for surface measurements. Oriented core in the Williams Fork coal at a depth of 2199 m provided a face cleat azimuth of N25W. Core taken at 2334 m had a face cleat of N23W.

One may conclude from these examples that coal cleat orientation can often be predicted on the basis of surface (airphoto or field) measurements. These predictions can be used to help site wells along fracture zones, and to lay out well patterns to best drain a coal gas reservoir. Note, however, that although one can generally predict which fracture trends will exist in the subsurface, it is not always possible to predict which of the surface joint sets will be the dominant (face) cleat in the coal horizon.

SECONDARY RECOVERY AND WATERFLOODS

Remote sensing can be used to help evaluate waterflood or other secondary recovery operations where there is the likelihood of fractures controlling the flow of injected fluids or leading to premature breakthroughs along the waterflood front. In some cases pulse tests have shown no fluid flow between injection wells and producing wells on opposite sides of airphoto lineaments. The assumption is that the fracture zone acts either as an impermeable barrier to fluid flow, or that the fracture zone acts as a conduit that carries fluids along the zone and prevents fluid flow across the zone.

Cottonwood Creek Field Case History

An example of using airphotos to evaluate a secondary recovery program in a fractured reservoir can be seen at the Cottonwood Creek field in the Bighorn basin, Wyoming (Fig. 8.16). Cottonwood Creek is a stratigraphic trap in the west-dipping Permian Phosphoria Formation at depths from 1538 to 3077 m. The trap is the result of a porous dolomite grading updip into an impermeable shale-anhydrite facies [12]. The Phosphoria is approximately 100 m thick in the field, and porosity averages 11%, with permeability averaging 13 md. Engineers first suspected a fractured reservoir when a gas injection program initiated in 1958 resulted in rapid movement of gas to producing wells and oil production declined. A water injection program a year later had similar

EXPLOITATION REMOTE SENSING 247

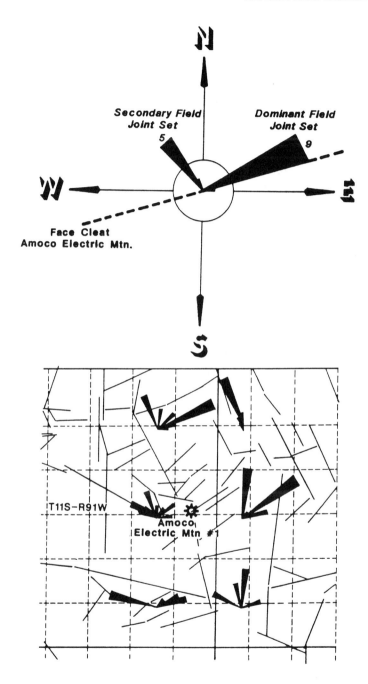

Figure 8.14 Airphoto interpretation with rose diagrams for a ten square kilometer grid, and field and oriented core measurements of fractures at the Amoco Electric Mountain #1, Delta County, Colorado. The number of field-measured joints in the dominant sets are noted on the rose. Face cleat in coals (1,477–1,485 m depth) corresponds to the dominant outcrop fracture set, and to one of several major photogeologic fracture trends around the well.

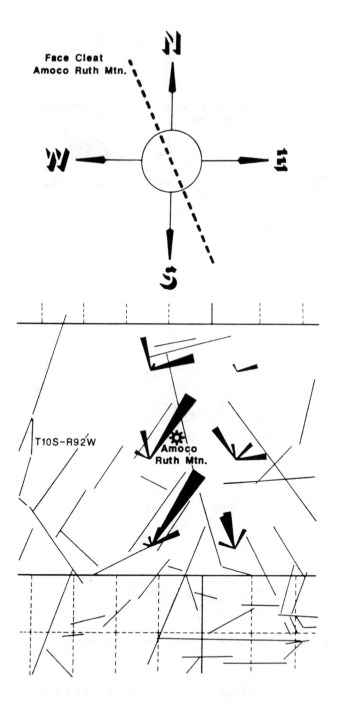

Figure 8.15 Airphoto interpretation with rose diagrams for a ten square kilometer grid, and field and oriented core measurements of fractures at the Amoco Ruth Mountain #1, Mesa County, Colorado. Face cleat in the coal (2,199 and 2,334 m depth) matches the secondary photogeologic fracture set. No outcrops were available for fracture measurements within several kilometers of the well.

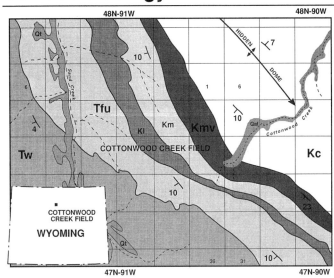

Figure 8.16 Surface geologic map of the Cottonwood Creek field area, eastern Bighorn basin, Wyoming. The field is a stratigraphic trap downdip from Hidden Dome.

results. Jointing proved to be a disadvantage in this case because the injected gas and fluid led to premature breakthrough along probable fracture zones.

Remote sensing was used to map fracture zones and determine if surface lineaments had any relation to subsurface reservoir properties. The idea was to use this information to plan the drilling of new wells. Lineaments were mapped on color-infrared stereo airphotos at a scale of 1:58,000 (National High Altitude Program photos). The lineaments were then digitized and the density of fractures was contoured in kilometers of fractures per 10 square kilometer area (Fig. 8.17).

The map of total fracture density did not correlate well with production data. Bedding was dipping west-southwest in this area, and many interpreted north-northwest lineaments may have been related to the strike of bedding. The next step was to determine if a correlation existed between production and fractures with a specific trend. Indeed, the map of northeast fracture density showed zones of more intense fracturing that corresponded to zones of increased total porosity-feet and high cumulative production (Figs. 8.18, 8.19). The results of this exercise allowed production engineers to take into account the role of fracturing when planning future secondary recovery programs in this field.

EXTENDING KNOWN MINERAL DEPOSITS

As with oil fields, one may extend a known mineral trend and/or increase reserves by stepping outward from existing deposits. The usual technique is to map the characteristics associated with the known occurrence, usually a specific rock type,

Northeast Lineament (Fracture) Density & Well Distribution Cottonwood Creek Field

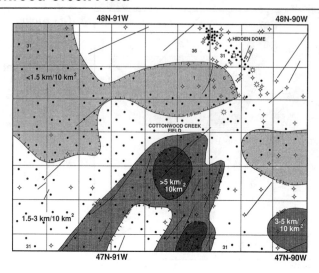

Figure 8.17 Map showing the distribution of wells in the Cottonwood Creek field superimposed on the northeast fracture density contours (kilometers of fractures per 10 kilometer square grid) and northeast interpreted fractures.

Porosity - Feet & NE Fracture Density

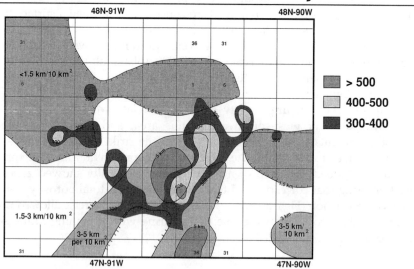

Figure 8.18 Contour map of porosity-feet, derived from well data, superimposed on the northeast fracture density contours. Porosity-feet is a measure of the reservoir volume above a specified porosity cutoff, in this case between 7 and 8.5% (e.g., 100 feet of 8% porosity yields 8 porosity feet). Note the northeast alignment of the high porosity zone.

Cumulative Production (BBLS) & NE Fracture Density

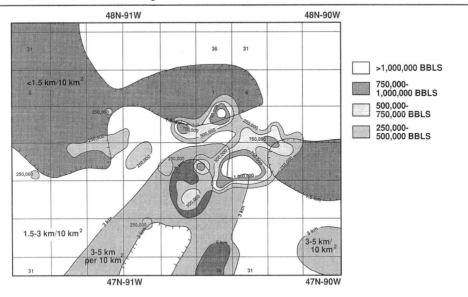

Figure 8.19 Contour map of cumulative production to 1986 (in barrels) superimposed on the northeast fracture density contours. Note the northeast alignment of high cumulative production.

alteration assemblage, or fault/vein trend, then look for the same characteristics in the surrounding countryside. The following example is taken from work by Agar and Villanueva [13].

Huancavelica Mining District Case History

The Andes of south-central Peru contain high-relief topography that makes it difficult for field parties to map on the surface. The high, arid, vegetation-free terrain is, however, well-suited for airborne remote sensing. Cia. De Minas Buenaventura S.A. operates several mines in the Huancavelica area approximately 250 km southeast of Lima. The northeastern part of the district contains Paleozoic basement exposed in the cores of domes and the Triassic-Jurassic Pucará group (including carbonates) and Cretaceous-Tertiary sediments in flanking synclines. The southwestern part of the district contains Tertiary volcanic sequences overlying the older rocks. The volcanics are intruded by penecontemporaneous andesites and dacites. Most of the mineral deposits are associated with extensively altered volcanics, although lead-zinc-silver mineralization occurs within the Pucará group. Field mapping in this area is difficult, expensive, and perhaps not as reliable as it could be due to restricted access.

Sixty-three channel hyperspectral data were acquired using the Digital Airborne Imaging Spectrometer (DIAS) operated by Geophysical and Environmental Research (GER). Imagery was acquired over 5000 square kilometers south of Huancavelica during October of 1996. Ground resolution varied between 10 and 14 m. The instrument has

27 channels in the visible and near infrared between 0.41 and 1.048 microns; 30 channels in the short-wave infrared from 1.40 to 2.42 microns; and 6 channels in the thermal infrared between 8.95 and 10.98 microns.

Ground spectra were gathered over hydrothermally altered areas prior to the overflights. These control points contain spectra of soil and fresh and weathered rocks. Weathered surfaces were found to contain spectra similar to, but less intense than the fresh samples. Likewise, spectral curves acquired during the survey correlated well with spectra from the USGS spectral library (Fig. 8.20) and ground control points (Fig. 8.21).

Images were evaluated for areas of hydrothermal alteration, and targets were chosen and ranked based on the presence of argillic, propylitic, iron oxide, and siliceous

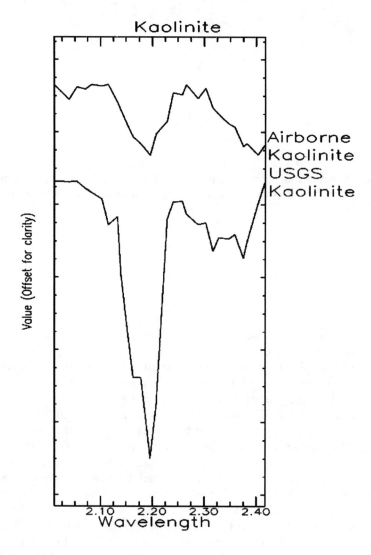

Figure 8.20 GER DIAS airborne spectra versus USGS spectral library. (a) Kaolinite; (b) Opal.

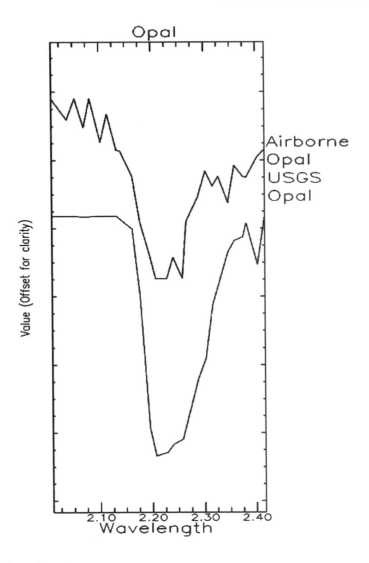

Figure 8.20 (continued)

alteration as well as structure (fig. 8.22). All of the known mines and deposits contain the alteration minerals that were found to occur within these targets. Some targets, such as Arcopunco, were on unclaimed land and were quickly claimed after the initial overflight and before the work continued. Evaluation of other areas was put on hold because of a lack of alteration. Narrow, vein-like targets in the Pucará limestone may represent Mississippi Valley-type mineralization. Areas considered most prospective for epithermal gold were subjected to further remote sensing processing.

Mineral distribution maps were created in high priority areas such as Arcopunco. A group of key alteration minerals was selected based on prior work. Images were generated for these reference minerals showing their distribution and where they are most intense (Fig. 8.22). These then became the end-members for spectral unmixing

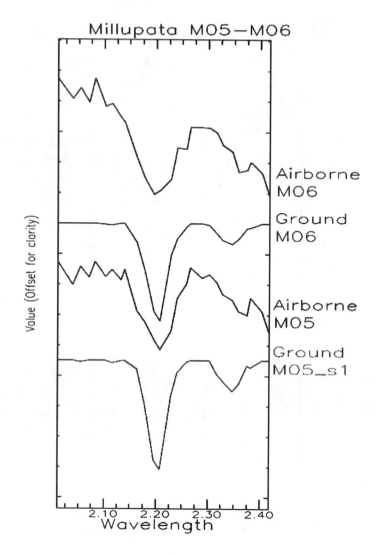

Figure 8.21 GER DIAS airborne spectra versus ground spectra of the same area, samples M05 and M06, Millupata area.

[14,15]. Spectral unmixing is a method of looking for combinations of these end members. The images produced by unmixing provide an estimate of the abundance of each mineral. These image maps were used to locate altered targets. Ground spectra were collected that confirmed the accuracy of the images.

In summary, alteration at producing mines was characterized by acquiring ground spectra. These spectral curves were used to calibrate and as control points for an airborne survey that then mapped similar zones in the adjacent region. These mineral alteration maps were used to direct the filing of claims and to prioritize the evaluation of properties already held. Epithermal gold targets were mapped, and a new class of target, the Mississippi Valley lead-zinc-silver association, was proposed.

EXPLOITATION REMOTE SENSING 255

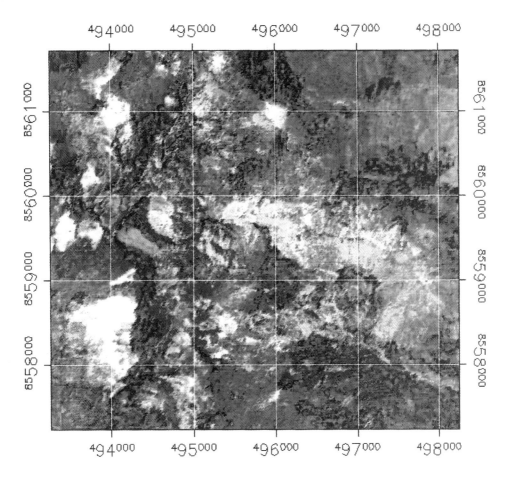

Figure 8.22 Hyperspectral alteration image of the Huancavelica area. Colors are designed to enhance propylitic alteration (chlorite, epidote, calcite), shown as white areas. Processing by A.G.A.R.S.S. Pty. Ltd. in conjunction with Cia. Buenaventura. (See color plate.)

MINE SAFETY ISSUES

There are a number of mining-related safety issues that can be addressed and possibly alleviated using remote sensing. Among these are subsidence over active and abandoned mines, slope stability in open pit mines, and roof falls in underground mines.

Mine subsidence can be recognized as topographically low areas that generally are not located along streams and do not have drainage developed into them because of their young age (see Subsidence Case History in Chapter 12). They may be filled with water or moist (darker) soil. They can be round, elongate, or branching, and can range in size from less than an hectare to several square kilometers. A series of depressions adjacent to tailings is a good clue that a mine has collapsed. Factors affecting the size and shape of surface collapse over underground mines include tectonic and residual stresses, faults and foliation, the width and height of the excavation, presence of groundwater, and the strength and thickness of the overburden [16].

Just as surface subsidence can result from geologic discontinuities (weaknesses such as faults), so, too, mine roof falls may result from the same factors. Lineament (interpreted fault) concentrations have been related to coal mine roof falls in Virginia [17]. Vertical decompression of 7000 psi was triggered by underground mine excavation across a fault in metasediments at approximately 107 m depth near Idaho Springs, Colorado: stresses determined by overcoring were unequal in opposite walls of faults, differing by a factor of two [16].

In general, fracture length (or the length of a zone of fracture segments) is more significant than density, as pervasive and apparently random short fractures suggest a surficial effect, whereas longer fractures are likely to penetrate deeper into the subsurface. As previously stated, a series of short faults along a trend, or a set of en echelon faults, may in fact be surface manifestations of a single large fracture at depth.

Slope stability in open pit mines (or highway roadcuts) can be adversely affected by the presence of joints and faults [18–21]. The least damaging are fractures perpendicular to the strike of the slope, whereas the most unstable conditions occur when fractures are parallel to the slope and may be undercut by excavation. Interpretation of faults and joint orientations and locations from high-resolution satellite images and/or medium to low-altitude airphotos can help direct remediation by rock bolting and terracing and thereby reduce the risk of catastrophic slope failure.

REFERENCES

1. R.F. Broadhead, in E.A. Beaumont and N.H. Foster (eds), *Structural Traps 1, Treatise of Petroleum Geology Atlas of Oil and Gas Fields* (Am. Assn. Pet. Geol., Tulsa, 1990), pp. 213–232.
2. R.H. Snyder and M. Craft, *Gulf Coast Assn. of Geol. Soc. Trans.* **27** (GCAGS, 1977): 376–443.
3. W.E. Galloway, T.E. Ewing, C.M. Garrett, N. Tyler,, and D.G. Bebout, *Atlas of major Texas oil reservoirs* (Texas Bur. Econ. Geol., Austin, 1983), pp. 41–42.
4. J.M. Parker and P.D. Hess, *Oil and Gas Jour.* (Oct. 13, 1980): 210–216.
5. R. Roundtree, *Western Oil Reporter* (Feb. 1984): 31–40.
6. D.L. Sawatzky and G.L. Raines, in D.W. O'Leary and J.L. Earle (eds.), *Proc. of the 3rd Intl. Conf. on Basement Tectonics* (Basement Tectonics Committee, Denver, 1981), pp. 91–100.
7. R.C. Nolen-Hoeksema and J.H. Howard, *Bull. Am. Assn. Pet. Geol.* **71** (1987): 958–966.
8. C.M. Tremain, D.L. Boreck, and B.S. Kelso, *New Mexico Geol. Soc. Guidebook, 32nd Field Conf.* (1981), pp. 241–248.
9. G.J. Bell, J.C. Seccombe, K.C. Rakop, and A.H. Jones, *Society of Petroleum Engineers* **SPE 14445** (1985): 1–11.
10. F.T.C. Ting, *Jour. of Pressure Vessel Tech., Trans. of Am. Soc. Mech. Engrs.* (1977): 624–626.
11. D.L. Boreck and M.T. Strever, *Colorado Geol. Survey Open-file Rept.* **80–5** (1980), 11 p.
12. R.W. Willingham and J.A. McCaleb, *Soc. Petrol. Engrs. Paper* **SPE 1770** (1967), 12 p.
13. R.A. Agar and R. Villanueva, *Proc. of the 12th Intl. Conf. Applied Geologic Rem. Sens.* (ERIM, Ann Arbor, 1997), pp. I-13–I-17.
14. J.W. Boardman, *Proc. IGARSS '89, 12th Thematic Canadian Symp. Rem. Sens.* **4** (ERIM, Ann Arbor, 1989), pp. 2062–2069.
15. J.W. Boardman and F.A. Kruse, *Proc. 10th Thematic Conf. Geologic Rem. Sens.* **I** (ERIM, Ann Arbor, 1994), pp. I-407–I-418.
16. F.T. Lee, J.F. Abel, Jr., and T.C. Nichols, Jr, *U.S. Geol. Survey Prof. Paper* **965** (1976), 47 p.

17. R.C. Milici, T.M. Gathright II, B.W. Miller, and R. Gwin, *Virginia Div. Mines Res. Rept.* **ARC-CO-7232-80-1-302-0206** (1982), 101 p.
18. R.D. Call, PhD Dissertation, Univ. Arizona, Tucson (1972), 201 p.
19. M.S. Ness, MSc Thesis, Colorado School of Mines, Golden (1982).
20. J.P. Savely, MSc Thesis, University of Arizona, Tucson (1972), 134 p.
21. P.M. Montazer, MSc Thesis, Colorado School of Mines, Golden (1978): 202 p.

ADDITIONAL READING

D. Greenbaum, S.H. Marsh, E.A. O'Connor, D.G. Tragheim, and A.J.W. McDonald, *Proc. of the 10th Thematic Conf. Geologic Rem. Sens.* **II**, (ERIM, Ann Arbor, 1997), pp. 115–126.

X. Huang, B. Zhang, D. Liu, J. Di, B. Rui, J. Fang, P. Chen, Z. Wang, C. Shan, Y. Li, and F. Xu, *Proc. of the 10th Thematic Conf. Geologic Rem. Sens.* **II**, (ERIM, Ann Arbor, 1997), pp. 138–143.

G.L. Prost, Soc. Pet. Eng. Preprint 35289 (1996): 25–32.

QUESTIONS TO CONSIDER

1. Why should subsurface fractures (faults and/or joints) be expressed at the surface? Would you expect to see the trace of the fracture itself? If so, when? If not, why not?
2. Would you expect the length of a fracture zone to be related to its depth of origin? Why? Could zones of short fractures be related to one master fault at depth? How might they be expressed at the surface?
3. What kind of fractured reservoir would you not expect to see evidence for at the surface?
4. If you know that your ore deposit terminates at a fault, what would you look for to locate the faulted-off portion of the body? What other techniques might you consider using?
5. Coals produce increasing amounts of methane as they are dewatered. Residents living near a coalbed methane field have begun to notice methane in their well water. What could remote sensing tell you about how the methane got there?
6. While interpreting an image that has been processed for alteration associated with porphyry copper you notice a cluster of color anomalies that look similar to alteration at the only known copper deposit in the region. What do you do?

9
Hydrology

CHAPTER OVERVIEW

The price of commodities goes up and down, as anyone in the oil and mining industries can attest. Yet water is one commodity that is in increasing demand just about everywhere. Whereas a company drilling an oil well may be able to use brackish or saline water to mix their drilling mud, fresh potable water is most desirable and not always available. Remote sensing can help locate sources of water, and in some cases one may be able to infer whether it is fresh or brackish. It may also be important to monitor surface water for changes in salinity, temperature, or turbidity, both onshore and offshore. Likewise, flood control can be assisted by monitoring snowpack and the extent of flooding over time. Both floods and normal runoff can cause erosion, as can wave action along coastlines. The progress of erosion, which can cost millions of dollars in property damage annually, can be observed using remote sensor systems. Whereas most geothermal resources in North America have been identified, many small thermal springs may not have been catalogued, and in other parts of the world economic thermal resources may never have been mapped. Thermal imagery provides a unique opportunity to map geothermal systems.

LOCATING SOURCES OF WATER

There are a number of ways one can use remote sensing to help look for sources of water. The simplest procedure is to use airphotos, either color or black/white, to locate dense vegetation in the bottom of gullies or dry creekbeds. Plants will have roots deep enough to tap into shallow or perched groundwater. Certainly if springs occur locally, these will show up on airphotos or satellite imagery as dark (moist) soil tones associated with vegetation. Healthy plants that appear bright green on color photography will be especially easy to locate on color-infrared imagery or photography because of their red color against an arid background of rocks and soil (blacks, browns, blues, greens).

Groundwater is often dammed against faults, forming alignments of springs that can be seen on imagery (Fig. 5.59). Near-surface groundwater often appears on near-infrared imagery as areas containing more vigorous vegetation in the summer and fall, and as areas of early leafout in the spring. In general, one is more likely to encounter groundwater close to the surface in topographically low areas, especially if they occur along a fault, and contain more dense or healthier vegetation than the surrounding areas. Detecting groundwater by mapping lineaments (fault zones) has been shown to be an effective means of ensuring a fresh water supply, particularly in crystalline terrains [1]. Some faults have a greater capacity for groundwater storage than others: in Guangzhou, China, northwest-trending transtensional faults contain the bulk of the groundwater in fractured granites, whereas northeast-oriented transpressive faults are for the most part barren of water [2].

A slightly more innovative way to look for groundwater is to use thermal imagery. The effects of solar illumination will overpower any signal derived from evaporative cooling in areas with moderate or high relief. In areas with low relief moist soils will appear cooler than surrounding dry areas. On pre-dawn imagery the effects of differential solar heating are subdued. Areas of near-surface water appear cool due to evaporative cooling, and areas of surface water will appear to be warmer than background as a result of the higher thermal inertia of water, that is, the capacity of water to retain heat longer than rocks and soils (Fig. 9.1).

Abandoned or ancient buried river channels may also be sources of groundwater. These channels may be covered by thin sand sheets or alluvium and may have a slight topographic depression or a slightly different texture at the surface. Thermal imagery can detect the telltale dendritic pattern of cooler soils over the old channels, or in some cases radar has been credited with detecting dark patterns of paleochannels in "hyperarid" regions by penetrating the sand cover to depths of one or two meters (Fig. 9.2) [3].

In arid climates water is likely to be brackish to saline if white crusts or light colored areas are seen on soils or in low spots (evaporites, including salt and gypsum, or alkali deposits). In temperate regions plant density will decrease, and plant growth will be less vigorous where near surface water is not fresh. Few plants (with the exception of mangroves, some palms, and coastal grasses and shrubs) can take large or even moderate amounts of salts in their water and will become stunted or even die as

Figure 9.1 Thermal image of surface water (warm = white) and moist ground (cool = dark), near the junction of Tincup Creek and the Salt River, Bonneville County, Idaho. Predawn image flown at 2,100 m altitude by Mars, Inc., Phoenix.

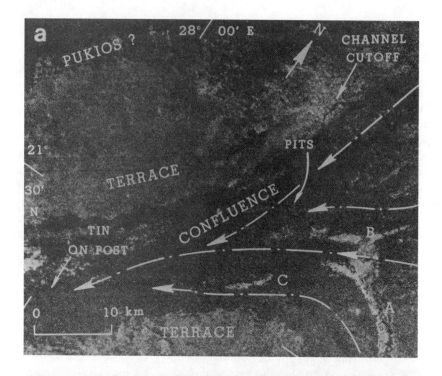

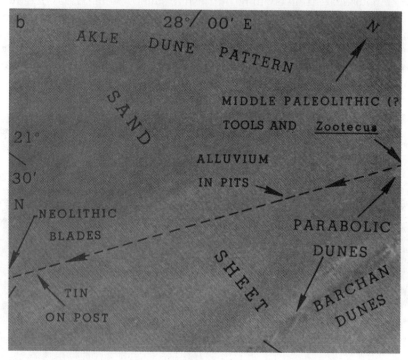

Figure 9.2 SIR-A radar image (a) of part of the Selima sand sheet, northwest Sudan, showing a buried river channel. Landsat image (b) of the same area shows no sign of the channel. From McCauley, et al. [3].

salinity increases. In south Florida marine salt water is encroaching on municipal and agricultural fresh water wells. The South Florida Water District is using satellite imagery to find the boundary between salt and fresh water by observing vegetation, like mangroves and saltgrass, that only grow near salt water (Fig. 9.3) [4].

Figure 9.3 Mosaic of Landsat images over south Florida showing agriculture and natural vegetation patterns [4]. Processed by EROS Data Center, Sioux Falls. (See color plate.)

262 REMOTE SENSING FOR GEOLOGISTS

In some areas water is as valuable or more so than oil. As part of an application for an exploration permit in Bahrain, Amoco was asked if sources of fresh water could be located as part of a remote sensing project. Day and night Landsat Thematic Mapper thermal images were processed to determine if sources of water could be located. Spatial resolution is 120 by 120 meters, and thermal resolution is estimated at about 2° C. Nighttime imagery is acquired around 9:30 p.m. Thermal and visible imagery was interpreted for evidence of faulting. Onshore and offshore areas were examined and it was determined that nighttime imagery showed cool spots off the east coast of the island near probable faults (Fig. 9.4). It was felt that these cool areas might indicate

Figure 9.4 Cool, possibly fresh water (blue) in warmer Persian Gulf waters (green to red) off the east coast of Bahrain. An abrupt break in temperatures suggests a northwest-southeast-trending fault. Landsat TM band 6 nighttime image. Image processed by Amoco Production Company. (See color plate.)

freshwater subsea springs. Groundwater would be cooler than the warm waters of the shallow Persian Gulf, and would rise to the surface due to the lower density of fresh water. Although there are unconfirmed reports of pearl divers encountering fresh water emanations offshore Bahrain, it is not known whether the cool spots mapped in this study were ever tested for fresh water.

In a similar study, Nelson, et al. [5] used thermal imagery and sampling to pinpoint small springs at Red Eagle Lake, Montana. Imagery was acquired at 6:00 a.m. from a helicopter using a thermal infrared video sensor angled approximately 45° from horizontal and a ground resolution of about 4 by 4 meters. One spring, estimated to be flowing 450 liters/minute, showed anomalous cool temperatures with up to 3° C variation from lake temperature as far as 10.5 meters from the spring.

MONITORING SURFACE WATERS

Short wavelengths penetrate water farther than long wavelengths, and blue light penetrates farthest. In clear water, light has been shown to penetrate between 20 and 30 meters (see Chapter 1, Water-covered Areas). The response at the sensor is a function of the water depth, suspended material (turbidity), and the color of the bottom material (e.g., white sand will reflect better than sea grass), and to some extent the incidence angle and surface roughness. Work by Lyon and Hutchinson [6] developed a radiometric model for evaluating shallow water bodies such as rivers using airborne scanners. A computer program simulated the flow of electromagnetic energy from the atmosphere through water, reflecting off the bottom and back to the surface. The program addressed the influence of the water, water surface effects, and the atmosphere on solar radiation transfer through time and space. Inputs included the solar constant, extinction coefficients, and bottom type reflectance. Water depths could be predicted to depths of 3.7 meters. Interestingly, blue wavelengths gave poor results due to scattering of these wavelengths in water. Blue-green, green, and red wavelengths showed the best correlation with water depths.

Shallow water appears light gray on panchromatic film, black on black/white infrared film, light green to light blue on color film, and light blue on color infrared film. Deep water appears dark gray on panchromatic film, black on black/white infrared film, dark blue to dark green on color film, and black on color infrared film. Because light is reflected from near-surface suspended particles, turbid water has the same appearance as shallow water. Radar wavelengths (microwaves) are either reflected by rough water or are reflected away from the antenna (no returns) on smooth water (Fig. 9.5). Thermal imagery generally senses surface water as warmer than background soil and rock during the night and cooler than background during the day. The temperature of surface waters can be monitored using thermal sensors (see Geothermal Resources, this chapter).

Turbidity is usually associated with wind, in shallow lakes, or with flowing water, particularly flood waters or Spring runoff. Rivers entering lakes or the ocean generally have higher turbidity than the body of water they are flowing into (e.g., Puyallup River flowing into Puget Sound, Fig. 3.13). Lakes generally show a turbidity gradient from inlet to outlet that appears as light blue to black on color infrared images (Lake Cabora Bassa, Mozambique, Fig. 9.6). Increased salinity can have the same effect as turbidity, increasing the reflectance of the water body (Figs. 9.7, 11.13).

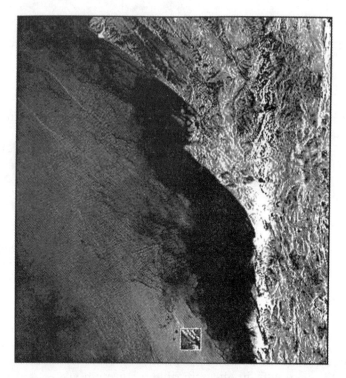

Figure 9.5 ERS-1 radar image of the Angola coastline showing rough and calm seas. Smooth sea surfaces can be due to wind shadows along the coast [7]. (From *World Oil*, July 1999.) Inset is a high resolution image showing internal waves.

FLOOD CONTROL

Surface waters can be monitored for flood prediction and public safety, irrigation planning, fisheries, and for recreational activities. The main remote sensing contributions include monitoring snowpack and monitoring runoff.

Satellite imagery is particularly well suited for mapping the extent of snowpack over time (Fig. 9.8). Although it may not be possible to remotely map the thickness of snow, if ground control points exist it should be possible to extrapolate thicknesses over the area of known cover. Additional information on slope steepness and azimuth (north-facing slopes retain more snow in the northern hemisphere) derived from a digital terrain model should further refine the calculations. After several years of imagery a historical database can be compiled showing the volume of water associated with a given extent and thickness of snow. Flood control reservoir managers will need this information to plan when to lower water levels in anticipation of higher than normal runoff. Farmers, particularly those cultivating rich bottom lands, will want to know the chances of flooding in any given year prior to planting. Downstream fisheries can use information on water salinity and turbidity related to lower or higher than usual river volumes.

Late snows combined with long or heavy rains and rapidly warming temperatures can cause catastrophic runoff and flooding. Some floods are purely the result of long

Figure 9.6 Light blue indicates turbidity on this color infrared image TM image of Lake Cabora Bassa, Mozambique. Image processed by EOSAT. (Landsat^{TM} images courtesy of Space Imaging, LLC, Denver, Colorado.) (See color plate.)

periods of intense rainfall. Monitoring floods may not do much for those affected by the floodwaters, but it can show which areas are prone to flooding (Fig. 9.9). This will allow those areas to be set off limits for development, or at least permit prudent planning for development of the floodplain. Insurance companies will be interested in the limits of flooding for setting rates, and cities and counties will need this information for preparing storm sewers and drainage projects that will minimize

Figure 9.7 Light blue water indicates increased salinity in evaporation ponds of the Dead Sea, Israel and Jordan. Landsat TM color infrared image processed by EOSAT. (Landsat™ images courtesy of Space Imaging, LLC, Denver, Colorado.) (See color plate.)

damage and control runoff. In the U.S., the National Flood Insurance Act of 1968 established the Federal Emergency Management Agency (FEMA) to manage a national flood insurance program. As part of the program the act required identification of all floodplains and flood risk areas. FEMA has prepared a series of Flood Insurance Rate Maps that are used to establish premiums and to help with flood relief operations. Input to these maps comes from topographic maps and historical flood records, including multitemporal remote sensing images [8].

Figure 9.8 Snowpack is shown as blue in this Landsat TM 5-4-3 (RGB) image of the Altun Mountains, China. Image processed by EOSAT. (Landsat™ images courtesy of Space Imaging, LLC, Denver, Colorado.) (See color plate.)

EROSION

Erosion is a function of several factors, including precipitation, currents, slope, dip of strata, vegetation cover, and surface materials. Imagery, whether airphoto, radar, or satellite-derived, can contribute to mapping of erosion hazards. Digital terrain models can provide a quantitative measure of slope, whereas image interpretation gives only a qualitative measure (i.e., a slope is more or less prone to fail). Mapping of scours and

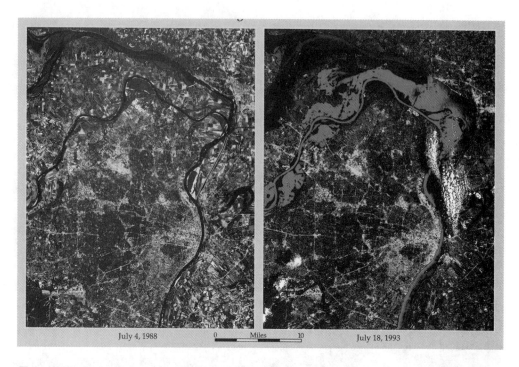

Figure 9.9 Landsat images of the St. Louis area showing the confluence of the Missouri and Mississippi rivers. The first, taken on July 4, 1988, shows normal flow; the second, from July 18, 1993, shows the rivers in flood stage. Images processed by EOSAT. (Landsat™ images courtesy of Space Imaging, LLC, Denver, Colorado.) (See color plate.)

cutbanks along rivers, and of headlands and barrier bars along a coastline, can also furnish information on where, historically, erosion has occurred. Effort should be taken not to build roads, dams, and other structures where they are likely to be undercut by running water or waves, where the slope is greater than the angle of repose for saturated soils, or where the dip of bedding is parallel to the slope.

An example of an erosion mapping project was described by Šúri, et al. [9] for the Slovak Republic. Both potential and actual soil erosion risk were evaluated on a regional scale (1:500,000) using five criteria: (1) erosional potential of rainfall, (2) potential soil erosion risk, (3) actual soil erosion risk, (4) slope, and (5) land cover. Remote sensing was able to contribute to mapping of soil erodability (soil texture and type), slope, and land cover (vegetation density, structures, soil conservation practices). Potential erosion risk indicates the tendency of an area toward erosion. Actual erosion risk takes into account land cover and conservation practices. They found that the loamy and silty soils developed on the strongly dissected slopes of the Carpathians put 75% of the country in the high soil erosion risk category. The actual erosion risk in the forested mountains is lower due to the plant cover.

Another approach was used by Rao and Adinarayana [10] to assess erosional potential in the Kukadi River watershed of Maharashtra, India. They feel that soil erosion is a function of three factors: (1) energy provided by falling rain, runoff, and topographic slope, (2) soil resistance, and (3) vegetation cover. Satellite imagery was used to map land use and to modify drainage maps derived from topographic mapping. This information was then combined with soil maps derived from field observations and lab measurements to develop a database that describes the relative erosion status for different watersheds in a quantitative way.

A study of coastal erosion was undertaken at Chesapeak Bay by the Maryland Geological Survey [11]. First they mapped the extent of shoreline change by examining historical maps and airphotos. Then they calculated the rates of erosion along transects perpendicular to the coast. They combined this information with sediment type and topographic relief to determine the volume of sediment eroded. This knowledge proved helpful for determining set-backs for construction and flood insurance, calculating the volume of sediment available for silting up other areas of the bay, assessing turbidity and nutrients in bay fisheries, and determining the factors that contribute to coastal erosion.

GEOTHERMAL RESOURCES

Lists of thermal springs exist for most developed areas [12]. For those areas where such lists do not exist there are several indicators of geothermal/hydrothermal systems. Among these are evidence of recent volcanism, alteration of bedrock and soils associated with springs, and indications of hot or warm water. Geothermal sources can be used for power generation, as at Geysers, California. They can be used for health and tourism, as at Bath, England. They can also be used as an indication that an area is volcanically active or has high near-surface heat flow (thus, may be suitable for generation of hydrocarbons in recent sediments, but may not be appropriate for hydrocarbon preservation).

Perhaps the easiest and most direct means of locating hot water is to use thermal sensors. These sensors do not measure temperature directly (see Chapter 3, Airborne Thermal Scanners), so it may be necessary to obtain ground control measurements to calibrate the imagery if temperatures are required. Thermal infrared scanners flown under pre-dawn conditions usually show shallow water bodies as warmer than the background soil and rocks because of the ability of water to retain heat (high thermal inertia). Flowing water, however, will appear cooler than point sources of warm water, which are a clue to warm or hot springs (Fig. 9.10). One should be careful not to map black shales, basalts, or other dark rock as heat sources. Other clues to thermal areas include areas of water in otherwise ice-covered rivers or lakes on winter imagery; ground fog over rivers and streams on cool, still mornings; and mineral deposits such as tufa, travertine, or siliceous sinter. Sulfur-secreting bacteria are common in some thermal springs and are green in color. Green algae thrives in water between 49 and 60° C; orange and red algae can be found in water between 60 and 71° C; white algae has been found in water as hot as 85° C. Although none of these criteria are unequivocal, all are indications of hot springs in some areas. Finally, recent cinder cones or lava flows, and epithermal alteration (iron staining, argillization) of outcrops and soils, especially if near springs, provide evidence for a near surface origin of heat [13]. These features can also be associated with regional near-vertical (linear) fault systems that tap deep crustal heat sources. All of these features can be observed and mapped on airphotos and satellite imagery.

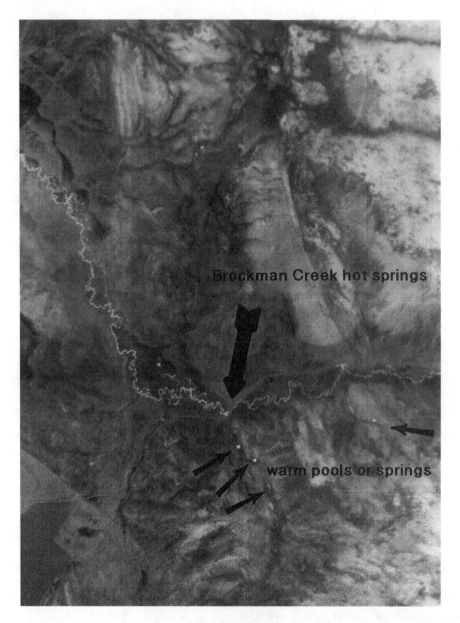

Figure 9.10 Predawn airborne thermal image of Brockman Creek hot springs, Idaho (large arrow) and numerous warm spots in nearby streams (small arrows). Image processed by Mars, Inc., Phoenix.

REFERENCES

1. M. Dalati, *Proc. of the 12th Intl. Conf. Applied Geologic Rem. Sens.* **I**, (ERIM, Ann Arbor, 1997), pp. 376–383.
2. L. Dongtian, and C. Yunchang, *Proc. of the 10th Thematic. Conf. Geologic Rem. Sens.* **II**, (ERIM, Ann Arbor, 1994), pp. 568–575.

3. J.F. McCauley, G.G. Schaber, C.S. Breed, M.J. Grolier, C.V. Haines, B. Issawi, C. Elachi, and R. Blom, *Science* **218** (1982): 1004–1020.
4. H. Hough, *Earth Observation Mag.* (Sept. 1995): 22–24.
5. J.E. Nelson, S.E. Silliman, and R.J. Wasowski, *Civil Engineering Applications of Remote Sensing and Geographic Information Systems* (Am. Soc. Civil Eng., New York, 1991), pp. 179–188.
6. J.G. Lyon and W.S. Hutchinson, *Photogram. Eng. & Rem. Sens.* **61** (1995): 161–166.
7. H. Hajji, *World Oil* **220** (July 1999): 39–40.
8. S. Grzeda, *Earth Observation Mag.* (April 1996): 22–25.
9. M. Šúri, T. Cebecauer, J. Feranec, and E. Fulajtár, *Proc. of the 12th Intl. Conf. Applied Geologic Rem. Sens.* **I**, (ERIM, Ann Arbor, 1997), pp. 315–322.
10. K.G. Rao and J. Adinarayana, *Asian-Pacific Remote Sensing Jour.* **7** (1994): 147–154.
11. L. Hennessee, R.T. Kerhin, J.J. Isoldi, R.A. Gast, M.S. Robertson, *Proc. of the 12th Intl. Conf. Applied Geologic Rem. Sens.* **I** (ERIM, Ann Arbor, 1997), pp. 309–314.
12. G.W. Berry, P.J. Grimm, and J.A. Ikelman, *U.S. Dept. Commerce Key to Geophysical Records Document* **12** (NOAA, 1980), 59 p.
13. S. Camacho, L. del Rio, L. Sanchez, and J. Gonzalez, *Proc. of the 16th Intl. Symp. On Rem. Sens. of Environ.* **I** (ERIM, Ann Arbor, 1982), pp. 145–153.

ADDITIONAL READING

S.O.H. El Khidir, A.I. Al Biely, and A.H.A. Mohamed, *Proc. of the 12th Intl. Conf. Applied Geologic Rem. Sens.* **I**, (ERIM, Ann Arbor, 1997), pp. 348–353.
K.R. Knapp, K.M. Morgan, N. Donovan, A. Busbey, and N. Kresic, *Proc. of the 10th Thematic Conf. Geologic Rem. Sens.* **II**, (ERIM, Ann Arbor, 1994): 155–160.
N. Kresic, *Proc. of the 10th Thematic Conf. Geologic Rem. Sens.* **II**, (ERIM, Ann Arbor, 1994), pp. 161–167.
G.A. Waring, R.R. Blankenship, and R. Bentall, U.S. Geol. Survey Prof. Paper **492** (1965): 383 p.
W.S. Warner, *Proc. of the 12th Intl. Conf. Applied Geologic Rem. Sens.* **I**, (ERIM, Ann Arbor, 1997), pp. 286–293.
A.J. Wiley, G.D. Louderback, F.L. Ransome, F.E. Bonner, H.T. Cory, and F.H. Fowler, Report of the Commission to Investigate the Causes Leading to the Failure of the St. Francis Dam near Saugus, California (Calif. State Printing Office, 1928), 79 p.

QUESTIONS TO CONSIDER

1. How do plants respond to shallow groundwater in arid and temperate climates?
2. What thermal response would you expect from groundwater ponded against a fault on daytime thermal imagery?
3. What environmental factors can cause a misleading interpretation of remote sensing water depth mapping?
4. How can moving water contribute to erosion? What would indicate erosion-prone areas?
5. How would you organize a program to explore for geothermal resources?

10

Logistics and Engineering

CHAPTER OVERVIEW

Logistics and engineering application are among the most important and at the same time most overlooked functions of geologic remote sensing. Many engineers, trained to be conservative in their approach to problem solving, are either unaware of the capabilities of remote sensing technology, or are hesitant to use new, often untried techniques for mapping and measuring environmental factors. This chapter will review some of the applications that have been used successfully by corporate engineering groups.

Engineering activities covered by remote sensing can be lumped broadly into three categories:

1) reconnaissance mapping, including locating old tailings, well sites, seismic lines, facilities, roads, and pipelines; and finding sources of raw materials such as aggregate for construction.
2) siting, i.e., determining the safest, most cost-effective sites for plants, storage facilities, pipelines, roads, and port facilities.
3) transport, i.e., finding the shortest and least expensive access to a drill site, determining the cheapest way to get product out of an area, determining ice movements and weather patterns that might interfere with truck or tanker traffic, and locating dangerous shoals.

Each of these topics will be covered in turn.

USING APPROPRIATE BASE MAPS

Base maps are most often topographic maps showing access roads, mines, or well locations, mine dumps, tailings, gathering systems, storage tanks, and facilities. A map that is more than a couple of years old is probably out of date: new wells are drilled, new roads are built, new pipe laid, new tailings ponds are filled. An airphoto enlarged to 1:25,000, or satellite image at the same scale or 1:50,000 can be acquired annually at a nominal cost and integrated into a geographic information system to provide up-to-date maps of development activities. An archive of such photos can be useful as well if there is ever a question of environmental liability regarding spills, mine drainage, mud pits, and pipeline or tailings pond leaks.

If maps are needed for purposes of surveying or construction, it is possible to generate orthophotos by eliminating tilt and radial distortion. Digital imagery can be rectified and projected to any map projection, datum, and spheroid, then plotted with a superimposed Universal Transverse Mercator (UTM) or latitude-longitude grid. Topographic maps can be generated from airphoto or satellite stereopairs: SPOT Pan and some radar imagery can be used to generate topographic maps with five meter contours or better. The shuttle radar mission will provide near global elevation data with 10–15 m vertical accuracy (chapter 3, Shuttle Imaging Radar). Generation of topographic maps usually requires that ground control points are surveyed and that elevation control points are provided. Recent advances in global positioning system (GPS) technology have made the acquisition of control points not only quick and inexpensive, but also accurate to plus or minus 10 meters (non-differential) and to within 1 meter (differential). It is necessary to bear in mind that control points must be readily identifiable on the imagery. Road intersections, bridges, large buildings, the point at which a tributary enters a river or a river enters a lake make good control points. These points cannot assist in removing image distortion unless they can be accurately located on the imagery. The usefulness of geographic information systems depends on accurate and up-to-date base maps.

GEOGRAPHIC INFORMATION SYSTEMS

Geographic information systems (GIS) have become the way to manage data for all large engineering projects. The GIS is simply a relational database that allows the user to call up and integrate several types of data based on their geographic location. There are three attributes that make the GIS useful: (1) the ability to automatically locate an object based on a given set of coordinates, (2) the ability to relate an object to other nearby objects, and (3) the ability to add descriptive text or tabular data linked to the object of interest. Generally speaking, remote sensing imagery or topographic maps are used as the base for overlaying graphical information on physical or administrative features. Physical information includes geology, roads or tanker lanes, river drainage networks, oil wells, mine pits and tailings, pipelines, and power lines. Administrative features include land use, wildlife habitat, and rights of way and land ownership. By specifying an area on the base map, one can be hot linked to all types of information related to that area, including historical photos, diagrams, video clips, and text such as lists of wells, soil samples, engineering reports, or legal land descriptions. The GIS allows the user to change scales for greater or lesser detail, and permits the interpreter or database manager to update all levels of the database instantly. It also gives the user instant access to various levels of data detail, which can be critical in the rapid and accurate analysis of diverse information. The user is able to point to a map feature and ask "what are you?" One is also able to query the feature, for example, "show me all the properties crossed by a pipeline right-of-way from point 'a' to point 'b'," or "show me all the areas where bedrock is shale, dip is greater than 30 degrees, rainfall is greater than 40 cm/year, and vegetation cover is sparse." It should be evident that this type of system can be very useful for engineering purposes. Two examples are given to illustrate applications of a GIS to engineering remote sensing.

Case History: Mine Management Using a Geographic Information System

Highland Valley Copper (HVC) has the largest base metal mine in northeast British Columbia, Canada, shipping about 320 million pounds of copper in 1996 [1]. They are using GIS technology to make and update maps on a daily basis. These maps support exploration for new ore bodies, the mining process, and restoration of the land. The company holds about 5,900 hectares of disturbed land, including two large open pits, facilities, dumps, and a tailings pond (Fig. 10.1). Local landowners include the provincial government, forest companies, and private farmers and ranchers.

Figure 10.1 Revegetation is underway at the Highmont and Lornex pits and dumps and tailings piles, British Columbia. A GIS is coordinating the process [1]. (See color plate.)

In the past, maps were made individually in a CAD system and were difficult to maintain, update, or scale up or down. Information existed on more than one map or computer file, and updates were not made consistently across all versions. With the introduction of a GIS, the engineers at HVC were able to take all existing maps and generate them in any coordinate system and scale. They also added baseline data from aerial photography and topographic maps at 1:5,000. This information is available to all the staff, from engineers to geologists and environmentalists, immediately after it is brought into the GIS environment.

During the mining process it is important to monitor slope stability. Pit crews expect to be able to record field observations tied to global positioning systems; show rates of ground movement; the orientation, size, offset, and amount of movement on joints and faults; changes in hydrology (water table, springs); and the relationship of these features to buildings and other structures using the GIS. Reclamation costs HVC about $2 million annually. About 1,300 hectares have been reclaimed. Reclamation involves re-sloping and contouring the land, capping waste dumps with soil, preparing the soil to seed, seeding and planting, then fertilizing and maintaining the area for up to four years. The GIS allows them to prepare maps of the reclamation status quickly and accurately. In addition to reclamation, the GIS shows the land ownership situation in Highland Valley. There is an ongoing need to document and update leases, rights-of-way, and water rights, among other items.

Case History: Pipeline Routing Using Remote Sensing and a GIS

Engineers at Bechtel planned a 700 km pipeline to bring oil from the Tengiz field in Kazakhstan on the Caspian Sea to Novorossiysk in Russia on the Black Sea [2]. The completed pipeline would be capable of delivering 1.5 million barrels of oil per day. A least-cost analysis was performed incorporating pipeline length, topography, geology, land use, and stream, wetland, road, and railroad crossings. Landsat TM and SPOT imagery was used as a base on which to display results and for land cover mapping.

Most of the pipeline route passes through low relief terrain along the northern end of the Caucasus Mountains. Tectonic activity began in the Caucasus in Miocene time, and most intense uplift and folding occurred in the Pliocene and Quaternary. Many faults are still seismically active. Although the shortest and generally least costly pipeline is a straight line, factors to be considered included population centers, environmentally sensitive areas, and technical issues such as slope steepness and stability, geology, and river crossings. Rugged topography requires more pump stations, whereas hard rocks require blasting and river crossings require bridges.

Maps and aerial photography are needed to evaluate these factors and make routing decisions. Airphotos were not available, so 1:100,000 and 1:200,000 topographic maps and TM and SPOT imagery were used. A supervised classification was performed on the satellite imagery to generate a land use map (Fig. 10.2). Topographic, drainage, geologic, and infrastructure (roads, cities) data were also input as separate layers to the GIS. A slope map was derived from the topographic elevation information. The least-cost analysis was performed by assigning costs factors associated with crossing of steep vs. gentle slopes, streams, wetlands, roads, railroads, outcrops, agricultural land, and urban and industrial areas. From this it was possible to generate a "weighted cost surface," that is, a surface analogous to a topographic surface with peaks and troughs

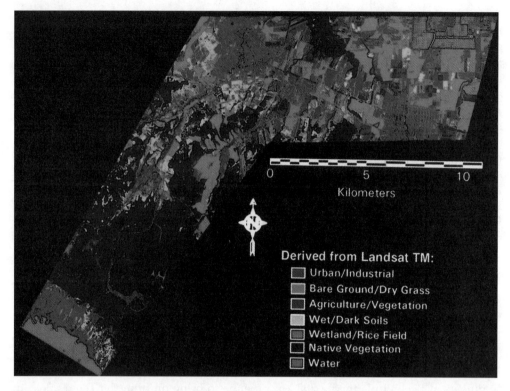

Figure 10.2 Land use map of the Caspian pipeline corridor derived from TM bands 3, 4, and 5. Image processed by Bechtel [2]. (See color plate.)

that correspond to high and low costs, respectively (Fig. 10.3). The least cost path could then be traced. Although the straight line path was shorter, the least cost path was 14% less expensive to build.

RECONNAISSANCE

Before going into a new area, particularly in a foreign country, it is important to know what the infrastructure is, that is, where production occurs, where pipelines and storage facilities exist, and where the roads, powerlines, refineries, and smelters are. This type of information is often not available, either because it is not recorded or not volunteered by the host country. One can learn, for example, something about the amount of oil produced in an area by counting the number of oil storage tanks (Fig. 10.4). Whether or not a pipeline exists is a crucial factor in calculating the economics of a specific region. If there are signs of a well having been drilled in an area, yet there is no record of it, this would suggest that an area previously thought untested may not be worth testing, or a least rates a higher exploration risk category. Old tailings can tell where mining activity occurred in the past: the lack of modern roads might tell you that nothing has happened there since the price of gold went over $32 per once.

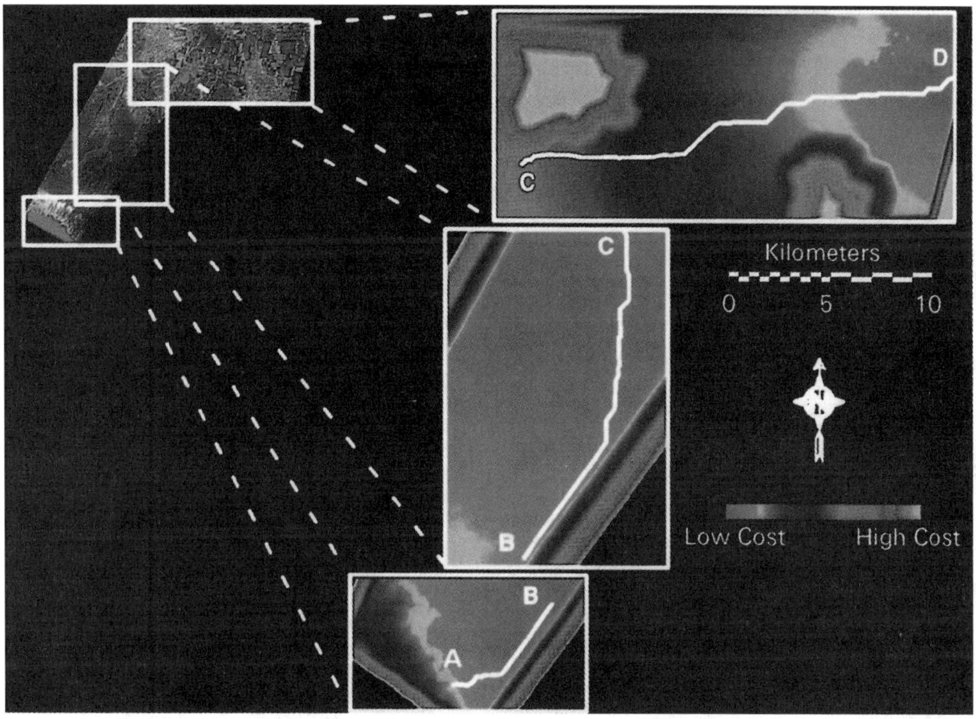

Figure 10.3 Cumulative cost surface and least cost path generated from a GIS database for the Caspian pipeline. Image processing by Bechtel [2]. (See color plate.)

When a new discovery is announced it is often not accurately located on existing maps. Detailed satellite images can show not only exactly where the discovery is, but can also provide information such as whether a well was drilled on surface structure or if a deposit is localized by faulting. If there is no obvious surface structure, this suggests a certain sophistication of the seismic (or other subsurface prospecting) techniques employed to locate the discovery. If there is surface structure, one can still learn something about the kind of structures that produce in an area, and perhaps use this information to explore for similar structures in the surrounding basin. If an ore body is located along a shear zone, it may be possible to explore along that zone, or to look for features analogous to the zone in the adjacent region. A lack of surface alteration suggests the use of potential fields methods or geochemical sampling to locate the deposit.

The number of wells in a field, and the degree to which they are interconnected by road and pipeline networks, can provide important information to companies evaluating partnership deals (Fig. 10.5). Likewise, the extent to which an area has prospect pits can reveal the breadth of the area thought to be mineralized.

A costly part of many seismic programs involves permitting and cutting paths for the survey crew and equipment. In many parts of the world it is fairly evident where previous seismic lines were cut because the vegetation has not completely recovered, even from surveys many years earlier. Such is the case in Somalia, where seismic lines

Figure 10.4 Oil storage tanks, southern Kharg Island, Iran. SPOT P image. North is at the top. Image processing by Amoco Production Company.

cut through thorn bushes 10 to 15 years before imagery was acquired are still visible as straight lines across the landscape (Fig. 10.6).

Gravel is often an important resource for building drill pads and artificial islands in shallow water offshore areas. The cost of transporting gravel is a great incentive for locating local sources. A quick airphoto or satellite survey of an area can often locate gravel bars in nearby rivers, glacial outwash, or coastal gravel banks that are excellent sources of aggregate (Fig. 10.7) [3].

Water is often necessary for drilling mud, for mining and smelting operations, to sustain field camps, and to contain dust in construction sites. Imagery can assist in locating the best surface water or shallow groundwater (Chapter 9).

Figure 10.5 Oil field showing wells, pipelines, and roads in Sharjah, United Arab Emirates. SPOT P image, processed by Amoco Production Company. North is at the top.

SITE SELECTION

For any engineering application the first step must be site selection and evaluation. Many factors go into selecting the site for plants, storage facilities, pipeline routes, and roads. Among them are construction suitability of the surface material, slope stability, environmental sensitivity, and proximity to transportation networks. The goal of a site selection program is to evaluate surface materials to determine whether they are permeable, cohesive, and stable; to evaluate soils, their thickness, and suitability for agriculture or other competing uses; to evaluate the local drainage, bedrock, and land use. Remote sensing, whether satellite-derived, airphoto-based, or some other combination of sensors, provides an economic and rapid means of obtaining this information [4]. Maps can be generated showing surface materials, surface geology and soils, potential hazards, hydrographic drainage networks, plant cover, and land use. An interpreter can then begin to predict the impact on the local environment, and plan a surface sampling program to confirm the observations.

Image analysis provides a means for evaluating any region on earth quickly and at low cost. Certain types of information are needed for site selection whether the plan is to build on the surface or tunnel through the earth, drill into it, or excavate and mine it. Among these are data on the surface geology including the soils, rocks, and structure. Important questions that require answers include the thickness, permeability, stability, and cohesiveness of soils; the type of subsurface

Figure 10.6 Seismic lines in Somalia northwest of Mogadishu. SPOT P image processed by Amoco Production Company. North is at the top.

and surface rocks, their attitude, and whether they are fractured. Local sources of sand and gravel are often important for construction. Local hazards, such as near-surface caverns, landslides, or active seismic zones, are of paramount importance (Chapter 12).

LOGISTICS AND ENGINE[

Figure 10.7 Sandbars and gravel in the Irrawady River near Prome, Myanmar. This SPOT P image was processed by Amoco Production Company. North is at the top.

Surface Materials

Facilities, roads, conveyor belts, powerlines, and pipelines should not be built in areas with potential hazards such as unstable slopes, river flooding, subsidence, and fluvial or coastal erosion. It is sometimes only possible to determine the extent of such hazards by obtaining multitemporal imagery and checking, for example, the extent of flooding over a period of a decade or more. Overgrown landslides and areas prone to grass or brushfires can often be detected by reviewing photos acquired over many years.

Generally speaking one prefers to build on a solid, stationary surface. Bedrock is good in the sense that it is usually stable. One must determine, however, that bedding or foliation is not parallel to or undercut by the slope of the surface, as it may then serve to initiate a detachment or slide. Soils are generally easy to work with as far as excavating or shaping the surface, but are subject to settling and uneven compaction, and thus may require special building techniques and later monitoring.

Different soils have different engineering properties that must be taken into account when planning site development. Sandy soils will have good water infiltration and thus minimize the need to account for storm water runoff. On the other hand, spills can quickly find their way to the water table in sandy material. Sandy soils are less prone to mass downslope movements than clay-rich soils, which can fail on slopes as shallow as one or two degrees. The ability to slide downslope as a result of gravity is enhanced by the types of particles and amount of water in the soil. Large, irregular-shaped particles (sands, conglomerates) generally have more resistance to sliding than fine-grained, flat particles (clays). All materials have less resistance to sliding when saturated with water, which tends to behave as a lubricant [5].

It should be possible to interpret sandy versus clay-rich soils on the basis of infiltration and relative drainage spacing, the steepness of drainage channels, and the type of vegetation growing in an area. Sandy soils have better drainage and thus a more widely spaced drainage network than clay-rich soils. They also tend to have V-shaped gullies, whereas silt and clay-rich soils tend to have U-shaped gullies. Clay-rich soils have grasses and shallow-rooted trees and shrubs, whereas sandy soils have deep-rooted plant communities (see Chapter 4). Areas with near-surface groundwater will appear as darker tones (moist) and have more verdant vegetation in all soil types.

Subsidence

Some projects are in areas of active subsidence, either due to sinkholes forming in a karst environment or as a result of groundwater or hydrocarbon withdrawal from the subsurface or from abandoned subsurface mining activity (see Chapter 8, Mine Safety Issues; Chapter 12, Subsidence Case History). In such regions it is necessary to avoid areas prone to subsidence when planing pipeline routes or facilities sites. Subsidence related to withdrawal of liquids from the subsurface can lead to undrained depressions or large open soil cracks. The probability of locating mine-related collapse depressions is greater if one can identify tailings in the vicinity. It is fairly simple to recognize areas that have collapsed; it is difficult, if not impossible, to predict areas that will collapse unless one has a map of old workings.

Surface Slope

For the most part, one tries to avoid steep slopes when building facilities or roads, since site preparation costs go up with the amount of earth that needs to be moved or leveled. Steep slopes are inherently less stable than gentle slopes, requiring more work to make the road, pipeline, etc., safe. Relative slope is easy to observe on imagery. With

Permafrost

Knowing the distribution of permafrost is extremely important to anyone wishing to build in the arctic, or shoot a seismic program. Its presence or absence dictates the construction methods to be used, or how a seismic program is laid out. One can sometimes determine the presence of permafrost by observing geomorphic features such as thermokarst lakes, patterned ground, ice wedge polygons, earth hummocks, solifluction sheets, and pingos (Fig. 10.8) [6].

Expanding Clays

As anyone who has built a house on montmorillonite-rich soils can attest, such soils are not the desired foundation for a stable home. Cracks suddenly appear, and walls shift positions. Likewise, roads built on expanding clays without the proper bed

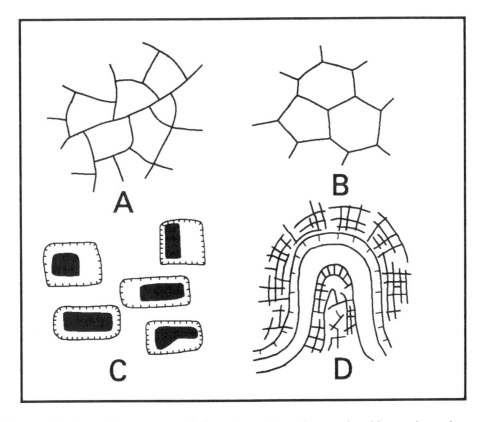

Figure 10.8 Permafrost patterns. A. Irregular patterned ground and ice wedge polygons. B. Polygonal patterned ground and ice polygons. C. Thermokarst (rectangular) lakes. D. Radial patterned ground around a river channel. Modified after Mollard and Janes, Fig. 7–5 [6].

preparation will quickly become uneven and possibly dangerous. The alternating swelling and contracting associated with wetting and drying often causes such clay-rich soils to form cracked and hummocky surfaces. In some cases they appear as badlands. It should also be possible to map changes in surface elevation over time with a high degree of accuracy (on the order of 1 cm), and this may help determine the existence of unsuitable soils. Radar interferometry is an ideal candidate for this type of mapping. Gabriel, et al. [7] has shown the ability to map the expansion of water-absorbing clays in the Imperial Valley of California using Seasat radar data to an accuracy of 1 cm over 10 m pixels.

Often the final site for a facility or route of a road or pipeline will be determined by a series of tradeoffs. For example, when construction costs to build a pipeline through the arctic run several million dollars a mile, one will necessarily want to take the shortest, most direct route from point A to point B. Yet, one is also obligated to avoid lakes and river crossings that require special construction techniques that can run up the costs even further. Then there are environmental factors to consider: environmental sensitivity, fish or mammal migration routes, and endangered habitats. These considerations are illustrated by the process involved in evaluation of a nuclear power plant site in India (see Kakrapar Power Plant Case History, this chapter).

TRANSPORTATION

One of the first things a drilling engineer asks in a new area is "how do we get to the drill site?" Easy access to a wildcat well site is important not only to keep costs down, but to be able to drill at the exact spot pinpointed by surface mapping, seismic, and other surveys. Good airphotos or high-resolution satellite imagery can assist the roadbuilding effort by pointing out shortcuts, steep and shallow grades, landslide-prone areas, etc.

In some areas it may be possible to gain access by barging supplies upriver. In that case it would be helpful to have an up-to-date map of the river channel and sandbars (Fig. 10.7). Imagery is more accurate than most topographic maps, and it is certainly easier to obtain up-to-the-minute imagery than to have a new set of base maps produced. Likewise, it may be possible to transport oil by barge down river instead of bringing it out by truck or building a pipeline.

It is becoming increasingly important to know the patterns of ice movement in the arctic as a result of increased drilling activity and the possibility of future ice-breaking oil tankers. Offshore Alaska, the North Sea, offshore Siberia, and the Canadian arctic are all areas of exploration interest. Loading facilities and offshore well platforms will have to be able to withstand bottom scour from sea ice. All of these require a thorough knowledge of the timing, direction, and rate of movement of ice floes. An excellent source of information is the historical archive of satellite images collected over the past 20 years. By looking at multi-temporal imagery from a single season, one can determine the rate and direction of movement of individual floes. By looking at images from multiple years, one can begin to map the average and maximum extent of the pack ice, and the usual location and concentration of icebergs at various times of year. Daily airborne radar has already been used extensively by the Canadians and Russians to map ice movements around offshore drill ships. With the advent of radar satellites, this monitoring of sea ice should become routine and much less expensive (Fig. 10.9).

LOGISTICS AND ENGINEERING 285

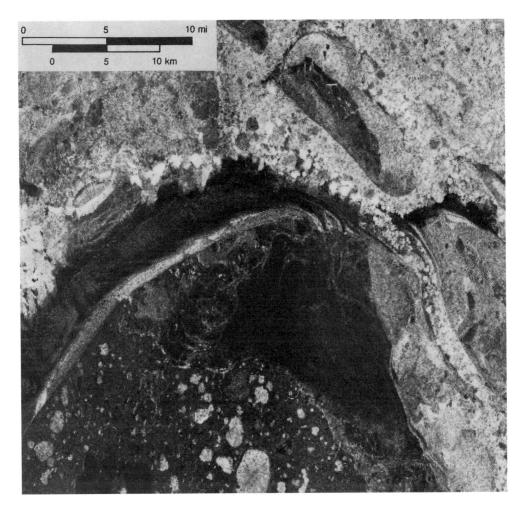

Figure 10.9 ERS-1 radar image of sea ice in the Kara Strait, offshore Siberia. The land area, bottom and center, has numerous frozen lakes. The ice offshore has different textures depending on its age. Image processing by Amoco Production Company. North is at the top.

A final transportation issue is water-depth mapping. It is necessary to know the location of shoals in and around harbors and transportation routes. Many of these are mapped on Admiralty Charts or National Ocean Service Hydrographic bases. Still, in some areas the shoals shift through time, and in other cases they were not mapped accurately. Airborne water-penetration film (Chapter 1), the Coastal Zone Color Scanner, and Landsat TM all have excellent water-penetration capabilities. It is fairly simple to generate relative depth maps by density-slicing, or using various film densities to indicate areas that are increasingly shallow (Fig. 10.10, 12.9). It is slightly more difficult to generate good bathymetric maps, since several widely scattered control depths would be required. Also, as mentioned in Chapter 9, the apparent water depth is a function of the bottom reflectance and the amount of suspended sediment. If

North Cat Cay, Bahamas

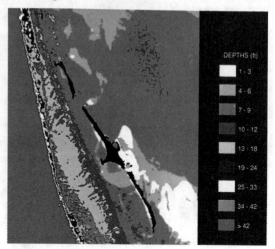

Contoured Depth Image (Depths in Feet)

Color-Coded Depth Image

Figure 10.10 Density-sliced and contoured bathymetric images of North Cat Cay, Bahamas. Landsat TM bands 1, 2, and 3 processed by ERIM, Ann Arbor. (See color plate.)

the bottom has both light and dark areas, such as is caused by sea grasses or changing rock types, this will affect the apparent depth. Suspended particulates will also reflect light as if they were the bottom. One can usually distinguish the pattern of sediment plumes from bathymetry patterns, however.

POWER PLANTS, PIPELINES, TUNNELS, AND DAMS

Construction of large engineering projects requires extensive planning. Such planning must take into account the least disturbance to the natural environment, the existing road, rail, and power networks, sources of water for cooling or construction, other resources in the area (e.g., agriculture, forestry, fisheries), population centers, seismicity, slope stability, and other factors. Power plants, in particular, may require access to railroads for fuel and water for cooling. They will be concerned with seismic activity that could disrupt operations [8]. Pipeline routes prefer gentle, stable slopes, the least possible amount of bedrock, and few river crossings (see Pipeline Routing Case History, this chapter). Tunnels must be concerned with seismicity and the lithology and amount of fracturing in wall rocks. Dam construction will require local sources of sand and gravel for construction. Safe construction of dams must account for strike and dip of rock in which the dam is anchored, fracturing of wall rocks, seismicity, porosity and permeability of the country rock, annual variations in runoff from snowpacks and rainfall, erosion and sedimentation of the drainage basin upstream from the reservoir, and drowning of agricultural land and other resources. Image analysis can be a primary source of information to answer all of these concerns. An example power plant in Gujarat, India.

Kakrapar Atomic Power Project Site Selection Case History

Satellite images and multi-scale aerial photographs were used to maps that addressed surface geology, geomorphology, seismicity, wa land use, forest and mineral resources, availability of raw materials for cons rail access, and environmental, health, and safety factors associated with the siting an atomic power plant in Gujarat [9]. The eastern half of the area is covered by Deccan basalts, whereas the western part of the area is mostly recent sediments. Dolerite and gabbro dikes of various ages intruded the area along ENE trends.

Regional scale satellite images were used to select a region of interest. This was followed by candidate site selection using finer resolution satellite images and airphotos to scales of 1:50,000. Detailed analysis was performed using airphotos at scales of 1:25,000 with support from ground mapping and sampling.

Geomorphology was used to classify stable slopes, areas of rugged topography, and locations of construction materials. Surface water was mapped in lakes and drainages, and probable sources of near-surface groundwater were located. Land use was classified as cultivated, urban, forest, and "waste" land. Existing roads were mapped, and locations for new roads and power lines were evaluated. Forest cover was mapped to prevent degradation of an already depleting resource. Seismic activity was investigated to determine the location, frequency, and magnitude of earthquakes in the area. Sites closer than five kilometers to active faults were not acceptable. Active faults are defined as having movement in the last 35,000 years, or multiple movements in the past million years. Image analysis, including low sun-angle photography, radar, and thermal imagery, was used to locate lineaments, recent fault scarps, shutter ridges, sag ponds, and aligned hot springs. A network of micro-seismic monitoring stations was established, and historical seismic activity was investigated. Trenching is not mentioned, but may have been used to determine the age of the last active faulting by dating of disturbed and non-disturbed soil horizons. In summary, several remote sensing systems were used to determine the suitability of the site for a power plant, and to determine the probable impact on the environment around the plant.

The number of engineering applications for remote sensing are as limitless as the resources and imagination of those who are willing to test them. As new problems arise, or new technologies become available, the number of applications will increase accordingly.

REFERENCES

1. V. Speed, *Earth Observation Mag.* (October 1996): 22–25.
2. S.C. Feldman, R.E. Pelletier, E. Walser, J.C. Smoot, and D. Ahl, *Remote Sensing Environ.* **53** (1995): 123–131.
3. I.P. Jolliffe and A.G. McLellan, *Geographical Magazine* **52** (1980), 3 p.
4. J.N. Rinker and R.E. Frost, *Computing in Civil Engineering* (1996): 359–371.
5. D.P. Krynine and W.R. Judd, *Principles of Engineering Geology and Geotechnics* (McGraw-Hill Book Co., New York, 1957), 730 p.

J.D. Mollard and J.R. Janes, *Airphoto Interpretation and the Canadian Landscape* (Energy, Mines and Resources, Canada, 1984), pp. 117–136.
7. A.K. Gabriel, R.M. Goldstein, and H.A. Zebker, *Jour. Geophys. Res.* **94 B7** (1989): 9183–9191.
8. H. Lee, R. Wightman, and R. Whittle, *Civil Engineering Applications of Remote Sensing and Geographic Information Systems* (Am Soc. Civil Engr., New York, 1991), pp. 199–208.
9. V.J. Katti, Y.R. Satya Saradhi, S.N. Kak, D.C. Banerjee, and R. Kaul, *Int. Jour. Remote Sensing* **14** (1993): 3291–3298.

ADDITIONAL READING

P.J. Beaven, Remote Sensing for Highway Engineering, Proc. of a Post-Graduate School (Univ. Dundee, 1984): 129–137.
J.M. Ellis, *Proc. of 11th Thematic Conf. on Rem. Sens.* **1** (ERIM, Ann Arbor, 1996), pp. 167–174.
J.M. Ellis, *Proc. of 12th International Conf. on Applied Geologic Rem. Sens.* **1** (ERIM, Ann Arbor, 1997), pp. 75–81.
C. Fischer and D. Hermsmeyer, *Proc. of the 6th Intl. Conf. on Tailings and Mine Waste* (A.A. Balkema, Rotterdam, 1999): 77–86.
T.W. Foresman and R.C. Brown, Jr., *1st Thematic Conf. Remote Sensing of Arid and Semi-Arid Lands* (ERIM, Ann Arbor, 1982), pp. 497–501.
C.A. Gachet, Fighting floods with Remote Sensing, ERMapper Mag. 4 (Earth Resources Mapping, 2000): 28–29.
K.C. Hardcastle, *Photogram. Engr. and Rem. Sens.* **51** (1995): 739–748.
R.W. Kiefer (ed.), *Civil Engineering Applications of Remote Sensing* (Am. Soc. Civil Eng., 1980), 195 p.
F. Kuehn, T.V.V. King, B. Hoerig, D.C. Peters (eds.) Remote Sensing for Site Characterization (Fed. Inst. For Geoscience and Natural Resources, Springer, Berlin, 2000): 252 p.
C.A. Legg, Remote Sensing and Geographic Information Systems: Geologic Mapping, Mineral Exploration and Mining (Ellis Horwood, Chichester, 1992), 256 p.
P.C. Leveque, *Memoires de la Societe Geologique de France, Nouvelle Serie* **157** (1990): 339–348.
D. Li and J. Li, *Proc. of the 10th Thematic Conf. on Geol. Rem. Sens.* **II** (ERIM, Ann Arbor, 1994), pp. 558–567.
J.G. Lyon, Remote Sensing and Geographic Information Systems in Hydrology: Environmental Hydrology (Lewis Publishers, Boca Raton, 1995): 337–368.
W. Murphy and M.H.K. Bulmer, *Proc. of the 10th Thematic Conf. on Geol. Rem. Sens.* **1** (ERIM, Ann Arbor, 1994), pp. 341–351.
D.C. Peters and M.A. Smith, *Proc. 10th Intl. Conf. on Ground Control in Mining* (West Virginia Univ., 1991): 184–193.
O. Rouzeau, J.M. Coudert, and J. Fairfield, *Proc. of the 10th Thematic Conf. on Geol. Rem. Sens.* **1** (ERIM, Ann Arbor, 1994), pp. 193–202.
J.D. Russell, *Civil Engineering Applications of Remote Sensing and Geographic Information Systems* (Am. Soc. Civil Eng., New York, 1991), pp. 209–219.
S.G. Ungar, *Remote Sensing Environ.* **26** (1988): 51–73.
M. Wyvill, *Earth Observation Mag.* (April 1992): 46–48.

QUESTIONS TO CONSIDER

1. What is the purpose of a geographic information system?
2. How can a GIS help plan a seismic program?

3. How can remote sensing help determine the best site for a hydroelectric project? The best route for a new interstate highway?
4. An aqueduct is being planned across California. What factors should be considered in planning the route? How can remote sensing help map these features?
5. What would you suggest is an appropriate base map (image type, scale) for a mine expansion project in China? What would be an appropriate base map for an oil exploration program in the Canadian arctic? For laying a natural gas pipeline across Los Angeles?
6. What are your main concerns regarding nuclear waste disposal? How could remote sensing minimize those concerns?

Section 4
Environmental Remote Sensing

INTRODUCTION

The purpose of these chapters is to show ways that remote sensing technology might be applied to help solve problems encountered by corporate, governmental, and private environmental organizations. Applications can include mapping the pattern of storm water runoff from plant sites or tailings, planning for spill containment, detecting onshore or offshore spills or pipeline leaks, and mapping environmentally sensitive areas such as wetlands or migratory bird nesting areas. Other uses for remote sensing include investigating historical archives to locate photos of old drill sites to determine the leakage from mud pits or impact on local surface drainage, monitoring changes in an oil field, mine, smelter or refinery over a period of years, or mapping the extent of hot water outflow from power plants or refineries. Environmental hazards, from landslides to earthquakes to volcanic eruptions, can be identified and monitored using remote sensing technologies. Legal questions can be raised or resolved, saving untold expense in court. Finally, the public can be educated and landowners placated by the appropriate use of imagery.

Environmental remote sensing is becoming a muture field, and has been associated with urban planning and the development of Geographic Information System databases. Remote sensing has served as both a photo-documentation tool as well as an up-to-the-minute base map for planning urban or industrial development, and for mapping and outlining environmentally sensitive areas. The primary environmental uses for remote sensing in the petroleum and mining industries are in preventing or minimizing damage to sensitive development sites or plant locations, in spill contingency planning, in monitoring pipelines, tailings, or platforms for leaks, tracking spills and leaks, and documenting past use (or misuse) of a parcel of land.

11

Environmental Baselines and Monitoring

CHAPTER OVERVIEW

Environmental remote sensing is concerned with the establishment of baseline maps through a process of mapping various components of the surface such as vegetation, soil, water and drainage networks, and man-made disturbances. During this process it should be possible to map animal habitats through a combination of slope, water, and vegetation mapping and field observations. Mine reclamation projects can be assisted by periodic monitoring. Surface and groundwater can be observed for evidence of pollution, areas of recharge and discharge, and for man-made input (e.g., power plant cooling water, fertilizers, acid mine water). Air pollution, although more difficult to monitor, can under certain circumstances be seen as a result of fire, smelter plumes, or even urban smog. All of these maps are derived in total or in part from the effective interpretation of various types of imagery.

ENVIRONMENTAL AUDITS

Perhaps the most suitable application for remote sensing technology, the monitoring of environments, is also the most routine and best-demonstrated application. The simplest use of this technique is in an environmental audit that examines a property prior to leasing or prior to divesting in order to determine whether surface damage exists (or existed) and requires remediation. Environmental audits are becoming an essential part of assuring compliance with environmental regulations and ensuring the exercise of "due diligence" during real estate transfers. Lending institutions are beginning to require this type of investigation to minimize their liabilities in future pollution cases. In the U.S. the 1986 Superfund Amendment and Reauthorization Act (SARA) established the "innocent purchaser" defense to environmental liability. This defense is applicable if the property owner exercises "all appropriate inquiry into the previous ownership and uses of the property" and conducts "appropriate inspections" at the time of acquisition [1]. Resource extraction companies, too, should carefully evaluate leases, especially in foreign areas, to be sure they are aware of existing environmental damage that they may be required to clean up at some later time.

The first stage of an environmental audit typically consists of an initial assessment of site conditions and the documentation of existing or potential environmental problems. This involves determining previous surface operations, noting present land use, observing surrounding land use, doing a site inspection, and interviewing those familiar with the site. Historical photos or images should be reviewed for any evidence of abnormal vegetation growth, trenching or landfills, surface structures, and any signs of contamination (spills, leaks, or dumps). Phase two of the audit involves confirming the presence of contamination by sampling soil, soil gas, water, plants, etc. Multispectral and hyperspectral scanner images can be processed to enhance anomalous soil tones that may reveal the extent of pollution [2].

HABITAT MAPPING

Other uses for remote sensing include mapping environmentally sensitive areas prior to building roads, pipelines, or facilities in order to preserve the unique character of the area as well as meet legal requirements. Many areas require that mining or well sites be returned to their original condition after drilling, and that the impact of access roads be minimized. This requires documentation of the original condition of the area, which is readily accomplished with archival and recent airphotos. Other areas such as wetlands require that drainage is not interrupted and that discharges are free of pollutants and/or particulates and are at ambient temperature. Airphotos or high-resolution satellite images provide accurate and detailed base maps for survey teams and construction crews, and can improve the efficiency of cleanup and restoration efforts. Radar images can be used to map areas such as rainforests that are under cloud cover much of the time.

Increasingly companies must take into account the impact of their facilities on neighboring urban areas. Many facilities were originally built in more or less isolated areas. These areas subsequently attracted development, and people living near these plants often complain about the effects of the facility, whether it be noise, smell, or discharges. These factors should be investigated at the earliest possible date, both to guide local development and to lessen the impact of an industrial site on neighboring communities by taking into account wind patterns, surface runoff, and development patterns.

The technique employed to evaluate an area or establish an environmental baseline (conditions at a point in time) usually involves generating or purchasing standard color or color-infrared pictures and then interpreting them for land surface relief, surface water, surface development and/or disturbance, and mapping the distribution, density, vigor, and types of vegetation cover. Several dates and seasonal cycles should be analyzed, if available. The combination of topography, water, and cover type determines the habitat for wildlife (Fig. 11.1). If certain combinations of parameters are known, for example, that elk range requires a specific combination of vegetation, slope, and slope azimuth (aspect), then it should be possible to characterize the habitat at a known location (the control point or "training site"), and automatically map (classify) all areas with statistically similar spectral characteristics using the unique reflectance of the control area [3, 4]. Spectral classes are a function of discreet wavelengths reflecting off single or mixed cover types and the percent cover or vegetation density. This technique, known as a "supervised" classification, is commonly used to map habitats and ecosystems once a set of control points has been established by field surveys. Prior to field surveys it is possible to generate "unsupervised" classification maps by providing a finite number of "classes" and instructing a computer to perform multivariate analyses that lump all surface features into the assigned number of

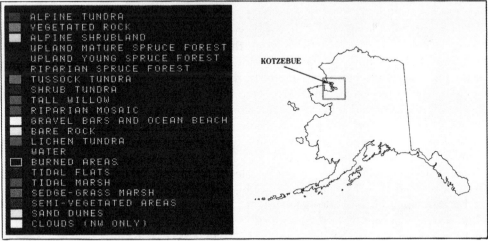

Figure 11.1 Caribou and grizzly bear habitat classification combining digital Landsat data and ground measurements near Kotzebue, Alaska. Processed by the General Electric Digital Image Analysis Lab. (See color plate.)

categories on the basis of similar reflectance characteristics. These categories then require field verification to determine their unique composition [5].

Simple images can also be used as base maps for field work, construction programs or for regional orientation purposes. Latitude-longitude grids and other survey information can be annotated on the images, thus providing highly accurate and up-to-date maps (Fig. 11.2). One can go further still and generate topographic maps of areas that have stereo overlap of airphotos or satellite images (see Chapter 10). Some companies will generate topographic maps from radar imagery.

Classification techniques are often used in conjunction with a GIS to categorize habitats on the basis of more than just imagery. GIS databases can be co-registered to an image and can be used to provide information concerning road networks, land ownership, well locations, drainage patterns, sample sites, rights of way, and tailings or dumps. Conversely, the image can be used to update the GIS databases. Using a GIS with the appropriate databases, it should be a simple task to identify, for example, all south-facing slopes less than 15° without forest cover on private land and having good road access.

SURFACE DISTURBANCE AND CHANGE DETECTION MAPPING

Mapping of surface disturbance can be achieved using standard photointerpretation techniques with color, color infrared, or black-white photography or imagery. Unlike habitat mapping, the objects being mapped are generally man-made (roads, building sites, storage tanks, or mud pits, etc.) and therefore are usually distinct from background. Sometimes we are interested in mapping changes due to flooding, landslides, coastal erosion, or due to some other natural cause. There are cases where dumps, mud pits, pipelines, and other features have been buried and forgotten, leaving only subtle clues to their existence. These clues include features such as linear bulldozer scars over landfills, abrupt changes in the natural contour of surface slopes, and geometric patterns of vegetation over a site [6]. These are often easier to see from the air than from the ground because the contrast with background slopes and vegetation is more apparent. Multi-temporal images are useful for observing change, particularly man-made change in an area. Photos of areas such as Love Canal taken over a period of years readily show where old developments were and where new developments have occurred [1]. Multispectral airborne video has proven to be particularly effective in monitoring hazardous waste sites over a period of time and should have widespread application to many environmental monitoring activities [7].

Another technique for locating buried wastes or pipelines is ground-penetrating radar. This can be used as a reconnaissance tool by running regional survey lines, or can be used to pinpoint a site once airphotos locate a disturbed area. Ground-penetrating radar (GPR) provides output that appears much like a seismic line (Fig. 11.3). The latest GPR can be processed to show either vertical or horizontal slices of the near surface. Ground-penetrating radar uses a single, low-power broadband impulse of 1 to 6 nanoseconds to transmit, followed by a 20,000 nanosecond recording interval. A common frequency is centered at 120 MHz (250 cm), although antennas from 80 to 900 MHz (3.33 to 375 cm) are available. Higher frequencies provide better resolution, but do not penetrate as deep. Penetration depths depend on soil type and moisture content. Maximum depths for low frequencies are on the order of 50 m in dry, sandy soil [8]. Most systems with bands centered at 100 MHz penetrate up to 20 m. Effective penetration depth is usually determined by soil moisture.

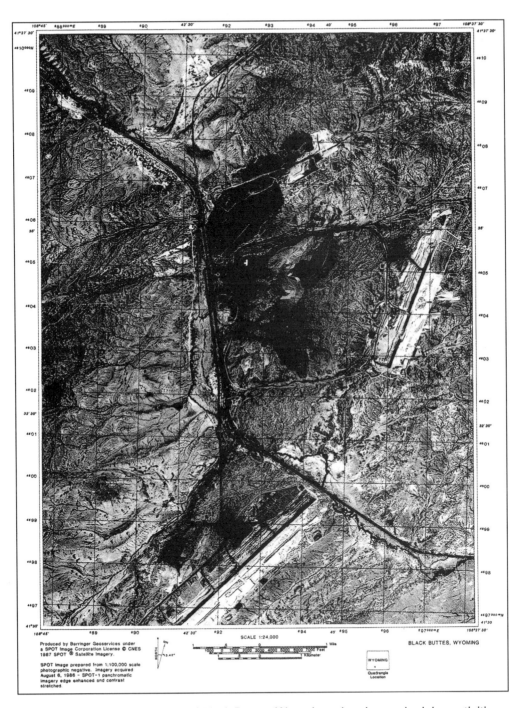

Figure 11.2 SPOT image map of Black Buttes, Wyoming, showing coal mining activities. Generated by Barringer Geoservices, Denver. (See color plate.)

GPR can detect a variety of materials, including the water table, bedrock-alluvium interface, clay or sand lenses, pipeline leaks (gas or liquid), and wood or metallic

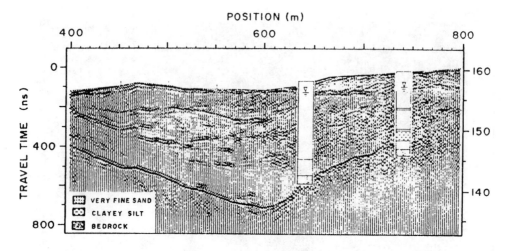

Figure 11.3 Ground-penetrating radar reveals bedrock under 20 m of sandy fill. From Davis and Annan [8.].

objects, depending on the contrast with background [9–11]. The primary factor controlling the amplitude of the reflector is the contrast between electrical properties of the target and surrounding materials. Metallic targets such as oil drums are strong reflectors under most conditions; non-metallic targets are generally weak reflectors. Filled-in trenches have been detected in the near surface, and law enforcement agencies have even used GPR to search for bodies [12]. No target is detectable if covered by a thin layer of moist clay. Saline groundwater also limits the depth of penetration. Clay is both a good reflector and strong absorber of radar waves, and is the least favorable medium for detecting buried objects. Dry sand and gravel, on the other hand, are excellent radar transmitters.

An advantage of GPR over geophysical methods such as electromagnetic terrain conductivity measurements is that GPR is not significantly influenced by surface cultural features. One limitation of conventional GPR usage is that the antenna must be towed along the ground. Rough or uneven ground can make this difficult. Newer techniques have begun to overcome this problem, but the GPR technique is only effective under a narrow range of conditions. GPR is particularly limited with regard to sensitivity, resolution, and penetration. The data generated can be difficult to interpret where soils have a wide variety of electrical properties.

Classification techniques, described earlier under Habitat Mapping, can also be used to locate cultural features. One must be careful, however, that the computer does not classify, for example, a rock outcrop or concrete building the same as a rock quarry since all have similar spectral characteristics.

We are often interested in learning whether surface disturbance is increasing due to development, illegal dumping, etc., or whether it is decreasing as a result of successful remediation efforts. Change can be detected using image subtraction. Photos or satellite images acquired at different times must be registered and, if necessary, resampled so that the resolution cell or picture elements (pixels) of each are the same size. Once this is accomplished, any changes that exceed a given threshold can be detected by simply subtracting the reflectance at one time from another. Those areas that show change are assigned an arbitrary color and are then superimposed on the more recent image (Fig. 11.4).

A fast and relatively inexpensive way to monitor leaks, spills, and illegal trespass, dumping, or mining on one's lease is to fly an aerial survey at given intervals (semiannual, monthly) over all oil fields, facilities, pipelines, and even undeveloped land. These photos can be rapidly evaluated in real time or in the office, and areas with indications of tampering, etc., can be quickly visited in the field. Many oil companies now routinely fly periodic thermal infrared surveys along buried oil pipelines to detect leaks before they become a threat to the environment. Such techniques are rapid and relatively inexpensive, and provide excellent documentation of current and past surface conditions.

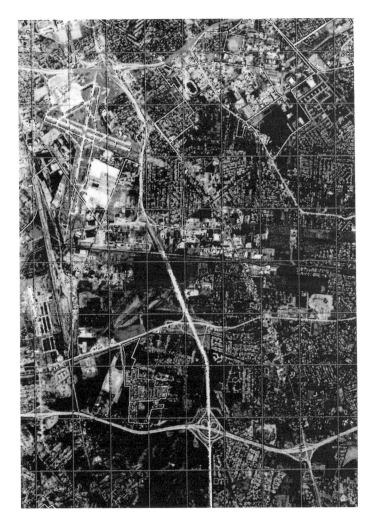

Figure 11.4 Change detection, Louisville, Kentucky. Areas that show changes (airport runways, quarries) between 1988 and 1991 are shown in red (bright changes) and cyan (dark changes) on this SPOT Pan image [13]. Image processing by Earth Satellite Corporation. (See color plate.)

COAL MINE FIRES

Coal mine fires are rare, but once started they can be devastating. They can be started by lightning or grass or forest fires that contact an outcrop, or can be accidentally started by man. Some fires have been burning for years, and they are difficult if not impossible to extinguish. They not only destroy a potentially valuable resource, but can destroy a landscape and make it uninhabitable. In 1962 a coal bed was ignited by a fire in a dump on the outskirts of Centralia, Pennsylvania. As the fire worked its way along the underground seams, smoke began to rise out of new cracks in the ground, and in some areas the ground became hot to the touch. Local fire departments and the U.S. Bureau of Mines tried unsuccessfully to pump water into the ground, dig trenches, and use explosives. Sensors showed temperatures close to 525° C less than 10 meters below the surface. Soon homes had unhealthy levels of carbon dioxide, and holes began opening in people's yards. In the 1980s it cost the government $42 million to evacuate the town [14]. The following example shows how coal fires can be monitored using remote sensing techniques [15].

Jharia, India, Coal Fire Case History

Fire, probably started by spontaneous combustion, was reported at the Jharia coal field, India, in 1916. Since that time, an estimated 37,000,000 tonnes of high-grade coking coal have burned, and another 1,864,000,000 tonnes are at risk unless the fire can be extinguished. Surface installations have been damaged, and the local environment has been degraded. The intensity and extent of the fires must be determined in order to plan how to extinguish them.

A study was undertaken in the Jeenagora colliery of the eastern Jharia coal field (Fig. 11.5). The surface is covered by sand and overburden consisting of the Permian Baraklar Formation. Eighteen coal seams dip about 6° west and lie at depths of a few meters to 80 meters.

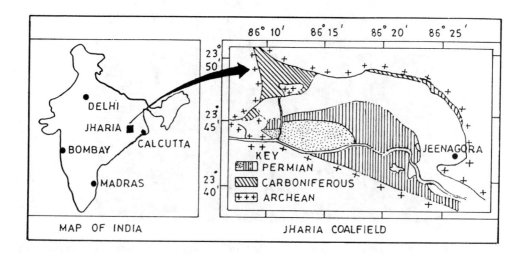

Figure 11.5 Location map, Jeenagora colliery, Jharia coal field, India [15].

A combination of airborne thermal infrared imagery and a magnetic survey were acquired to help determine the extent and intensity of the fires. Predawn imagery was acquired in the 8.5–13.0 micron range with a ground resolution of 1.5 meters at nadir and a thermal sensitivity about 1° C. Two preset black bodies at 10° C and 65° C were used to calibrate the system. Isotherm maps of the surface were generated from relative radiance of the thermal imagery. Simultaneously, a magnetic survey consisting of 17 profiles between 50 and 130 meters apart were acquired both over the known fire zones and over areas known to be unaffected by fires. The accuracy of the instrument was on the order of ±10 gammas. Laboratory work has shown that magnetic susceptibility of sediments is dramatically affected by heating. For example, the magnetic susceptibility of feldspathic sandstone and soils from the fire-affected zone was measured at less than 100×10^{-5} (SI), whereas the same units from areas unaffected by fire ranged from 100×10^{-5} to 200×10^{-5} (SI).

Density slicing and color coding of the thermal imagery in conjunction with some surface temperature measurements showed where the surface has been heated by conduction or convection processes. Temperatures varied mainly from 5° C to 15° C, with some readings as high as 45° C (Fig. 11.6). However, the technique is limited by the ground conditions (surface cover, relief, disturbances).

Magnetic data was converted to vertical intensity and contoured, then smoothed (Fig. 11.7). The areas affected by fires contain clinkers that have magnetic susceptibilities ranging from –1000 gammas to +2000 gammas. These highs and lows are felt to be influenced by the location of rooms and pillars in the coal as well as the location of fractures that allow groundwater and air movement to the coals. Unburned areas show almost no magnetic character, for example, south of Underground Barrier-1.

The combination of thermal infrared imagery and ground magnetic surveys appears to be an effective way to monitor the location and intensity of underground coal fires.

MINE RECLAMATION

Many areas contain abandoned mines. Some are small and the danger consists of animals or children falling into them. These mines can be mapped using pre-dawn thermal imagery, since the mine entrance will appear cool, even when filled with rubble or soil. Other mines are associated with large areas of disturbed ground, unstable slopes, groundwater seepage, and tailings piles or dumps that pose a danger of debris slides or heavy metal contamination in surface and groundwater. Satellite images and airphotos, combined with a geographic information system, can help plan efficient, cost-effective remediation and revegetation of mine tailings and dumps [16]. The following discussion describes an attempt to map mine waste by mapping the distribution of characteristic minerals [17].

Couer d'Alene, Idaho, Mine Reclamation Case History

A project to map the distribution of abandoned mines in the Couer d'Alene district of northern Idaho used AVIRIS hyperspectral imagery to locate tailings and dumps (Figs. 11.8, 11.9). The purpose was to identify areas prone to high concentrations of acid water discharge and heavy metal contamination.

The Couer d'Alene district contains structurally complex sediments of the Precambrian Belt Supergroup, including argillites, quartzites, siltstones, and dolomites.

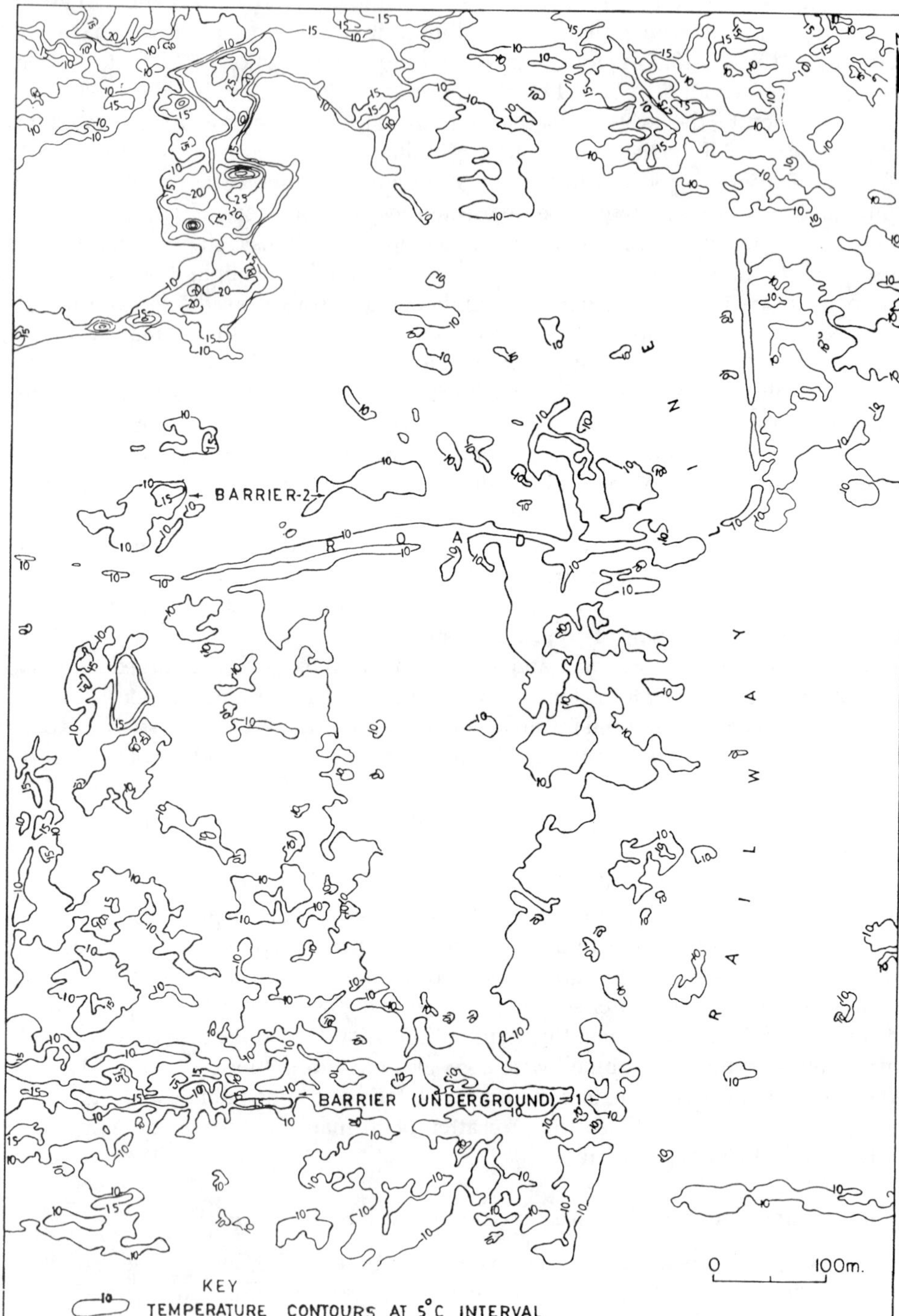

Figure 11.6 Surface temperatures, derived from thermal imagery, over the underground coal fire, Jharia coal field [15].

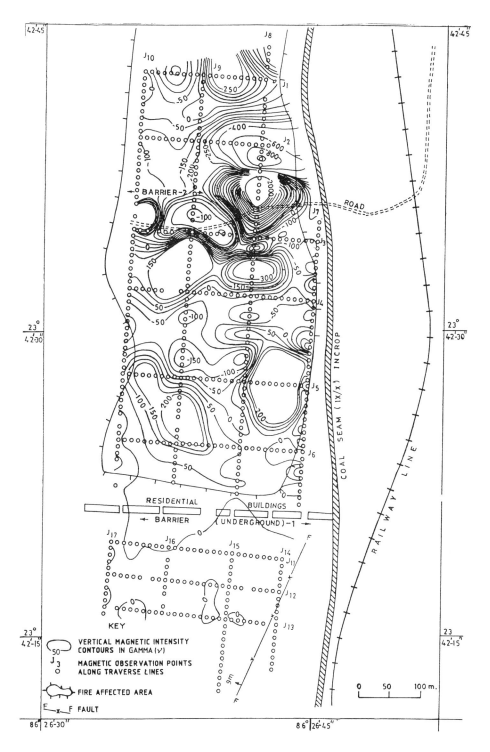

Figure 11.7 Magnetic field over underground coal fire, Jharia coal field [15]. Highs and lows are influenced by the location of rooms and pillars as well as fractures.

304 REMOTE SENSING FOR GEOLOGISTS

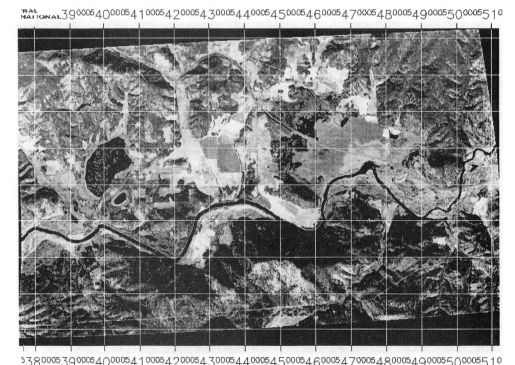

Figure 11.8 AVIRIS hyperspectral image over the lateral lakes of the Couer d'Alene River, Idaho [17]. Rectified false color image by Spectral International and Borstad Associates. (See color plate.)

Lead-zinc-silver mineralization is hosted in the clastic units as vein-type deposits containing sphalerite, galena, and tetrahedrite. The main toxic metals of concern include lead, zinc, and cadmium.

The approach used in this project was to learn as much as possible about the geology and history of the district, then examine the area during a field reconnaissance and document the results. Sites were located using a GPS unit. They were photographed, rock samples were collected and analyzed using a portable X-ray fluorescence instrument, and rock, soil, vegetation, and solar spectra were obtained using a field spectrometer. This was followed by acquisition of airborne imagery, lab analysis of samples collected in the field, and generation of a spectral library. Image processing was followed by the generation of mineral maps and a final field check of the maps.

The high resolution field spectrometer allowed the image processor to chose "end member" spectra for use in spectral unmixing algorithms. The "pure pixel end member method" [18] was used. Raw data was calibrated to relative reflectance, then the imagery was warped to GPS control points, and a combination of channels was chosen to create false color images that highlighted the distribution of iron minerals, micas, and clays (muscovite and illite) (Figs. 11.10, 11.11). Dumps are characterized by clay minerals (altered rock), whereas stream bottoms contain iron coated cobbles concentrated where mine drainage enters the stream (Fig. 11.10). X-ray diffraction

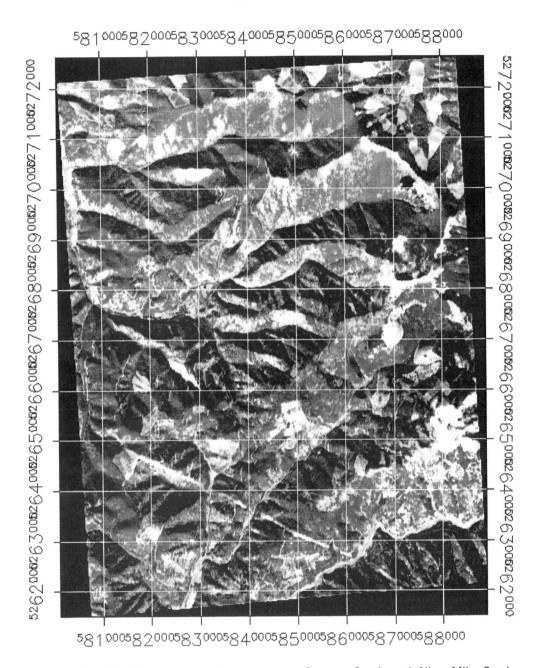

Figure 11.9 AVIRIS hyperspectral image over Canyon Creek and Nine Mile Creek drainages, Couer d'Alene mining district, Idaho. Rectified false color image by Spectral International and Borstad Associates [17]. (See color plate.)

work prior to image acquisition showed that tailings in the Mission Flats area are composed primarily of siderite, degraded muscovite, and silica. Pixels containing these minerals were easily identified using spectra extracted from the AVIRIS data. Field

Figure 11.10 Classification image of the lateral lakes of the Couer d'Alene River area, Idaho [17]. Tailings are indicated by the 2.2 micron absorption band color coded as yellow (best match to siderite and mica) to blue to magenta. Processing by Spectral International and Borstad Associates. (See color plate.)

mapping following image processing suggests there is a 90% match between minerals (and thus mine waste) mapped on imagery and actual minerals on the ground.

WATER POLLUTION

Rivers, Lakes, and Ponds

Surface water is relatively easy to monitor because of the uniform background that clean water presents on imagery. Features that can be monitored include surface slicks and emulsions, suspended particulates, changes in water depth, clarity, algal blooms, effluent, and thermal (hot or cold) discharges and mixing zones.

Equipment used to monitor surface water includes black/white, color, or color-infrared cameras and scanners, ultraviolet scanners, airborne laser fluorosensors, radar, and thermal scanners. Black/white, color, and color-infrared imagery can show sediment plumes due to riverbank or coastal erosion, algal blooms resulting from excess runoff of fertilizers or leaks in sewage treatment plants, and water clarity changes due to a variation in salinity, for example (Figs. 11.12, 11.13). Ultraviolet scanners and laser fluorosensors can reveal the presence of naturally occurring or man-made surface slicks and many types of organic effluent. Radar can reveal the surface

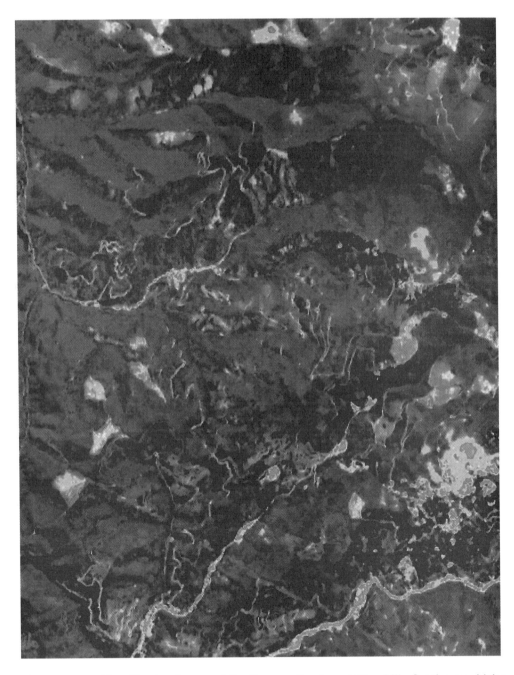

Figure 11.11 Classification image of the Canyon Creek and Nine Mile Creek area, Idaho [17]. Processing emphasizes iron minerals (red is best match; yellow moderate match). Processing by Spectral International and Borstad Associates. (See color plate.)

roughness that might show where discharge is occurring. Thermal scanners can detect oil slicks (Fig. 11.14) as well as hot water discharges from industrial sites such as refineries or power plants (Fig. 11.15). Hyperspectral scanners can detect mineral

Figure 11.12 Algal blooms (red) in the Adriatic Sea near Ravenna, Italy. Landsat TM infrared image from 9 July 1989. Processed by Geospace, Inc., Bad Ischl, Austria. (See color plate.)

precipitates on a riverbed resulting from mine discharges (see Couer d'Alene case history, this chapter).

Groundwater

Most remote sensing technology examines only the surface microlayer. It is not possible to determine the depth to the water table, or to locate water table contamination using conventional imaging remote sensing technology. Similarly, there are no conventional imaging techniques that will locate or track hydrocarbon accumulations on the groundwater surface. It is possible, however, to map areas of surface moisture, springs, fracture zones, and surface cover related to groundwater.

ENVIRONMENTAL BASELINES AND MONITORING 309

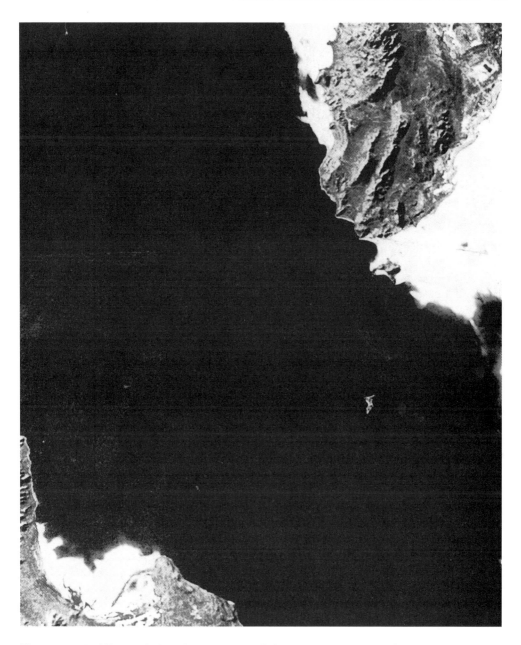

Figure 11.13 Water clarity changes as salinity increases south of the railroad trestle crossing Great Salt Lake, Utah. Landsat MSS image processed by Earth Satellite Corp. (See color plate.)

In Chapter 9 under "Locating Sources of Water" we discussed using satellite imagery and airphotos to locate springs, seeps, and near-surface water. Thermal sensors can detect warm or cool layers beneath a veneer (several centimeters) of sand [19]. Radar flown from aircraft or satellites can penetrate up to 2 m under ideal

310 REMOTE SENSING FOR GEOLOGISTS

Figure 11.14 Thermal image of an oil slick from an offloading tanker. Image acquired by Clyde Surveys, Maidenhead, U.K.

conditions, whereas ground-penetrating radar, a nonimaging system, can detect the water table at a maximum depth of about 50 m.

Soil moisture variations appear as changes in soil tone on black/white, color, or color-infrared imagery: moist soil is generally darker, and dry soil is generally lighter in appearance [1]. The expression of moist soil and surface water on thermal imagery

ENVIRONMENTAL BASELINES AND MONITORING 311

Figure 11.15 Hot water discharged into tidelands. Ebb tide above; flood tide below. Colors indicate temperature differences of 0.5° C. Survey by Clyde Surveys. (See color plate.)

varies depending on the time of day and amount of moisture (Fig. 9.1). Moist soil generally appears cooler than dry background soils due to evaporative cooling. Thermal imagery has been used to map near-surface groundwater in fractures and could be used to map leaks in pipelines or dams [19,20].

Airborne radar will be most sensitive to changes in moisture content if an area is relatively flat and has a uniform, homogeneous surface. This is because radar waves react to changes in the electrical properties of materials. Areas of moist soil generally have a brighter appearance (higher backscatter coefficient) on radar images [19].

Ground-penetrating radar surveys reportedly can detect the water table as well as contacts between fresh water and saline water or water and other liquids such as gasoline [9,10]. This is caused by an electrical conductivity contrast between subsurface fluids and appears similar to an acoustic impedance contrast on seismic lines.

An example of using remote sensing as part of an evaluation of groundwater pollution can be seen in a study done by the Baylor University Geology Department [21]. The Lower Cretaceous Paluxy aquifer in north-central Texas provides water for both domestic and agricultural purposes. The vulnerability of the aquifer to pollution was assessed using a geographic information system to integrate depth to the water table, net recharge, aquifer type, soil type, topography, impact of the vadose zone, and conductivity. These criteria were weighted based on their importance in determining regional groundwater quality. Land use and land cover derived from satellite imagery was combined with the other criteria to determine the final groundwater pollution potential of the area.

Marine, Estuarine, And Wetlands Environments

Remote sensing of marine waters, estuaries, and wetlands is similar in most respects to sensing rivers, lakes, and ponds. It is often easier and less costly to use satellite imagery than airborne sensors because of the economies of scale involved in surveying large areas. Spills, on the other hand, require detailed daily coverage by a variety of sensors, and satellites do not provide the most effective sensors nor the daily coverage necessary to identify the spill, determine the volume of the spill, or track the spill.

Estuaries, wetlands, and deltas are environmentally sensitive areas that may require periodic monitoring, especially if oil fields or chemical facilities are located nearby. For the purpose of monitoring or inventorying changes in these areas, it is appropriate to use color or color infrared satellite imagery or aerial photographs. These can be purchased at a variety of scales and resolutions, depending on the detail required.

Some states require that the amount of particulates discharged along with water from refinery cooling ponds be below a specified level. Algae growing in the ponds are considered particulates, and the ponds must be periodically drained, cleaned, and painted with special algae-inhibiting paint. It should be possible to monitor blue-green algae growing in ponds by taking advantage of their chlorophyll reflectance peak at 0.7–0.75 microns (Fig. 11.12). Algae may be distinguished from suspended sediment by subtracting the reflectance at 0.665 microns from the peak reflectance at 0.710 microns [22].

The Fraunhofer Line Discriminator (FLD) can detect oil refinery waste, sewage effluents, feedlot effluent, sludge (wet or dry), algal blooms, and phosphate processing effluents by measuring the luminescence of these materials [23]. Fraunhofer lines are absorption bands in the solar spectrum. Using the sun as a source, the FLD measures luminescence in these absorption bands under daylight conditions.

Marine spills can occur as a result of flushing of bilges, tankers running aground or colliding, platform blowouts, ships colliding with platforms, pipeline leaks, etc. (Fig. 11.14). In addition, natural seeps exist in many areas, often slow but steady, in other cases intermittent and triggered by earthquakes or reservoir pressure changes. It is virtually impossible to tell the difference between natural and man-made slicks unless one can pinpoint the source and identify the type of hydrocarbon. Most manmade slicks originate at a point and spread outward from it; many natural seeps appear dispersed over large areas and are visible over long periods of time [24].

Operator's permits for power plants and refineries often require monitoring of surface temperatures of cooling water discharges. Thermal imagery can be used to monitor these thermal plumes (Fig. 11.15). Thermal imagery has been used since the mid-1970s to detect acid mine drainage into fresh water bodies. Aircraft imagery can produce high-resolution imagery at frequent intervals, but costs are relatively high. It has been shown that Landsat TM thermal imagery has sufficient thermal and spatial resolution to map the cooling effluent from power plants such as the Diablo Canyon nuclear facility [25]. When radiosonde measurements are used to adjust atmospheric models (such as LOWTRAN 6), the surface radiances can be converted to temperatures that are within $0.6°$ C of ground truth.

In any spill or leak situation there are three questions that must be answered as soon as possible.

1. Where is the spill located?
2. How much has been spilled?
3. Which direction and how fast is the spill moving?

Most companies will have contingency plans in place. These plans have modeled oceanic current movements and wind directions at various times of year. They should also have emergency manpower and monitoring equipment (buoys, surveillance aircraft) on standby.

Daily repetitive flying with a suite of sensors that include radar, ultraviolet, and thermal scanners is the best way to answer the questions posed above. Radar can reveal variations of surface roughness, particularly offshore, and anything on the surface, including drilling platforms, ships, and effluent. Radar imagery will pinpoint a slick from high altitudes in any weather and at any time of day or night under normal conditions. Oil slicks appear on radar images as smooth areas due to the suppression of small surface waves (Fig. 11.16) [24,26]. Radar will not detect spills if there are storm waves, or if the sea surface is smooth due to a wind shadow, for instance. An ultraviolet (UV) scanner or a FLD can identify the slick at lower altitudes as an organic substance, and the thermal scanner can characterize the thickness of the spill. In general, slicks less than 300 microns thick appear cool because the emissivity of oil (0.94–0.97) is less than that of water (0.99). Slicks thicker than 500 microns act as an absorber of solar radiation and tend to appear warmer than background [27]. However, if an oil has a high concentration of volatiles these will evaporate quickly and make the oil appear cool. All of these sensors, as well as

314 REMOTE SENSING FOR GEOLOGISTS

Figure 11.16 Radarsat image of an oil spill at St. Ann's Head, Wales, taken 22 February 1996 [26]. Oil appears as dark streaks along the coast. Processing by Radarsat, Inc.

color or color-infrared images, photos, or video can provide the areal extent and movement direction of the spill and, when combined with inertial navigation systems or global positioning systems fixed to buoys dropped into the slick, they can pinpoint the location [28,29].

Repetitive coverage, combined with accurate positioning, will provide movement vectors of a slick and allow the emplacement of booms and skimmers, the application

ENVIRONMENTAL BASELINES AND MONITORING 315

of dispersants, and optimize oil-collection for in situ burning. In high risk, heavy traffic areas it would be wise to know the movement of currents during various seasons so that in the event of a spill local conditions can be quickly taken into account. Certain types of imagery, particularly true color or thermal infrared, can assist in mapping currents if there are suspended sediments (near the coast) or if there is a temperature contrast between warm and cool currents (Fig. 11.17).

For some purposes it is important to know where the coastline is. While this sounds obvious and trivial, there are areas where the mapped coastline can change by tens of meters or even kilometers depending on the tide. The same sensors mentioned above can be used to map the extent of high and low tides for the building of jetties and marinas, breakwaters, for erosion control along beaches, for siting property developments, and for coastal heavy mineral and placer mining.

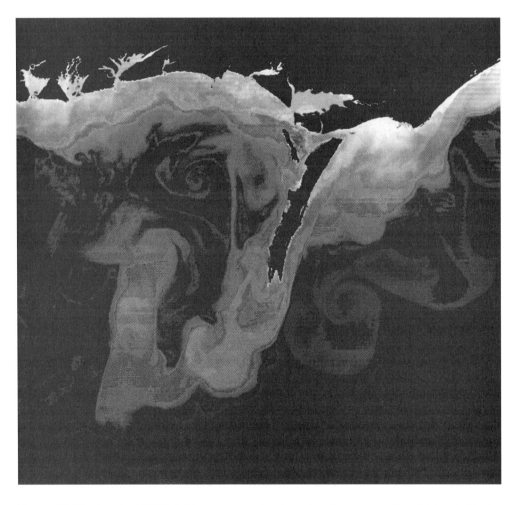

Figure 11.17 Landsat TM 3-2-1 image reveals current patterns near Kayak Island, Alaska. The currents are visible because light is reflected off of suspended sediment. Image processed by the University of Alaska, Fairbanks. (See color plate.)

AIR POLLUTION

Black/white, color, or infrared images can detect smoke plumes, and color photos are sensitive to atmospheric haze such as vehicle-derived smog (Fig. 11.18). These images can show, for example, those areas most subject to acid rain as a result of coal-burning mills, copper smelters, or power plants. None of the sensors mentioned so far can, however, detect or identify clear gaseous emissions such as volatile organic compounds from factories or plants. At least one company in the late 1970s used satellite images to map "fuzzy" or "hazy" tonal anomalies associated with naturally occurring photochemical smog over hydrocarbon accumulations that leaked methane at the surface. It is not known whether this technique was ultimately successful, but there are several atmospheric conditions, such as ground fog or wind-blown dust, that could thwart such an approach to exploration.

In the United States, the Clean Air Act Amendments (CAAA) of 1990 and Clean Air Act of 1992 require increased ambient air monitoring by industry. As an example of such monitoring, a thermal infrared video camera was used to record near source plumes from a turbine stack at an oil gathering center at Prudhoe Bay. Because the ambient temperatures were very cold, and the plume was hot (270–331° C), this method of monitoring plumes was shown to be more effective and more accurate than standard plume rise models such as the Industrial Source Complex model [30]. There are some nonimaging instruments available and others in development that might detect and possibly identify fugitive emissions [31]. Nondispersive infrared spectrometry (NDIR) can be used to detect carbon monoxide and carbon dioxide in vehicle exhaust. An infrared source (heating element) is placed on one side of a road, and an IR detector is placed on the other side. When a vehicle passes through the IR beam, the instrument measures the strength of the absorption for carbon monoxide at 4.3 microns and for carbon dioxide at 4.6 microns [32–34]. The ratio of carbon monoxide to carbon dioxide can be used to calculate emissions in terms of percent or grams per gallon. Similar absorption bands exist for hydrocarbon vapors such as benzene, toluene, xylene, and other aromatics. The signal to noise ratio is the major problem in development of these hydrocarbon detectors.

A related technology, Fourier transform infrared spectroscopy (FTIR) is being used by several companies to detect both hydrocarbon and nitrous oxide emissions. A portable version is being developed to detect airborne ethyl ether, an by-product of illegal drug production.

LIDAR (light detecting and ranging) has been used to measure movement and concentration of urban air pollution and to determine the composition of emissions near industrial plants. The technique consists of projecting a short laser pulse followed by reception of reflected or re-emitted radiation from atmospheric constituents such as molecules, aerosols, clouds, or dust. The incident radiation interacts with these constituents and causes a change in intensity and wavelength. There are several types of lidar systems. The most commonly used systems for pollution detection include atmospheric backscatter lidar (ABL), differential absorption lidar (DIAL), and fluorescence lidar (FL).

Atmospheric backscatter lidar is the most common type, consisting of a nontunable high-power pulsed laser. The system transmits at one wavelength and detects changes in the backscatter due to aerosols and dust in the atmosphere. This is used to track

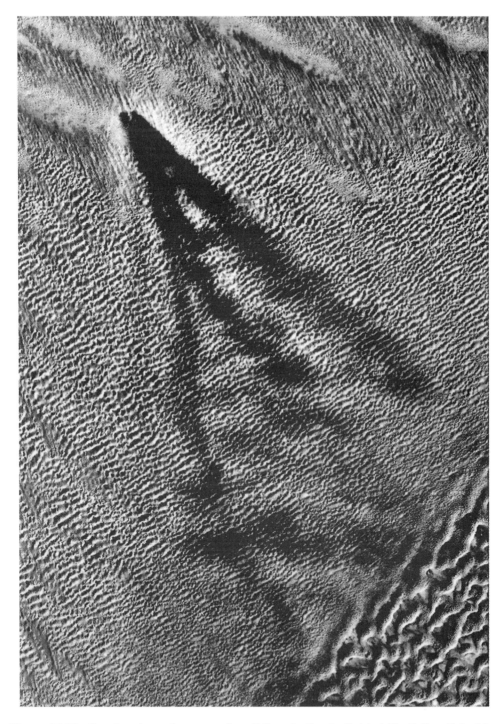

Figure 11.18 Smoke plume from an oil well flare in the Ar Rab' al Khali, Saudi Arabia. Although well flares are not that common, this Landsat MSS image dramatically demonstrates the ability of satellite imagery to reveal smoke plumes and some types of industrial stack emissions. Image processed by ERIM, Ann Arbor. (See color plate.)

turbid effluent and gas plumes, and may be applicable to the monitoring of stack emissions from coal burning plants, smelters, and refineries [35].

DIAL measures the concentration of molecular species in the atmosphere by transmitting in two wavelengths, only one of which is absorbed. The difference in the intensity of the returns is measured. Alternatively, the laser is aimed at a mirror up to 2 km away and the return beam shows absorption by molecules along its path. The amount of aborption is a function of gas density and path length. Molecules studied using DIAL include SO_2, NH_3, O_3, CO, CO_2, HCl, NO, N_2H_4, NO_2, and SF_6 [35]. Methane, ethane, propane, and water all have absorption spectra near 3.3 microns and can be detected at concentrations as low as 0.1 ppm [36, 37].

Fluorescence lidar uses two wavelengths and a spectrometer to separate the wavelength-shifted fluorescence from the strong atmospheric (Rayleigh) backscatter. The laser is tuned to the absorption band of the molecule of interest, and reradiated fluorescence is detected by spectral filtering of the returned radiation. Fluorescence is greater in the ultraviolet than in the IR, but for some applications this limits the effectiveness of the system because the detector is overwhelmed by normal solar background radiation. Thus, the system works best at night and when tuned to wavelengths less than one micron. So far it has been used to detect atmospheric trace metals including Na, K, Li, and Ca, and the hydroxyl ion [35].

None of these instruments are traditional remote sensing systems in the sense that they do not generate map-like images. They are included here because traditional remote sensing techniques have not been developed to monitor most atmospheric pollutants. These optical systems all share certain advantages and limitations. Advantages include:

(1) No sampling is required.
(2) Path-averaged measurements are more representative than "grab" samples.
(3) Exposure of the analyst to toxic gases is limited.

Among the limitations are:

(1) Instrument costs are high.
(2) The possibility of spectral interference from non-pollutants (water vapor, dust).
(3) Spectral data bases are required and must be updated.
(4) The minimum resolution may not be adequate for regulatory purposes.

REFERENCES

1. J.G. Lyon, *Photogram. Engr. and Rem. Sens.* **53** (1987): 515–519.
2. J.D. Gustin and L.A. Neal, *Houston Geol. Soc. Bull* (Dec. 1990): 17–21.
3. T.P. Huber and K.E. Casler, *Int. Jour. Remote Sensing* **11** (1990): 907–912.
4. J. Koltun, *Earth Observation Mag.* **2** (1993): 32–40.
5. D.L. Isaacson, D.A. Leckenby, and C.J. Alexander, *Jour. of Applied Photographic Engr.* **8 (1982): 51–57.**
6. R.S. Williams, Jr., 1983, in R.N. Colwell, ed., *Manual of Remote Sensing* (American Society of Photogrammetry, Falls Church, 1983), 2nd ed., pp. 1667–1954.
7. S.E. Marsh, J.L. Walsh, C.T. Lee, and L.A. Graham, *Photogram. Engr. and Rem. Sens.* **57** (1991): 1221–1226.

8. J.L. Davis and A.P. Annan, *Geophys. Prospecting* **37** (1989): 531–551.
9. Battelle Labs, *Geophysical Survey Capabilities* (Pacific Northwest Labs, Richland, 1983), 17 p.
10. F.L. Graf, Using ground-penetrating radar to pinpoint pipeline leaks: *Matter Performance* **29** (April, 1990): 27–29.
11. D.R. Inkster, J.R. Rossiter, R. Goodman, M. Galbraith, and J.L. Davis, *Proc. of the 7th Thematic Conf. on Rem. Sens. for Explor. Geol.* (ERIM, Ann Arbor, 1989), pp. 127–140.
12. J. McCormick, *Newsweek* (June 14, 1999): 54–55.
13. C. Bynum and M. Wyvill, *Geo Info Systems* (June 1992): 28–29.
14. B. Bryson, *A Walk in the Woods* (Broadway Books, New York, 1998), pp. 180–184.
15. T.K. Bandyopadhyay, T.K. Mukherjee, and B.B.P. Shrivastava, *Proc. ERIM 10th Intl. Conf. Geologic Rem. Sens.* **II** (ERIM, Ann Arbor, 1994), pp. II-576–II-584.
16. R.G. Kremer, J. Zack, A. Davis, S. McLeod, and C. Byrns, *Tailings and Mine Waste'98* (Balkema, Rotterdam, 1998), pp. 789–793.
17. P.L. Hauff, N. Lindsay, D. Peters, G. Borstad, W. Peppin, L. Costick, and R. Glanzman, *Proc. of the 13th Intl. Conf. Applied Geologic Rem. Sens.* (ERIM, Ann Arbor, 1999).
18. J.W. Boardman, *Fourth JPL Airborne Geoscience Workshop* **93–26** *I* (Jet Propulsion Lab, Pasadena, 1993), pp. 11–14.
19. F.F. Sabins, Jr., *Remote Sensing Principles and Interpretation* (W.H. Freeman and Co., New York, 1987), pp. 198–199.
20. G.J. Weil and R.J. Graf, *Civil Engineering Applications of Remote Sensing and Geographic Information Systems* (Am. Soc. Civil Engr., New York, 1991), pp. 189–198.
21. T. Fritch, C. McKnight, and J. Arnold, *Proc. of the 13th Intl. Conf. Applied Geologic Rem. Sens.* (ERIM, Ann Arbor, 1997): 339.
22. G. Quibell, *Int. Jour. Remote Sensing* **12** (1991): 177–182.
23. R.D. Watson, M.E. Henry, A.F. Theisen, T.J. Donovan, and W.R. Hemphill, *Proc. of the American Chem. Soc. 4th Joint Conf. on Sensing of Environ. Pollutants* (1978), pp. 667–671.
24. J.E. Estes. R.E. Crippen, and J.L. Star, *Geology* **13** (1985): 282–284.
25. D.E. Gibbons, G.E. Wukelic, I.P. Leighton, and M.J. Doyle, *Photogram. Engr. and Rem. Sens.* **55** (1989): 903–909.
26. RADARSAT, *Earth Observation Mag.* (April 1996): 62.
27. R.H. Goodman, in A.E. Lodge, ed., *The Remote Sensing of Oil Slicks* (John Wiley and Sons, Chichester, 1989), pp. 39–65.
28. J.H. Hielm, in A.E. Lodge, ed., *The Remote Sensing of Oil Slicks* (John Wiley and Sons, Chichester, 1989), pp. 67–76.
29. R.J. Meyers, *Oil Spill Response Guide* (Noyes Data Corp., Park Ridge, N.J., 1989), pp. 46–69.
30. C. Rickel, B. Lamb, A. Guenther, and E. Allwine, *Atmospheric Environment* **24A** (1990): 2835–2838.
31. W.B. Grant, R.H. Kagann, and W.A. McClenny, *J. Air Waste Manag. Assoc.* **42** (1992): 18–30.
32. G.A. Bishop, J.R. Starkey, A. Ihlenfeldt, W.J. Williams, and D.H. Stedman, *Analytical Chemistry* **61** (1989): 671–677.
33. G.A. Bishop and D.H. Stedman, *Environ. Sci. Technol.* **24** (1990): 843–847.
34. D.R. Lawson, P.J. Groblicki, D.H. Stedman, G.A. Bishop, and P.L. Guenther, *J. Air Waste Manag. Assoc.* **40** (1990): 1096–1105.
35. D.M. Considine, ed., *Van Nostrand's Scientific Encyclopedia* (Van Nostrand Reinhold, New York, 1988), pp. 1678–1679.
36. K. Shirley, AAPG Explorer (Am. Assoc. Pet. Geol., August 2000), 3 p.
37. P.R. Wamsley, C.S. Weimer, J. Applegate, S.P. Beaton, and B.S. Beyer, Mobil system for open-path trace gas detection in the mid-infrared using a Raman-shifted Cr: LiSAF source: Proc. SPIE, v. 3757 (1999), 9 p.

ADDITIONAL READING

G.W. Green, R.M. Moxham, and A.H. Harvey, *Proc. of the 6th Intl. Symp. on Rem. Sens. of Environment* (ERIM, Ann Arbor, 1969), pp. 517–525.

F.B. Henderson III, Remote Sensing for Acid Mine Sites, Geotimes 45 (2000): 32–34.

T.K. Mukherjee, T.K. Bandyopadhyay, and S.K. Pande, *Photogram. Engr. and Rem. Sens.* **57** (1991): 1203–1207.

R.J. Slavecki, *Proc. of the 3rd Symp. on Rem. Sens. of Environment* (ERIM, Ann Arbor, 1964).

Working Party, *Quart. Jour. Eng. Geol.* **15** (Geological Society, 1982): 265–316.

QUESTIONS TO CONSIDER

1. You just found the perfect piece of land for a summer cabin in the hills. Why might you want to obtain historical airphotos over the property before you buy it?
2. If you have a choice between using a supervised or unsupervised classification, which would you use for (a) mapping grizzly habitat in the Arctic, (b) a mine reclamation study in Nevada? Why?
3. What sensors could be used to monitor buried pipelines for leaks? What about potential leakage around an earth-fill dam or tailings pond?
4. Your company has an operation on the shores of San Francisco Bay. In the past 24 hours it has been accused of causing a large fish kill in the bay. How might you use remote sensing to deny (or confirm) this allegation?
5. New England's acid rain has been blamed on Canadian smelters. How might remote sensing help determine whether this is true?
6. How might you distinguish between a natural seep and a man-made spill off the coast of California (or Louisiana, or the Ivory Coast)?

12

Environmental Hazards, Legal Aspects, and Public Relations

CHAPTER OVERVIEW

Building a nuclear power plant on an active fault or building an oil pipeline across an active or potential landslide are examples of the types of activities leading to disasters that are all too common, yet may often be prevented with the careful use of remote sensing imagery as a first step in a hazard assessment. The construction of roads, pipelines, or facilities within river floodplains costs industry billions of dollars each year. Drilling wells or building roads on landslides, and building facilities or pipelines across faults or in areas of active subsidence are also costly and often avoidable.

Detecting these potential hazards is fairly straightforward: telltale signs can be seen on most types of aircraft or satellite imagery, including black/white, color and color infrared, multispectral, hyperspectral, radar, and sometimes thermal data.

LANDSLIDES

Landslides are among the most costly and damaging of natural hazards, both in the destruction of life and property. Hazard assessment can be performed using geomorphic techniques as well as statistical methods that analyze multiple variables such as proximity to active faulting, slope azimuth and steepness, vegetation cover, soil and rock type, dip of bedding, rainfall amounts, snowmelt, forest and grassfire history, and historical landslide patterns [1,2]. Landslides are characterized by concave escarpments and generally lobe-shaped (in map view) hummocky topography (Fig. 12.1). Trees are often knocked down or tilted at unnatural angles. Many forest-covered slides have groves of aspen or other colonizer trees surrounded by the old-growth forest. Valleys are often dammed behind old debris flows, forming lakes, and rivers will often narrow where they are downcutting through old slides that have partially blocked their flow. The hummocky ground on landslides often has many small ponds with no well-developed drainage system. On gentle, grass-covered slopes that are water-saturated and prone to slump one often sees patterns of downslope soil creep revealed by ridges and depressions parallel to slope contours. As the surface particles creep downslope slightly faster than deeper soils, plants with deep root systems can be seen to tilt slightly downslope. Many of the nongeomorphic aspects of detecting and predicting landslides can be evaluated with geographic information

Figure 12.1 Color infrared airphoto of the Dugout Creek landslide south of Glenwood Springs, Colorado. This landslide is characterized by hummocky topography, numerous small lakes and poorly-developed drainage, and narrowing of the East Muddy Creek valley along the toe of the slide. Location of the Amoco Dugout Creek well is shown. (See color plate.)

systems that include a digital topographic database and surface materials database. Detailed stereo airphotos at scales from 1:15,000 are most useful for mapping landslide features, although satellite imagery can also be used. Global positioning systems can be used in conjunction with a GIS to monitor movements, as can interferometric radar imagery. The ideal map will contain information such as slope stability, bedrock versus soils and unconsolidated units, past slides and slumps, the most likely areas of failure, and the probable areas downslope that would be affected.

EARTHQUAKES

Many regions of the world have large populations living in tectonically active settings. Certainly this is true for the Pacific Rim; it is also true in areas considered tectonically quiet such as the coastal plain of Texas, which is prone to growth faulting. Sites subject to liquefaction during earthquakes are generally in flat, low-lying areas along a coastline that is either natural or consists of man-made landfills and is water saturated. Construction in places such as Kobe, Japan, or the San Francisco Bay area, whether for roads and bridges or high-rise offices and refineries, must take into consideration earthquakes and faulting (Fig. 12.2). Some areas have regulations concerning the siting of power plants that require, for example, that faults in the vicinity have not been active during the past 10,000 years or more (see Chapter 11, Kakrapar Power Plant Site Selection Case History). Ultimately this requires trenching and age dating the soils to determine the youngest strata cut by a fault. In the initial stages of this evaluation, however, it is convenient to use remote sensing imagery to map the indications of faulting in an area. There are a number of criteria used to recognize faults, including linear escarpments, offset drainage patterns, and abrupt changes in rock or soil types on imagery (Chapter 5).

Radar interferometry (see Chapter 3) provides an almost real-time method for monitoring creep along active faults. Rosen, et al. [3] used ERS-1 radar images taken 14 months apart to map creep along the San Andreas fault near Parkfield, California. The radar observed the fault from roughly 800 km. Offset along the fault has been estimated at 32 mm/yr. Any change in motion from parallel to the fault would indicate

Figure 12.2 Aerial view of the 1987 Whittier fault, Los Angeles basin. Photo by Rose, Newsweek.

an increase in strain across the fault. Although strain patterns related to a change from locked to creeping are not clear, a decrease in creep occurs from northwest to southeast along an 80 km segment of the fault. The relative slip in the northwest segment of the fault is observed to be about 10 mm over 14 months. This is a vector displacement in the radar look direction equivalent to 32 mm/yr along the fault.

A similar study examined nearly instantaneous ground surface displacements near Kobe, Japan [4]. A large and destructive earthquake struck Kobe on January 17, 1995, resulting in over 5000 deaths and 40,000 injured. The Japanese National Institute for Earth Science and Disaster Prevention, which is responsible for investigating such disasters, analyzed the deformation to better prepare for future occurrences. Two JERS-1 synthetic aperture radar images collected one and a half years apart were used, along with a digital elevation model, to measure surface deformation. The DEM was used to subtract topographic elevation information from the radar interferogram generated from the two images. The resulting differential interferogram revealed a maximum horizontal deformation of about 1.5 meters toward the southwest as a result of movement along a northeast-southwest fault in the area.

Earth Remote Sensing Data Center (ERSDAC) also analyzed this in the same area. The maximum vertical displacement occurred at the northern tip of Awaji Island, where the surface moved up nine color cycles, or about one meter (Figs. 12.3, 12.4).

FLOODS

River floodplains can be identified as those low-lying flat zones on either side of a river that extend outward until one encounters bluffs and higher ground (Figs. 9.9, 12.5). Floodplains often contain swampy depressions and abandoned meanders known as oxbow lakes. Construction in these areas must account for both periodic flooding and unstable soil due to near-surface water tables. Roads and railroad grades may need to be elevated. Facilities will often have to be raised on pilings, and wells should be drilled on artificial gravel islands. Shifting of the river channel and erosion along the banks must be taken into consideration. In cool temperate and arctic regions one must consider the effects of ice jam-induced flooding as well as the effects of river ice on pilings and abutments during spring breakup. The best way to determine the extent of areas affected by annual flooding is to look for indications of recent sandbars and/or a lack of mature shrubs and trees. Identification of zones affected by major floods requires examination of multitemporal imagery over a period of several years to several decades. These images will also show those locations where the channel is most prone to shifting, and provide an estimate of the direction and rate of the channel shifts.

VOLCANOS

Generally speaking, volcanism is a hazard of greater concern to urban planners than to the petroleum or mining industries. However, there are some parts of the world where volcanism has been and will continue to be a concern. Parts of Hawaii and Iceland, for example, often have active lava flows that destroy property. The slopes of Mt. Rainier, in Washington state, have been and are prone to mudflows when eruptions melt glaciers on the volcano. The outskirts of Mexico City, like ancient Pompeii, are vulnerable to ash flows from nearby Mt. Popocatepetl. In an area where explosive

Figure 12.3 JERS-1 OPS color infrared image of Osaka and northeast Awaji Island, Japan, taken 31 October 1992. Image processed by ERSDAC. (See color plate.)

eruptions are possible one should not locate mines or facilities on the slopes of active volcanoes, nor should they be located in low-lying areas that are prone to volcanic eruptions or debris flows. Such areas, and those affected by previous eruptions, can be readily identified on imagery by mapping lava and mud flows and ash falls. Steam eruptions and heightened seismic activity may be the best indications of an impending eruption, but it is also possible to monitor fissures and melting icepack using thermal imagery to detect increasing heat flow in the near surface. Radar interferometry has

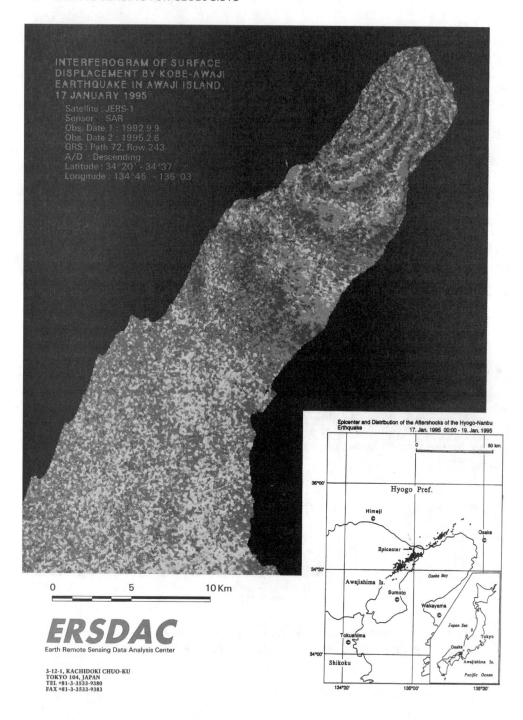

Figure 12.4 Radar interferogram derived from JERS-1 SAR over northeast Awaji Island, Japan. Displacement is a result of the Hyogo-Nanbu earthquake of 17 January 1995. Vertical displacement of about nine color cycles represents approximately one meter. Image processed by ERSDAC. (See color plate.)

Figure 12.5 Landsat MSS image showing a section of the Ob River, Russia. The floodplain (between arrows) is revealed by low-lying areas with sandbars and lacking trees. Image processing by Amoco Production Company.

been used to map deformation (subsidence, doming) around volcanoes in an effort to understand the stress regime before and after eruptions. In a study of Mt. Etna, Briole et al. [5] were able to map areas with a maximum subsidence of 47 mm/yr and attributed this to a combination of compaction of recent lava flows and relaxation of the substrate up to 3.5 years after an eruption. In a similar study of the area around Ito City, Japan, swelling and uplift up to four centimeters is indicated by JERS-1 radar interferometry and is attributed to magma moving at depths between 5 and 10 kilometers [6].

SUBSIDENCE

Subsidence can be natural or culturally induced. In low-lying carbonate platforms such as southern Florida, it is important to know where karst solution collapse has occurred and will occur. Karst collapse features tend to form along fracture zones due to groundwater movement along those zones. For this reason they often form narrow, linear caverns and subsequent collapse areas (see, e.g., Fig. 4.9). These depressions often have better-developed soils and near-surface groundwater, making them ideal areas for cultivation. They can often be detected prior to collapse by using thermal surveys to map areas where cool caverns appear as thermal anomalies [7]. Not only are caverns usually cooler than the surrounding rock due to evaporative cooling, but their temperature tends to be constant whereas the surrounding rocks and soils will warm up during the day and cool off at night. Production of groundwater or oil can also lead to subsidence, as at Long Beach, California, or Galveston Bay, Texas. This can be monitored using radar interferometry.

In areas such as the Silesian basin of Poland or along the foothills south and west of Denver, Colorado, it is important to know where coal mine collapse has occurred, and where it may occur in the future. It is possible to map areas that have subsided in the past, but it is nearly impossible to predict those areas that will subside in the future. Subsidence is indicated by topographically low areas distinct from natural drainage patterns, often containing ponds or indications of mining activity such as tailing piles or rectilinear depressions that contradict the natural slopes (see Chapters 8, 10).

Case History: Subsidence over Abandoned Coal Mines

The U.S. Bureau of Mines has been interested in mapping subsidence over abandoned mines since at least the early 1980s [8, 9]. Subsidence over mines can occur during mining, or may not happen until decades after a mine has been abandoned. An area of abandoned coal mines in northern and northeastern Colorado Springs, Colorado, was investigated using remote sensing techniques, primarily lineament analysis of satellite images and airphotos [10]. The subsidence hazard was mapped by interpreting the surface expression of geologic structures, chiefly faults and joint systems. Subsidence should occur where the mine roof is weakened by fractures. It is also possible to map actual subsidence using large-scale stereo airphotos and stereoplotters, airborne laser profiling, or radar interferometry.

Coal was mined in the project area starting in the 1870s and continued until 1957 (Fig. 12.6). The mines occur at depths from the surface to 150 meters, and most of the mine locations are known from old maps. Coals of the Upper Cretaceous Laramie Formation are up to 6 meters thick and dip 2° to 5° northeast. The mining method was room and pillar. Overburden consists of sandstones, siltstones, and claystones of the Cretaceous-Tertiary Dawson Formation, and scattered eolean deposits exist on the

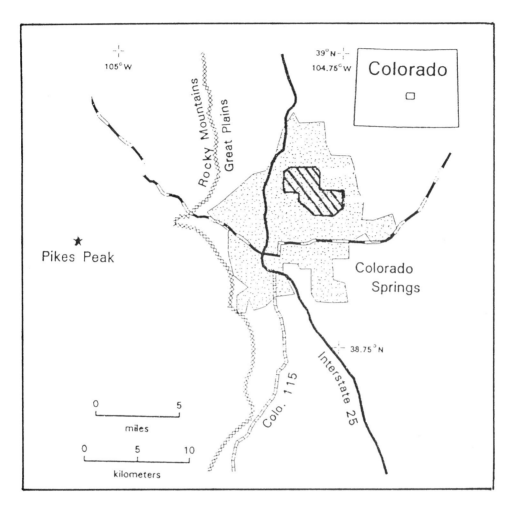

Figure 12.6 Location map showing area affected by subsidence over abandoned coal mines, Colorado Springs, Colorado [10].

surface. No faults had been previously mapped in the mined areas. Prior to 1950 most of the surface was used for agriculture and no one was concerned with subsidence. Since that time urban development has modified the surface through grading and the construction of roads, homes, and commercial buildings. Occasionally an old mine collapses and surface subsidence destroys property, as happened after heavier than usual spring rains in May, 1995.

Landsat TM (28 meter resolution), SPOT XS (20 meter resolution), and SPOT P (10 meter resolution) satellite images were acquired. In addition, high altitude color and black/white airphotos at scales of 1:58,000 and 1:80,000 were used, as were large-scale black/white airphotos taken in 1947 (1:23,600) and 1969 (1:28,000). The SPOT images and high altitude airphotos were not deemed helpful due to the clutter of cultural features. The coarser resolution of the Landsat TM made structural features more continuous and subdued the noise of the cultural features. It was felt that the larger fractures identified using Landsat would more likely extend to depth and affect subsidence. Large and subtle features are also more difficult to recognize on detailed airphotos because the faults often appear composed of multiple smaller features. The contrast of the photos and images was enhanced to maximize their spatial and spectral detail, and airphotos were then examined in stereo. Linear patterns of topography, vegetation, soil tones, and outcrops were interpreted as controlled by faults and/or joint zones. Interpreted structures were then compared to the location of known mines, known subsidence features, and published isopach and structure contour maps.

Factors controlling subsidence include type of mining, groundwater movement, and thickness and lithology of the overburden. For example, gaps in subsidence could be correlated to areas that are buried deeper or where thick sandstone channels lie between the mines and the surface. Although it could not be confirmed that the northwest-trending lineaments are indeed fractures, evidence exists that probable structures identified through this analysis have an effect on the stability of the mined areas. Previously unrecognized northeast and northwest fracture systems appear to influence the location of subsidence over abandoned coal mines (Fig. 12.7).

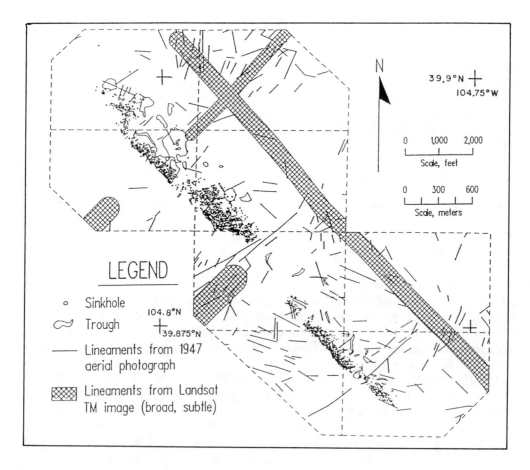

Figure 12.7 Fracture systems and subsidence features mapped during this project [10].

MARINE HAZARDS

Natural marine hazards include submarine landslides or slumps that can disrupt pipelines and platforms, shoals that can disrupt or impede shipping, and sea ice that can damage ships, platforms, and shallow subsea pipelines or facilities. Cultural hazards include collisions between ships or ships and platforms, and anchor damage to pipelines, for example. In shallow water it is possible to use visible light (preferably true color) images to examine the ocean surface or seafloor. This is because blue and green light penetrates water more than other wavelengths. Radar and satellite imagery is useful for mapping sea ice movements. In deeper water it is necessary to use side-scan sonar to image ocean-bottom hazards.

Submarine Slumps

Submarine slumps have cut trans-ocean phone cables and can cause damage to other man-made structures. They appear similar in surface pattern to landslides: arcuate escarpments, hummocky topography and lobate shape are characteristic. In areas of unstable slopes, such as on the Mississippi or Mackenzie River deltas, it is fairly easy to identify recent slumps (Fig. 12.8). These areas should be avoided if possible when planning pipelines, platforms, or subsea well completions.

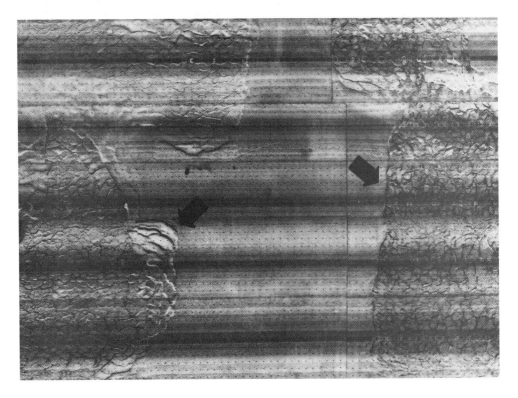

Figure 12.8 Sonar image of submarine slumps (arrows). Image by John Chance and Assoc., Lafayette.

332 REMOTE SENSING FOR GEOLOGISTS

Shoals

Shoals are best mapped using the visible blue or green light channels of Landsat or SPOT, or any true color photography (Fig. 12.9). These wavelengths penetrate

Figure 12.9 Landsat TM 7-4-2 image of shoals near Abu Dhabi City. Image generated by ERIM for the U.S. Geological Survey. From Photogrammetric Engineering and Remote Sensing (Sept. 1990). (See color plate.)

water up to 20–30 m, depending on the water clarity. Shoals have also been noted on radar imagery because currents moving through a channel cause the water to accumulate over the shoals, and this slight rise in the surface is detectable on radar imagery (also see Chapter 10). Sandidge and Holyer [11] used AVIRIS hyperspectral imagery and a neural network to map shallow bathymetry in Tampa Bay off the west coast of Florida and in the Florida Keys. The spectral radiance of coastal waters was determined to be a function of optical properties of the water column (scattering, absorption, and fluorescence of suspended sediment, organic matter, and chlorophyll), illumination conditions, and bottom type as well as depth. Hydrographic maps existed for control. Only 36 bands in the range 0.400 to 0.742 microns were used, as these have the sufficient water penetration required for mapping. An additional five bands between 0.867 and 1.203 microns were used for atmospheric corrections. Maps were generated to depths of six meters with a RMS error of 0.39 to 0.48 m. Maximum correlation of radiance to depth occurred at the band centered at 0.608 microns.

Sea Ice

Sea ice can be mapped using any visible or near-infrared sensor system, or using radar during night or Arctic/Antarctic winter. Imagery provides information on the average and maximum extent of sea ice, ice floe distribution and movements, and timing of ice formation and breakup. This information is necessary for supply ships, ore barges, and tanker traffic to avoid collisions and for safe operation of drill ships and platforms [12]. Side scan sonar may be necessary to identify areas with sea ice bottom scour so as to avoid them when planning subsea well completions and the laying of pipe. Sonar can also detect areas where pipelines have been uncovered by currents or damaged by shipping (Fig. 12.10).

Gas Vents and Craters

Gas vents can indicate unstable areas where natural gasses escape to the surface. Sometimes these vents occur as domes consisting of bottom mud, tar, and chemosynthetic communities of organisms. At other times there are craters where gas has escaped suddenly and violently. Gas bubbles and oily slicks may be visible on the sea surface. Not only is the gas potentially explosive, but rapid escape of large volumes of gas can pose hazards to ships by decreasing their buoyancy. Drill rigs prefer to avoid encountering shallow sub-bottom gas for the same reasons. These vents are often aligned along recent fault scarps. Side-scan sonar or sparker (shallow seismic) surveys may be the best ways to map venting gasses. Photography, scanned imagery, and radar can map associated surface slicks.

LEGAL ISSUES

Politics, scare tactics, and pseudoscience often determine how environmental issues are perceived by the public. Remote sensing offers those who use the technology one way to meet the growing list of environmental requirements in a cost-effective manner. Using imagery is not only cost effective, but can help avoid alarmist responses that exacerbate the perceived problems.

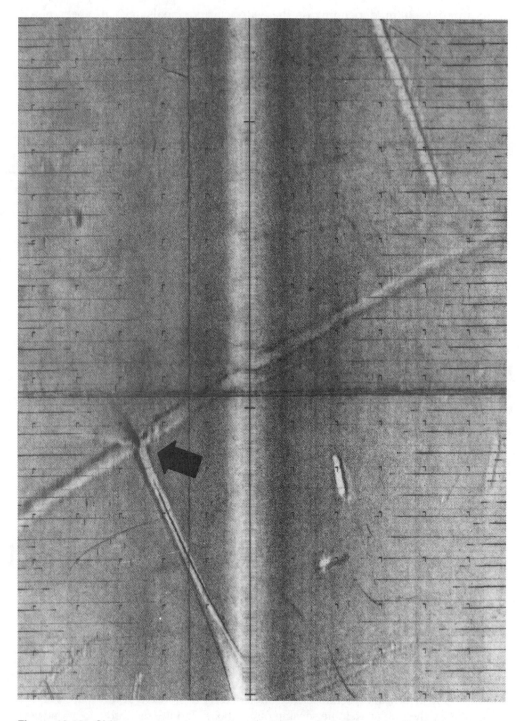

Figure 12.10 Side scan sonar image of a pipeline gas leak (arrow) in the Gulf of Mexico. Apparently the pipeline was snagged by a ship's anchor and pulled out of its trench. From John Chance and Assoc.

Regulations

In the United States the Federal Resource Conservation and Recovery Act (RCRA) outlines a system of permitting, record keeping, and waste management for companies that generate, transport, treat, store, or dispose of hazardous wastes. What makes these regulations a matter of concern to the mining and oil industries is that hazardous wastes have been defined to include certain tailings, drilling fluids, and produced water. This is having an impact on mining and all drilling operations for oil, gas, and geothermal exploration and production. At the very least, records will be required showing the disposition of dumps, tailings, mud pits and drilling fluids. Imagery provides such a record at relatively low cost.

The Comprehensive Environmental Response, Compensation and Liability Act of 1980 (CERCLA) was designed to help clean up hazardous waste sites where no responsible party was available to pay the cleanup costs. CERCLA imposed retroactive liability on parties responsible for hazardous sites. These responsible parties were defined as past and present owners or operators of a site, those responsible for disposal at a site, or those who selected a site. Those who purchase properties where contamination is later discovered routinely sue sellers under CERCLA. Courts have imposed liability on banks and other lenders who have participated in management of the polluted site. "Bankers are becoming increasingly concerned about lending money for a property that may contain hazardous waste. Banks can end up as defendants in lawsuits arising from deals involving real estate with hazardous waste" [13]. In many countries oil and mining companies are being asked to clean up environmental disasters that existed prior to their presence. These factors have dramatically increased the demand for environmental audits. As mentioned in Chapter 11, imagery is an essential part of any environmental audit, providing the historical evidence that is often lacking when the original property owners, or their records, are unavailable or uncooperative.

Remote sensing images can serve as point-in-time documentation of environmental conditions in an area. This type of documentation can be valuable either before or after an environmental problem occurs. Since most imagery is in the public domain, multitemporal records provide indisputable evidence of the surface condition of an area. Imagery can show the extent of environmental damage that existed on a lease before a company acquired property, thereby limiting liability [14]. Imagery can show the extent of damage, the timing of spills, leaks, trespass, or dumping, and perhaps provide clues as to the party involved.

Law Suits

In recent years there have been numerous cases involving contaminated groundwater and the oil and mining industries. Private landowners claim that past activity on or near their land has contaminated their surface or well water. The cause of action for lawsuits can be filed as a nuisance (seeping water, offensive stenches, leaking fumes or gasses), negligence (unreasonable use of the surface owners' land), gross negligence (reckless disregard of the consequences of one's conduct, say, in reinjecting drilling fluids), and possibly trespass (unauthorized entry, including cracks from a frac job or subsidence due to mining). It may be possible using historical imagery to show exactly where mud pits and tailings piles were, whether liners were used, whether there were crude or fuel spills around a well, the disposition of drilling fluids and produced water, or the extent of actual subsidence related to abandoned mines or hydrocarbon or water withdrawal.

In the case of tanker or pipelines spills, the liability lawsuits will be based on negligence that can be assigned in three categories:

1. The grounding, collision, or source of the break,
2. the containment, and
3. the cleanup.

It is not likely that there will be image documentation of the accident itself. Immediately afterward, however, it should be possible to use imagery to map the distribution and direction of movement of a spill so that it is possible to contain and clean the pollution using appropriate techniques and adequate resources (Fig. 12.11). In the damages portion of a lawsuit there is also a role for remote sensing. Most compensatory damage calculations depend in some way on the extent of damage to natural resources. It thus becomes essential to quantify the impact of a spill, broken tailings dam, or eroded mine dump. This means an environmental audit or damage assessment program. Such a program will measure the extent of the injury to resources, assess the impact such that full compensation will be provided, and determine how rehabilitation or restoration can be assured. Imagery can be used to help determine the extent of injury and guide rehabilitation efforts by monitoring the progress of the recovery program and comparing conditions to those existing before the accident.

Perhaps the simplest application of remote sensing is to show compliance with regulations. Thermal images of a refinery's hot water discharge, for example, will show whether or not the mixing area is within the U.S. Environmental Protection Agency guidelines (Fig. 11.15).

Images can be used in international disputes to demonstrate to a court, for example, that a nationalized lease is still producing, or to document illegal operations. Satellite imagery can clearly show flares from wells and piles of sulfur near the refinery where the oil is processed. This could help to determine the amount of compensation due to the former lease owner.

PUBLIC RELATIONS

Everyone loves to see an image of their hometown and all the familiar places nearby. The value of airphotos and satellite images as gifts should not be underestimated. A framed image of a city, county, or province, with company logo, will most likely be a much sought after item to be hung in the office of the governor, environmental agency head, minister of natural resources, or other important persons in the country where you are working. This relatively inexpensive gesture can go a long way towards generating favorable publicity and goodwill. Producing high quality imagery is also a wonderful and effective way to promote the high tech capabilities of a company through an annual report or media advertising.

Environmental groups need not necessarily be the natural enemies of resource extraction industries. A gift of imagery over migratory bird nesting areas or caribou migration routes could help generate a cooperative atmosphere when projects of mutual concern are being planned.

Figure 12.11 Landsat TM mosaic of Prince William Sound near Valdez, Alaska, showing the oil spill as of April 7, 1989. The oil mousse appears in the lower left as green filaments in this principal components image. Processing by KRS Remote Sensing. (See color plate.)

Companies are always trying to recruit the best students from universities. The goodwill and publicity generated by donating imagery to a geology or environmental sciences department, for example, could go a long way to making students aware of a company and impress them with its capabilities. At the very least it shows that a company is both cost-conscious and willing to use cutting-edge technology for solving exploration and environmental problems.

REFERENCES

1. J. McKean, S. Buechel, and L. Gaydos, *Photogram. Engr. and Rem. Sens.* **57** (1991): 1185–1193.
2. F. Mantovani, R. Soeters, and C.J. Van Western, *Geomorphology* **15** (1996): 213–225.

3. P. Rosen, C. Werner, E. Fielding, S. Hensley, S. Buckley, and P. Vincent, *Geophys. Res. Letters* **25** (1998): 825–828.
4. J. Ehrismann, B. Armour, M. van der Kooij, and H. Schwichow, *Earth Observation Mag.* (April, 1996): 26–29.
5. P. Briole, D. Massonnet, and C. Delacourt, *Geophys. Res. Letters* **24** (1997): 37–40.
6. S. Fujiwara, P.A. Rosen, M. Tobita, and M. Murakami, *Jour. Geophys. Res.* **103** (1998): B2–2411–2426.
7. C.M. Marino, G. Del Pero, A.M. Tonelli, 18th *Intl. Symp. On Rem. Sens. of Environ.* **III** (ERIM, Ann Arbor, 1984), pp. 1905–1914.
8. C.D. Elifrits and D.J. Barr, in C.J. Johannsen and J.L. Sanders (eds.), *Remote Sensing for Resource Management* (Soil Conservation Soc., 1982), pp. 347–361.
9. Dames and Moore, Colorado Springs subsidence investigation (1985).
10. D.C. Peters, K.E. Livo, and P.L. Hauff, *Environmental Geosciences* **3** (1996): 11–20.
11. J.C. Sandidge and R.J. Holyer, *Remote Sensing Environ.* **65** (1998): 341–352.
12. V. Kostilainen, *Offshore Technology Conference* (1990): 217–224.
13. R. Bivens, *The Houston Chronicle* (November 18, 1990).
14. P. O'Donovan, in A.E. Lodge, ed., *The Remote Sensing of Oil Slicks* (John Wiley and Sons, Chichester, 1989), pp. 105–113.

ADDITIONAL READING

P.N. Bierwirth, T.J. Lee, and R.V. Burne, *Photogramm. Eng. Remote Sens.* **59** (1993): 331–338.
R.G. Burdick, L.E. Snyder, and W.F. Kimbrough, *U.S. Bureau Mines Rept. Invest.* **9050** (1986), 27 p.
M.A. Khan, Y.H. Fadlallah, and K.G. Al-Hinai, *Intl. Jour. Remote Sensing* **13** (1992): 605–614.
F. Kuehn, G. Trembich, B. Hoerig, *Proc. of the 12th Intl. Conf. Applied Geologic Rem. Sens.* (ERIM, Ann Arbor, 1997), pp. I-425–I-432.

QUESTIONS TO CONSIDER

1. What are some of the features that characterize both landslides and submarine slumps?
2. What kinds of imagery might be useful in mapping earthquake hazards? What would one expect to map?
3. How are floods monitored using remote sensing? What should you be aware of when planning a river crossing in the Arctic?
4. List some marine hazards that can be mapped using remote sensing. What instruments are best suited for each?
5. How would you map karst collapse features, and potential collapse features, in west Texas? In Qatar on the Persian Gulf?
6. You are about to take over a mining operation in a foreign country. How would you use remote sensing to minimize your future legal liabilities with respect to cleanup and environmental damage?

Closing Thoughts

Remote sensing is an integrative "science," involving technologies such as satellite multispectral sensors, radar, thermal, and hyperspectral technology and image processing. The interpretation of images, however, is not new. New applications are constantly emerging as industry struggles to succeed in an ever more competitive world. The role of remote sensing interpretation in mining, petroleum, ground water, engineering, and the environment is to further reduce exploration risk, engineering uncertainty, and environmental impact. The driving force behind remote sensing has been and will continue to be the cost-effective nature of the work. It is the least expensive way to quickly survey and evaluate large areas of the earth's surface.

The greatest advances of the past five years lie in the emerging applications of geographic information systems (GIS) and combining global positioning systems (GPS) technology with remote sensing imagery. GIS allows us to easily update existing maps using imagery and to better evaluate areas by integrating multiple diverse data sets. GPS provides the opportunity to make images into accurate base maps so that they can be easily registered and merged with other geographic databases. These two technologies have revolutionized remote sensing by making it more accurate and easier to use when analyzing multiple data sets. In addition, hyperspectral remote sensing has provided the ability to map an ever-greater suite of minerals, soils, and surface materials. Radar interferometry provides a means to measure small changes in the earth's surface related to earthquake deformation and subsidence over oil fields or abandoned mines, it also is a way to generate digital elevation models and make topographic maps. Recent radar and hyperspectral satellites are changing how we view our world.

What can we expect from remote sensing in the next five years? In the short term there will continue to be improvements in airborne and satellite multispectral scanners. Instruments with many narrow bandwidth channels will become even less expensive and more routinely used. Our planet may finally get full radar coverage. Sensors will have both better resolution and a global perspective. Private corporations are even now launching satellites that will provide one meter resolution, stereo coverage, and near real-time turnaround. Better resolution will enable geologists and engineers to focus on finer and finer details of the surface, whereas coarser resolution will be used by environmental scientists to chart worldwide changes in plant cover, ocean currents, atmospheric ozone, etc.

In the not too distant future geologists and environmental scientists will routinely visualize field areas by merging digital topography with imagery and displaying the three-dimensional representations of the surface on powerful desktop workstations.

This will allow someone interested, say, in viewing the structural fabric of Nepal, or locating access routes in the Amazon to actually "fly" over the terrain and view it from various perspectives using their computers. Although the technology exists today primarily in military applications, it is expensive and good digital terrain data does not exist for many parts of the world. This kind of pre-travel analysis will make trips more cost effective by minimizing the amount of time needed to get oriented on the ground.

Looking further into the future, sensors will continue to be refined and, combined with spectral libraries and unmixing algorithms, will allow us to generate ever more accurate and detailed maps of surface materials. There will always be a place for the image interpreter, however, both to determine the types of processing required and to explain the significance of what is being displayed. This is true whether one is searching for copper in Peru or for water on the moon. It is hoped that this text will be a useful basic reference for geologic image interpreters of the future.

Responses to "Questions to Consider"

The following are responses to the questions posed at the end of each chapter. It should be noted that in many cases there is no "correct" answer, but there may be a wide range of innovative and imaginative responses that will solve the given question or problem.

CHAPTER 1

1. You have been asked to determine if there are seeps in the Natuna Sea, Indonesia. This area frequently has cloud cover. What sensor systems should be considered?

    The primary sensor for this circumstance would be radar, either airborne or satellite. This is because radar has all weather capability and can detect oil seeps due to the damping of capillary waves (slick should appear darker, or smoother, than surrounding water).

2. It is important to know whether the stratigraphy in Somalia is similar to that in Yemen. Would color or black/white images be more useful? Would satellite images or airphotos be better? Give examples of what information each would provide to help make a usable photogeologic map.

    Color images are generally more useful in mapping and correlating stratigraphy since it is not always clear which gray tones correspond to a given unit when comparing two areas. For large areas such as Somalia and Yemen satellite images would probably be more useful, since they cover a larger area. Airphotos, however, will provide better resolution, but at greater cost. Black/white images might provide better structural information (less distraction due to lithologic colors), whereas color images would provide a color basis for correlation.

3. How would you go about finding and obtaining airphotos over Kenya, Siberia, or Peru?

    The first step would be to check the internet under the heading "airphotos Siberia," "geological survey Kenya," or "University of Lima remote sensing". Another possibility would be to contact a local source of data (e.g., EROS Data Center) and ask for their contacts in the country of interest. Still another approach would be to contact the American Society of Photogrammetry (or local society) and ask about companies that have flown or are capable of acquiring airphotos in the specific country.

4. What scale imagery would you use for a regional gold exploration program in Kazakstan? Why? What scale would you use to locate sources of groundwater in one township in Wyoming?

   A regional exploration program will attempt to cover a large area rapidly and focus in on likely targets based on alteration patterns and structures. A scale of 1:250,000 would be a good compromise between area coverage and resolution. A township in Wyoming comprises 36 square miles (92 km^2). Such an area could be easily covered by a 1:50,000 satellite image or airphotos (1:24,000 to 1:80,000) would probably work just fine.

5. What platform, sensor, and wavelength of imagery would you use to evaluate a mining claim in northern Mexico? In southern Mexico? Would they be different, and if so, why?

   Northern Mexico is arid, whereas southern Mexico is tropical. Low to moderate altitude true color or high resolution color infrared satellite images would work well in arid areas for mapping hydrothermal alteration patterns, as would hyperspectral imagery. In humid tropical areas the atmosphere scatters much of the blue wavelengths, making true color and in some cases even standard color infrared imagery less useful. In such cases near infrared photography or imagery (e.g., TM bands 7-5-4 as RGB) avoid much atmospheric scattering. Vegetation cover in the tropics could make alteration mapping less useful than structure mapping, in which case radar or black/white near infrared band (e.g. SPOT band 3), combined with some color infrared vegetation stress mapping, could help locate mineralized or altered areas.

6. You need an image of Mongolia next week to tell management what roads and other infrastructure exists. Cost is no object. What would you do? Would you handle it differently if you are on a tight budget?

   If cost is not a problem, contact an image processing company and place a "rush" order on one or more recent images of the area of interest (see Table 1.2). Since there is a time limit, existing satellite images are the best bet. Depending on the size of the area, high to moderate resolution (1–30 m pixels) should be adequate. On a tight budget one would look for similar imagery, but it would necessarily be "off-the-shelf," i.e., imagery already acquired and processed and in an archive.

CHAPTER 2

1. Your office has a great view, but does not have a light table. The imagery you are using contains dark areas which, you are convinced, contain details important to your project. How might you get around this problem?

   Tape your image to the window and use natural sunlight to illuminate the darker areas as you carry out your interpretation.

2. No imagery or geologic mapping exists in your area of interest, and you have been asked to quickly evaluate a broad, low-relief valley for evidence of faults that could localize springs. What other type(s) of data could you use?

   Topographic maps are often available when no imagery exists in an area. These maps can be evaluated for escarpments, offset streams, aligned ponds, and vegetation patterns indicative of springs along faults.

3. You have the option of ordering low altitude color imagery or high altitude black/white imagery for a project. What trade-offs are involved? How do you decide which to use?

   Low altitude color imagery will require more frames, since each frame will cover a smaller area. The color film will also cost more, but may provide more information on stratigraphy, alteration, or vegetation. Alternatively, the black/white imagery will be less expensive, cover more area (thus require fewer frames), and may be processed and delivered sooner. The imagery used will depend on how critical it is to get information that only color can provide, the budget available, and the turnaround time requirements.

4. You know your project will take you to the field, where conditions will be harsh. Do you want your imagery on film or paper, and how will you carry and protect it?

   Laminated paper images will survive better than unlaminated paper or film. It can be written on and erased without damaging the emulsion. Photos can be carried on a clipboard or in a cloth or leather pouch along with pens, colored pencils, erasers, scales, and other equipment.

CHAPTER 3

1. What geologic features are you more likely to see on daytime thermal imagery over Qatar (a low-relief peninsula with surface carbonates in the Persian Gulf) versus night thermal imagery?

   Any topographic features will be enhanced by differential solar heating of slopes facing the sun vs shadowed slopes on daytime thermal imagery. Cool spots are likely to be moist soils (evaporative cooling) and standing water (higher thermal inertia, so warm up slower than surrounding bedrock). Nighttime imagery tends to show changes in lithology (virtually none) and moisture (standing water warm, moist soils cool).

2. Which would be more useful for detailed geologic mapping in the humid countryside north of Rangoon: SPOT panchromatic 10 m data or SPOT multispectral 20 m data? Why?

   Although the 10 m Pan imagery has better resolution, one should consider that the 20 m multispectral data has both better spectral resolution and, more important, the near IR band effectively ignores atmospheric haze caused by humidity in the blue and green portions of the spectrum. The black-white near IR band may provide more detail than the panchromatic image with twice the spatial resolution.

3. What do you want to know prior to planning the optimum radar survey over Guatemala?

   Of major importance is the amount of relief in the area and the trend of geologic structures. The relief will determine the depression angle of the airborne SAR (high relief, steep depression angle; low relief, shallower depression angle). The geologic trend of structures will determine the orientation of flight lines: lines should be laid out approximately 30° from the strike of structures so that the beam hits the trends at 60° to strike.

4. What are the advantages and disadvantages of using Landsat TM vs. Seasat radar to map oil slicks offshore Ivory Coast?

   Landsat TM will see slicks as dark streaks or blobs on the ocean as a result of the darker color. Radar will also sense the slicks as areas darker than background due to the suppression of capillary waves (smoother surface). Seasat had a limited time in orbit, however, and may not have imagery available in the area of interest. Other satellite radar images (ERS-1 or -2, Radarsat, etc.) may be available. Radar can see slicks through the tropical cloud cover; Landsat cannot.

5. What are the advantages and disadvantages of using airborne multispectral/hyperspectral scanners vs. color airphotos in mineral exploration in Australia?

   The principle advantage is the ability to identify and map minerals and alteration suites using multispectral/hyperspectral imagery. The main disadvantage is the cost of acquiring and processing this type of imagery. Color photos can assist in mapping alteration, but cannot identify minerals. They are faster and less costly to acquire, however.

6. How is near-infrared imagery different from thermal infrared imagery?

   Near-infrared imagery is acquired using a camera and film, or a sensor that records reflected light in the range 0.7 to around 2.5 microns. Thermal sensors are crystal detectors that record emitted thermal radiation in the range 3–15 microns.

7. What wavelengths penetrate water farther, blue or red?
   Blue.

8. Which wavelengths penetrate the atmosphere with the least scattering, blue or red?
   Red. The sky is blue because of the scattering of blue wavelengths.

9. What is the relationship between microwave wavelength and penetration of surface materials and vegetation?

   Longer wavelengths penetrate farther than short wavelengths. No wavelength will penetrate more than one or two meters, and then only under hyperarid conditions in coarse sand or soil.

CHAPTER 4

1. How would you expect limestone to weather, and how would you recognize it in the arctic, in a temperate climate, and in the tropics? A sandstone? A serpentine?

   Limestone weathers to angular blocks in the arctic; is generally resistant to erosion (ledge-former) in temperate climates although karst topography and deranged drainage are common; and forms erosional remnants, karst topography, and terra rosa soils in the tropics. Sandstone and serpentine weather to angular blocks in the arctic; in temperate and humid climates sandstones generally form ridges whereas serpentine will weather to clay-filled depressions.

2. A thrust fault puts a Carboniferous limestone over a Cretaceous shale in Wyoming. How might one recognize the thrust contact?

   If units in the thrust sheet are deformed and units in the subthrust are flat, the fault can be recognized by the abrupt termination of bedding at the contact. The limestone will be more resistant to erosion than the shale and will form cliffs over gentle slopes. The fault will be at the break in slope.

3. Although jungle covers the surface, one can make out drainage patterns in the Amazon basin. What is suggested by alternating bands of dense and widely-spaced drainages?

   Alternating layers of sandstone and shale.

4. While examining Landsat TM imagery over pediment surfaces in the Altiplano of Chile you detect clusters of small yellow-green outcrops on the color-infrared composite image. What could these be telling you?

   Yellow-green outcrops on color-infrared imagery correspond to bright red outcrops on the ground. These could be gossans or iron-oxide alteration associated with mineral deposits. Alternatively, they could be red shales or sandstones with hematite or limonite cement. Check for other clues such as bleaching that could indicate argillic alteration.

5. You recognize badlands on an image of the San Juan basin in New Mexico. What lithology does this most likely represent? Outcrops of a dipping unit along the margin of the San Juan basin have pine trees growing on them, whereas the surrounding plains are covered in grass and sagebrush. What is the probable lithology of the ridges?

   Badlands generally form in shales due to the poor infiltration of water and easily eroded units. Ridges with pine trees suggest resistant sandstone with good infiltration of water. Limestone would also form ridges, but has good infiltration only where fractured.

6. In the same part of the San Juan basin walls of rock radiate outward from a large, almost circular and jagged outcrop of dark rock. Any idea what this could represent?

   The walls of rock radiating outward from a central core most likely are dikes extending outward from a volcanic stock. An example might be Ship Rock, New Mexico.

CHAPTER 5

1. You are exploring for giant oil fields with four-way closure. What surface features would you look for in the Amazon basin? In the Andes? On the flat plains of the Tibetan Plateau?

   The Amazon is essentially flat, and only drainage patterns will provide clues to structure. In the Andes one can map the strike and dip of units to identify opposing dips. In the arid, low-relief Tibetan Plateau one can map bands of light and dark strata and an occasional strike and dip to identify structures.

2. How might a fault appear at the surface? A mineralized fault?

   Faults can be expressed as juxtaposition of two different rock types, as linear escarpments, as adjacent streams consistently offset in the same sense, as offset bedding, as linear streams or rivers, as breaks in slope, as vegetation changes, as sag ponds, etc. A mineralized fault can be more resistant to erosion than the host rock if, for example, it is silicified. It can also form linear valleys or depressions when rocks are crushed and easily eroded along its length, or because alteration has weakened the host rock if, for example, argillic alteration occurs along the fault. A mineralized fault may be altered and adjacent rocks bleached or iron-stained.

3. Where would you drill for water if pre-dawn thermal imagery showed a cool zone up slope from a faulted pediment? Why?

   The cool zone probably represents groundwater ponded on the up slope side of the fault. Drill in the cool zone.

4. What does a linear fault trace tell about the inclination of the fault plane? What does a curved fault trace tell about the fault geometry?

   A linear fault trace generally means a fault is near-vertical. A curved fault trace can be either an inclined planar fault cutting across a topographic high or low, or it can indicate a listric fault (normal or thrust) that curves from near vertical at the surface to bedding-parallel at depth.

5. How can normal faults and strike-slip faults appear similar on imagery? How can one tell them apart?

   A normal fault that offsets dipping strata can appear similar to a strike-slip fault. To tell them apart look for areas where the strata changes dip, or look for a consistent offset of strata, drainage, or topographic features. The normal fault may be concave toward the downdropped side; strike-slip faults tend to be straight over long distances.

6. How can unconformities and thrust faults appear similar on imagery? How can one tell them apart?

   Bedding with different inclinations can come into contact at both unconform-ities and thrust faults. If one knows which way is up-section, bedding above an unconformity will be parallel to the unconformity surface, whereas bedding below a thrust flat will be parallel to the thrust. Unless one can see the thrust flat-ramp-flat geometry it may be impossible to tell the difference between an unconformity and a thrust ramp. Bedding above an unconformity is generally less deformed than bedding below it. Bedding above a thrust can be more deformed than bedding below it. Thrusts can repeat section: unconformities do not.

CHAPTER 6

1. Why should you be able to distinguish shales from carbonates in an arctic environment using radar?

   You should be able to distinguish these units by the texture of the erosional products: blocky for carbonates (high radar return) and smooth and moist (low radar return) for shales.

2. On what basis and criteria would you establish a stratigraphic framework when working in a new and unknown area?

    The most commonly used criteria are color, tone, and resistance to erosion (texture, drainage density, ledge vs slope-former). Sometimes it is possible to use vegetation patterns (e.g., sandstones having tree cover whereas shales have grass cover). Correlations are done by matching similar photogeologic sections when bedding cannot be followed continuously.

3. You are in charge of a program to find new diamond deposits. What sensors would you use, and what would you look for? What if you are looking for coal? For Mississippi Valley-type lead-zinc deposits?

    Hyperspectral imagery would best assist in mapping the minerals associated with diamond-bearing kimberlites (olivine, phlogopite mica, diopside, apatite, calcite, chlorite, serpentine) and Mississippi Valley-type deposits (calcite, dolomite, sphalerite, galena, chalcopyrite, pyrite, bornite, barite, and jasperoid). Locate coal using color airphotos or satellite imagery by looking for dark bands associated along strike with reddish clinker zones (areas of molten rock adjacent to burned coal beds).

4. Why could deeply buried channel or beach bars have some expression at the surface? How might they be expressed?

    Shale drape over a sand channel or barrier or beach bar can cause a slight structural high in overlying units and a subtle topographic high at the surface. If the channels and bars were originally controlled by structures such as faulting, then some evidence of the structural control may also be visible at the surface, such as aligned drainages.

5. What is the advantage, if any, of hyperspectral imagery over multispectral imagery? Are there any drawbacks?

    Hyperspectral imagery should more accurately show the distribution of minerals, as long as there is a full spectral library available and spectral unmixing is possible. It takes a bit more image processing due to the greater number of channels (usually over 100 vs 3 to 6 bands for multispectral), and the imagery may cost more to acquire than multispectral imagery.

6. What, if anything, prevents us from flying an instrument or suite of instruments and generating geologic maps automatically, i.e., without an interpreter?

    Although unsupervised classifications can map groups of surface materials that have similar reflectance or emmittance, an interpreter is still required to explain what the groups represent. Supervised classification, by definition, uses an interpreter to chose the classes based on some knowledge of the area or experience in similar areas. Some aspects of geologic maps, such as the presence of unconformities, inclined faults, folding, and the dip of units, cannot be mapped automatically without a real possibility of serious mapping errors.

CHAPTER 7

1. A salesman calls and tells you he has a remote sensing study that can help you locate gold deposits in central Asia. What are some of the questions you should ask about the study (other than "how much does it cost?")?

Some of the logical questions to ask include: Who did the study, and what are their qualifications? What instruments were used to acquire the data, i.e., what kind of data is it? What were the instrument parameters (altitude, resolution, time of day, season, wavelengths). How was the data processed? Why should the technique work? Is the study exclusive or available to others?

2. You are planning a remote sensing study of a frontier oil basin. What other kinds of data would you want to have to support your work?

   In addition to remote sensing imagery, it would help to have some ground control such as a stratigraphic column, and some idea of where key rock types are and what they look like (color, texture, resistance to erosion). It would help to know something about the structural style: this may be available from reports or maps in a library. Some idea of the subsurface, such as the depth of the basin (from gravity, magnetics, or seismic data), and major basement faults or changes in rock type (from gravity, magnetics or seismic data) would help with the interpretation. It would also help to know where other field discoveries are, where wells have already been drilled (both productive and dry), and where known seeps are located: these can all help focus attention on the most prospective areas. Some of these can be obtained through library work, whereas others may require a visit to the field.

3. Before deciding what to look for on imagery, what kind of "up-front" work can you do?

   Go to the library (public or university) and find maps and reports on the area of interest. Check the internet for anything that has been done and may be available on the area. Contact universities in or near the area of interest for mapping, research, dissertations and theses, and local expertise. Contact the geological survey or local geological society for information on what has been done and who the experts are in the area of interest. Find out what imagery exists. Find out what auxiliary data exists, including gravity, magnetics, seismic, and geochemistry.

4. What are some ways geologic structures can contribute to stratigraphic hydrocarbon traps?

   Structures such as faults and folds can control reef buildups by creating horsts or high blocks for reefs to root on. Faults can direct the course of stream channels, or can localize depocenters where sand deposition is greater than in adjacent areas. Growth faults can cause thickening of reservoir sands on the downthrown side, and can in some cases determine the shoreline and location of beach or barrier bars. Faults that uplift reservoir units can contribute to development of updip unconformity traps. Fault-controlled water gaps or sediment bypass zones can control the location of sand-rich depocenters.

5. What alteration minerals are easiest to map using multispectral/hyperspectral scanners in the visible-near infrared (to 1.7 microns) range? In the region from 2.0–2.5 microns?

   The easiest minerals to map in the visible and near infrared are iron oxides including limonite, goethite, hematite, and jarosite. These minerals have absorption bands near 0.4 and 0.85 microns, and a local peak around 0.7 microns. Clay minerals have an absorption band centered at 2.2 microns: carbonates, chlorite, epidote, and talc have absorption bands between 2.35 and 2.40 microns.

CHAPTER 8

1. Why should subsurface fractures (faults and/or joints) be expressed at the surface? Would you expect to see the trace of the fracture itself? Is so, when? If not, why not?

    Faults may be expressed at the surface if they broke through to the surface when formed, or if erosion has uncovered them. Joints may form at the surface due to flexural folding and propagate downward, or may form in the subsurface and be uncovered by erosion. You can see the trace of a fault or joint at the surface if they have been uncovered by erosion or formed at the surface. You would not expect to see the trace of a fault or joint that lies beneath an unconformity, or that dies upward into folding or in a ductile unit (salt, shale).

2. Would you expect the length of a fracture zone to be related to its depth of origin? Why? Could zones of short fractures be related to one master fault at depth? How might they be expressed at the surface?

    Empirically it has been demonstrated that longer fractures have deeper origins than short fractures. One might speculate that this is a function of the energy required to break rock: greater stresses derived from deep in the crust generate longer and greater displacement faults. Zones of short fractures can sometimes be linked at depth to one master fault. They may appear at the surface as a series of en echelon (shingled) faults or faults along a trend separated by short sections of unbroken rock.

3. What kind of fractured reservoir would you not expect to see evidence for at the surface?

    You would not expect to see joints formed due to overpressure, since overpressured zones are usually confined to strata below a ductile seal.

4. If you know that your ore deposit terminates at a fault, what would you look for to locate the faulted-off portion of the body? What other techniques might you consider using?

    Look for consistent offset in a specific direction. Try to determine the type of fault that displaced the body. Consider using aeromagnetics, geochemical sampling, or surface potential fields methods such as induced polarization or very low frequency electromagnetics.

5. Coals produce increasing amounts of methane as they are dewatered. Residents living near a coalbed methane field have begun to notice methane in their well water. What could remote sensing tell you about how the methane got there?

    Mapping a fault that passes through the vicinity of a water well would suggest that methane began migrating along the fault once the coals were dewatered and reservoir pressure was lowered.

6. While interpreting an image that has been processed for alteration associated with porphyry copper you notice a cluster of color anomalies that look similar to alteration at the only known copper deposit in the region. What do you do?

    Send in a field crew to stake a claim and sample the anomalies.

CHAPTER 9

1. How do plants respond to shallow groundwater in arid and temperate climates?

   They are generally more vigorous (greener in visible light; redder on color infrared film) than surrounding vegetation. They will leaf out sooner in the spring, and will drop their leaves later in autumn.

2. What thermal response would you expect from groundwater ponded against a fault on daytime thermal imagery?

   Groundwater is generally cool due to evaporative cooling both on daytime and nighttime imagery. The contrast with the surrounding rocks and soils will, however, be greater on daytime imagery.

3. What environmental factors can cause a misleading interpretation of remote sensing water depth mapping?

   Misleading water depths can be a result of dark versus light bottom (sediment, rock, or vegetation), suspended particulates, variations in water surface reflectance, and atmospheric conditions (cloud shadows, ground fog).

4. How can moving water contribute to erosion? What would indicate erosion-prone areas?

   Moving water can erode streambanks, undercut cliffs or levees, or can move beach sand offshore or along the coast. Sheetwash can move an enormous amount of material during flash floods. Erosion-prone areas are those with steep slopes (near the angle of repose), little or no vegetation cover, in sandy or coarse clastics, in water-saturated sediments, or in poorly consolidated or unconsolidated materials. Look for areas with evidence of previous erosion (e.g. channel scours) or landslides (escarpments, hummocky topography).

5. How would you organize a program to explore for geothermal resources?

   Under ideal conditions one would use both a thermal scanner acquiring imagery with one meter resolution and a hyperspectral scanner that can map hydrothermal alteration minerals with a 5–10 meter resolution. Fly pre-dawn thermal imagery with multiple ground calibration stations. Look for areas with evidence of recent volcanism.

CHAPTER 10

1. What is the purpose of a geographic information system?

   The GIS is a way to relate various data sets to a geographic location. They can locate an object, put it into context with surrounding features, and link the object to descriptive text, tables, and illustrations. They can be used, among other things, for guiding urban growth, revegetation of mine tailings, and development of oil fields or mines. A well-integrated system allows the manager to update all levels of data simultaneously and instantly and to present information at any scale and at varying levels of detail. The user should be able to pick a feature and answer questions such as "what are you?," "where are you?," and "when I change you, what else is affected?"

2. How can a GIS help plan a seismic program?

   A GIS can provide the planners of a seismic program with information regarding land ownership (for permitting), where the flat to gentle topography is as opposed to more rugged terrain (topographic input), where surface statics might be avoided (sand dunes, deep weathering), where environmentally sensitive areas are, where line layout may be difficult due to urban areas, agriculture, rivers and lakes, and where roads and tracks already exist for access and to minimize cutting new lines (derived from image interpretation).

3. How can remote sensing help determine the best site for a hydroelectric project? The best route for a new interstate highway?

   Plant siting can be facilitated by mapping present surface use, potential hazards, general geology, topography, and environmental sensitivity. By updating existing maps, remote sensing can help with engineering and environmental assessment of both hydroelectric and highway siting.

4. An aqueduct is being planned across California. What factors should be considered in planning the route? How can remote sensing help map these features?

   The route should be as straight as possible to save miles and lower costs. On the other hand, it should avoid river crossings, active faults, steep and/or unstable slopes, rocky areas that may require blasting, environmentally sensitive areas, and urban and agricultural areas. Most of these factors can be evaluated using image interpretation with topographic maps.

5. What would you suggest is an appropriate base map (image type, scale) for a mine expansion project in China? What would be an appropriate base map for an oil exploration program in the Canadian arctic? For laying a natural gas pipeline across Los Angeles?

   Mine mapping requires detailed information regardless of where you are. Use black/white or color airphotos at scales of 1:1200 to 1:12,000, as appropriate. A regional oil exploration program could use satellite imagery at scales from 1:100,000 to 1:500,000. Landsat, SPOT, or IRS color or false color imagery would be appropriate. Laying a pipeline in a crowded urban area requires great detail. Again, use black/white or color airphotos at scales of 1:6,000 to 1:24,000 depending on detail required.

6. What are your main concerns regarding nuclear waste disposal? How could remote sensing minimize those concerns?

   Nuclear waste disposal causes much anxiety in communities along the train or highway routes, and adjacent to the disposal sites. Routing wastes away from populated areas, and disposing wastes in seismically inactive sites would go a long way toward easing those concerns. Image interpretation can help direct traffic onto less used routes, and can help map faulting. Trenching is required to determine how recently the faults were active.

CHAPTER 11

1. You just found the perfect piece of land for a summer cabin in the hills. Why might you want to obtain historical airphotos over the property before you buy it?

You may not be aware (and the previous owner may not be aware) that this land was used as (take your pick): a toxic waste dump, an oil field slush pit, a burial ground, a fuel depot that leaked into groundwater, an air force bombing range with unexploded ordnance, a stockyard, and so forth.

2. If you have a choice between using a supervised or unsupervised classification, which would you use for (a) mapping grizzly habitat in the Arctic, (b) a mine reclamation study in Nevada? Why?

   It doesn't really matter what it is that one is mapping. If you have some idea of the classes of surface materials and where they are, and know how many classes you want in the final map, then a supervised classification will give a better result. If you have no information regarding the distribution of surface materials (soils, plants, rocks, water, etc.), then an unsupervised classification will assign similar reflectances to various groups and you will have to determine what, in fact, they represent by doing field checks.

3. What sensors could be used to monitor buried pipelines for leaks? What about potential leakage around an earth-fill dam or tailings pond?

   The imagery most sensitive to changes in temperature and moisture include thermal and radar. Thermal will show leaks as cool if the leak is high pressure gas, as warm if the pipe is carrying oil across the arctic. For pipelines or dams look for anomalous warm or cool spots depending on time of day or season. Radar will show anomalous moisture changes if the surface materials are essentially flat and uniform (all factors being equal).

4. Your company has an operation on the shores of San Francisco Bay. In the past 24 hours it has been accused of causing a large fish kill in the bay. How might you use remote sensing to deny (or confirm) this allegation?

   Find repetitive detailed imagery showing the area around your plant over several years and over the past few days. Order imagery to be flown immediately to show whether anything has changed regarding leaks, dumping, algal blooms, etc. You may be able to show that unusually high fertilizer runoff due to recent rains has caused algal blooms, oxygen deficiency, and the resulting fish kill. You may find that someone has been dumping illegally near your property. Or you may find that one of your settling ponds leaked, allowing cyanide into the bay. In any case, you will have documentation showing what changed.

5. New England's acid rain has been blamed on Canadian smelters. How might remote sensing help determine whether this is true?

   Monitor vegetation changes over a period of years downwind from the smelters. Map changes and sample from background into the anomalies for soil acidity and vegetation vigor. Map wind directions using smelter plume trends.

6. How might you distinguish between a natural seep and a man-made spill off the coast of California (or Louisiana, or the Ivory Coast)?

   It is difficult to be absolutely certain, but man-made spills generally have a point source at a ship or pipeline. Ships leave wakes that can be traced for several hours after the ship has passed through the area, and the leak will dissipate over time. A pipeline will continue to leak from the same spot. Natural seeps also have fixed locations, but tend to be intermittent, leaking only when tectonic conditions favor open fractures or when pressures reach a critical point. Man-made leaks also tend to be uniform in composition (all crude oil, or gasoline), whereas natural seeps generally have a range of compositions and may contain a range of expressions from heavy oil-water emulsions to light oil sheens.

CHAPTER 12

1. What are some of the features that characterize both landslides and submarine slumps?

    Escarpments, generally concave toward the downthrown side. Hummocky topography on the sliding block. The lobate shape of the slumped block. Thrusting and/or compression at the toe of the slump.

2. What kinds of imagery might be useful in mapping earthquake hazards? What would one expect to map?

    Any visible-near-infrared imagery or radar would be useful in mapping earthquake hazards. Scale should probably be intermediate (1:50,000 to 1:250,000) so one can map both local and regional features and still see sufficient detail. Features indicating an earthquake risk include evidence for recent faulting and steep and unstable slopes. Recent faulting is suggested by fresh escarpments, offset of cultural features (orchards, fence lines, roads), and recent changes that have not yet reached equilibrium (offset coastline, dammed streams). Unstable slopes are characterized by recent landslides. Radar interferometry could be useful in monitoring the types of surface displacements (uplift or subsidence, horizontal creep) that preceed a fault rupture. Areas with volcanism may indicate areas that are seismically active.

3. How are floods monitored using remote sensing? What should you be aware of when planning a river crossing in the arctic?

    Repetitive coverage of flood and non-flood conditions using either airphotos or satellite imagery, optical or radar, can be used to monitor flood conditions. In addition to normal flooding, in the arctic the formation of ice dams during breakup of river ice can cause extreme flooding in different areas each year. Look for evidence of high water (piles of driftwood, gravel bars) to determine the extent of flooding during planning for engineering projects.

4. List some marine hazards that can be mapped using remote sensing. What instruments are best suited for each?

    Marine hazards that can be detected and monitored using remote sensing include reefs and shoals, sea ice, unstable submarine slopes, and gas vents. Reefs and shoals are perhaps best detected using true color or color-infrared imagery due to the penetration of water by blue and green wavelengths. Although radar can detect shoals when there is a current, due to the water surface rising over the shallow areas, it is probably not a reliable way to map such features. Sea ice can be detected and monitored using either visible-near infrared sensors or radar. Radar, in fact, may be the more versatile instrument since it can penetrate clouds and works just as well during the long periods of Arctic darkness. Both unstable slopes and gas vents are best mapped using sonar (look for steep slopes and evidence of slumps, and craters and/or mud volcanos), although rising gas bubbles usually bring colder bottom waters to the surface and can thus be detected using thermal imagery. Gas bubbles may also bring entrained oil to the surface, causing a sheen that can be detected using visible-near infrared instruments.

5. How would you map karst collapse features, and potential collapse features, in west Texas? In Qatar on the Persian Gulf?

   One would map these features essentially the same way in both areas. Using optical sensors one would look for evidence of solution collapse along fracture trends (linear depressions filled with moist soil surrounded by carbonate bedrock). One can also look for deranged drainage (intermittent streams, dry wadis) that flows for a distance and stops. Using thermal sensors one could look for areas that are cooler than surrounding bedrock, especially during the day, due to evaporative cooling of moist soils.

6. You are about to take over a mining operation in a foreign country. How would you use remote sensing to minimize your future legal liabilities with respect to cleanup and environmental damage?

   Obtain historical photos and imagery showing the site since the beginning of operations and look for evidence of dumps, buried drums, pits, old facilities, ponds, tailings, and anything that might contain toxic materials. Have any suspicious areas sampled or unearthed and tested. Take images of the site at the present time and put wording in the purchase agreement that your company is not responsible for (or has limited liability for) cleanup of pre-existing conditions. Use the imagery to document the pre-existing conditions.

INDEX

Absorption bands, 177
Access planning, 229
Accommodation zones, 142
Acid rain, 316
ADEOS, 67
Advanced earth observing satellite (ADEOS), 67
Advanced solid state array spectro-radiometer (ASAS), 34
Advanced spaceborne thermal emission and reflection radiometer (ASTER), 64
Advanced very high resolution radiometer (AVHRR), 8
Age of folding, 165
Aggregate, locating sources of, 278
Aggregates, locating sources of, 229
Air pollution, 313
Airborne imaging scanner, (AIS), 35
Airborne radar, 40
Airborne scanners
    disadvantages of, 40
    hyperspectral, 35
    improvements in, 339
    multispectral, 33
        advantages of, 34, 40
    ranges of, 34
    spatial resolution of, 34
    thermal, 35
Airborne Visible/Infrared Imaging Spectrometer, *see* AVIRIS, 35
Airphotos, 8, 27
    fault patterns mapped by, 148
    forest canopy changes detected by, 197
    mosaics, 28
Algal blooms, 306, 312
Algal mounds, 204-205
Alignments, 142
Allanite, 85
Alluvial cover, topographic highs in, 105

Alluvial fans, 106, 169
    recognition of faults by offset of, 148
Alluvial plains, stream patterns in, 115
Alluvial ponding, 116
Alluvial terraces, 98
Alluvial terrain, fracture patterns and, 112
ALMAZ-1, 6, 59
Alteration, 8, 29, 85, 214
    hydrothermal, 210
Altimeters, laser, 70
Alunite, gold mining and, 210
Angular unconformities, 163, 167
Anhydrite, 84
Annular drainage, 119
Anomalies
    circular, 233
    color, 106
    linear tonal, 201
    thermal, 328
    tonal, 106, 108, 316
    topographic, diagenetic soil chemistry changes and, 108
    vegetation, 110, 201
Anticlinal fields, 108-109
Anticlines, 98
    buried or blind, 104
    definition of, 98
    lateral ramp, 157
    truncation, 162
Apatite, 216
Applications
    hydrocarbons, 191-193
    minerals, 193-197
Aragonite, 178
Arctic ice movements
    mapping of, 229
    patterns of, 284
Arcuate faults, 144
Arcuate fractures, 130

Arcuate joints, 130
Arcuate strike valleys, drainage and, 119
Argillic alteration, 87
Argillites, 301
Argillization, 85, 91
ASTER, 7, 35, 64
Asymmetric folds, 126-127
Atmospheric backscatter LIDAR (ABL), 316
Atmospheric haze, 24, 316
Atmospheric scattering, 24, 29, 45
Austin Chalk, oil recovery from, 233
AVHRR, 5, 8, 67
AVIRIS, 223
    characterization of mineral assemblages by, 216
    limitations of, 213
    locating tailings and dumps with, 301

Backlight, 20-21
Backscatter image, 187
Badlands, 78, 83-84, 284
Bald know, 201
Banded iron formations, 193
Bandwidth, 34
Barbed tributary or drainage, 138, 142, 146
Barrier bars, 268
Basalts, 80
Base maps, 23
    using appropriate, 272
Base metal deposits, 194
Base-to-height ratio, 43
Basin analysis, 76
    choosing imagery for, 4
Bathymetric images, 286
Bathymetric maps, difficulties in generating, 285
Bathymetry, hyperspectral imagery use to map, 333
Bauxite, 197
Bedrock-alluvium interface detection, 297
Benzene, detection of, 316
Beryl, 85

Bighorn basin, Wyoming, 246
Biotite, 86
Black body, 38
Black Warrior basin, coalbed gas resources in, 244
Black and white photography, 29 (*see also* photography)
    limitations of, 29
Bleaching, 29, 86, 89, 91, 109-110
Blind thrusts, 163
Bornite, 195
Braided stream segments, 123
Bravo Dome, as an example of paleotopography, 231
Breccias, 216
    mineralization in, 166
    silicified chert, 195
Buried folds, 104, 107, 112
Buried waste location, ground-penetrating radar use for, 296
Buttes, 93

C-band radar, 40
Calcite, 178, 216
Calderas, recognizing images of, 80
Cambrian salt, 85
Cameras, 27
Camera height, 28
Camouflage detection film, *see* color-infrared film
Cane Valley diatreme, 216
Carbon dioxide, detection of, 316
Carbon monoxide, detection of, 316
Carbonate cements, 109
Carbonate platforms, subsidence in, 328
Carbonate rocks, bauxite formation and, 197
Carbonates, 83, 186
    intruded, 86
    rhombic jointing in, 131
    stable, 196
Carbonatite, 77
Carbonitization, 85

Cassiterite, 196
Caverns, 280
Cenozoic age, 196
Chalcocite, 195
Chalcopyrite, 195
Change detection, 296
Channels, 34
Channel incision, 116, 121
Channel width-depth adjustment, 116
Chemical precipitates, 193-194
China, porphyry copper deposits in northeastern region of, 214
Chlorite, 87, 194, 219
Chromite, 196, 216
Chronostratigraphic framework for mapping, 184
Circular features, 214, 233
    natural, 127
Circular karst sinkholes, 128
Classification, 295-296
    maps, 213
    supervised, 294
    unsupervised, 294
Clay minerals, 89
Clay-rich soils, 282
Clays, expanding, 283
Cleat, 244, 248
    frequency, 246
Clinoform bedding, 169
Cloud cover, 9
Coal, 193
    cleat in, 244
    recognition and mapping of, 194
Coal cleat orientation, prediction of, 246
Coal fires, monitoring, 300
Coal mines, abandoned, subsidence over, 328
Coalbed methane, 234, 244
Coastal erosion, 269
    mapping changes due to, 296
    monitoring, 229
    sediment plumes due to, 306
Coastal zone color scanner (CSCS), 5, 285

Color, influence of, 24
Color anomalies, 106
Color vs. black/white imagery, 9
Color photography, 29 (*see also* photography)
Color-infrared film (CIR), 24, 29
Color-infrared photography, 31
Comb Ridge, Utah, 216
Compact airborne spectrographic imager (CASI), 5, 285
Compaction, 118
Complex dielectric constant, 40, 187
Compressed meanders, 116, 122
Compressive stresses, oil field formation due to, 208
Concentric cooling joints, 130
Concentric drainage, 119, 123
Conglomerates, 82
Conjugate shears or fractures, 153-154
Contact metamorphic halos, 79
Cooling joints, 130
Cooling ponds, refinery, 312
Copper, 195
Copper deposits, porphyry, 87
Core
    drilling program, planning of, 233
    evaluation of, 184
    oriented, 246
Correlation, 174
Cost, 4, 8, 24
Cottonwood Creek, 246
Coverage, 8
Crabbing, 28
Craters, 333
Creep, 83
Cretaceous Niobrara Formation, 112, 206
Cretaceous-Tertiary Dawson Formation, overburden of, 328
Crosscutting relationships, 169
Cuestas, 95, 97, 124
Cumulative production, 237, 249, 251
Cutbanks, 268

Dams, 286
Daxinganling basin, porphyry copper deposit mapping in, 214
DD-5, 7
Dead ice moraines, 112, 114, 118
Decollement, 84-85
Decorrelation stretch, 177
Deltas, 312
Dendritic network, 93
Denver Basin
    fractured reservoirs in, 206
    oil recovery from, 233
Depositional patterns, 230
Deranged, 83
Depression angle, 42
Detachment depth, 160, 163, 165, 167
Devonian limestone, gold mineralization in, 195
Diagenetic alteration, 108-109
Diamonds, fluvial placer deposits containing, 196
Diapirisim, 102
Diapirs, 98
    salt, 84, 105, 128
Diatremes, 128
    Cane Valley, 216
    exploration for, 216
    Moses Rock, 216
    Mule Ear, 216
Dickite
    ground spectra of, 212-213
    as an indicator of gold mineralization, 210
Dielectric constant, *see* Complex dielectric constant
Differential absorption LIDAR (DIAL), 316, 318
Differential compaction, 105, 112
Diffraction grating, 23
Digital airborne imaging spectrometer (DAIS), 251
Digital chart of the world (DCW), 222

Digital elevation model (DEM), 21, 43-44, 187, 190, 223, 264, 267
Digital terrain model, *see* Digital elevation model
Digital satellite imagery
    advantages of, 47
    disadvantages of, 48
Dikes, 130
    basaltic, 134, 141
    gabbro, 287
    ring, 80, 214, 221
        Virginia Dale, 216
Diopside, 216
Dip, 21
    direction of, 95
    estimating, 95
    geomorphic, 96
    recognizing, 93
    slope, 95-97, 101
    zones of rapid change, 112
Dip magnitude, determination of, 22
Dips, opposing, 103
Dip-slip faults, 136
Distortion, 22
    radial, 28, 273
    spectral, 28
Dolerite, 287
Dolomites, 84, 86, 178, 193, 195, 204, 216, 301
Domes, 98
    drainage and, 119
    granite exfoliation, 128
    salt, 110
Double drainage deflection, 123
Downlapping, Clinoform bedding as a clue to, 169
Down plunge projection, 159
Drainage
    consequent, 94
    dendritic, 94
    dip slopes and, 124
    distributary, 106

double deflection, 123
obsequent, 94
offset patterns of, 323
parallel drainage deflection, 124
radial, 232
radial and concentric (annular), 119, 232
texture, 116, 123
trellis pattern of, 95
tributary, 106
Drainage patterns, 25
folding revealed by, 115
Drainage systems, deposits in, 193
Drape, 231
folds, 105-106, 112, 115, 118, 128
Drape folds, 128
Drill core, hyperspectral techniques applied to, 184
Drilling direction, 241, 243
Due diligence, 293

Eagle Springs field, 200
Early Bird, 68
Earth tides, 129
Earthquakes, 323
Economic minerals, 193
Edge enhancement, 23
Effluent, 306
Electric Mountain #1, 246
Electromagnetic spectrum, xiv
Emissivity, 38
Emulsions, 306
Endlap, 21
End member minerals, 219, 253
En echelon folds, 149, 153
horsetail features of, 155
Engineering applications of remote sensing, 229
Engineering characteristics, 76
Entrapment, erosion and fault truncation as factors of, 200
Environmental audits, 293
Environmental baseline, establishing, 294

Environmental issues, regulations concerning, 335
Environmental monitoring, multispectral airborne video use in, 296
EO-1 Hyperion, 6, 7, 64
Eocene Sheep Pass Formation, 200
Eocene Wasatch Formation, 246
Epithermal deposits, 87, 269
EROS A1, 9, 64
EROS satellites, 69
Erosion, 93
entrapment caused by, 200
mapping, 267, 268
mapping progression of, 258
monitoring, 229
sediment plumes due to, 306
Erosional outliers, 156
Erosional remnant, 105-106
ERS satellites, 6, 57
ERTS, 49
Escarpment, 95
Estuaries, 312
Ethyl ether, detection of airborne, 316
Euphotic zone, 10
Evaporative cooling, 142, 259
Evaporites, 84, 102, 130, 259
Pennsylvanian Paradox, 130
Exaggeration
factor, 97-98
vertical, 28
Exfoliation domes, 128
Expanding clays, 283
Exploitation, oil field, 228
Exploitation applications of remote sensing, 228
Extending mineral deposits, 249
Extension faults, 169
fluid migration in, 166
Extrusive rocks, 80
Eye seperation, 28

Faceted spurs, 149

Facies changes, 96, 104, 230
    mapping details of, 4
Facilities sites, choosing, 229
False color film, *see* color-infrared film
Fatigue failure, 129
Fault bend folds, 162
Fault intersections, locating diatremes at, 221
Fault line scarp, 139
Fault truncation, entrapment caused by, 200
Faults, 128, 135 (*see also* lineaments)
    active, definition of, 287
    apparent strike-slip, 148
    arcuate, 144
    Bright Angel, 129
    criteria used to recognize, 323
    dip-slip, 136
    extension, 169
    fluid migration in, 166
    groundwater and, 142
    growth, 136, 143-144, 148
    horsetail, 155
    hydrothermal fluid migration in, 214
    listric, 144, 169
    listric normal, 136, 138, 147, 157
    mapping for evaluation of fracturing, 234
    mineralization and, 144, 146
    normal, 136, 138, 142
    northeast, mineral occurrence in, 214
    northwest, ore bodies in, 214
    photointerpretation of, 203, 207
    planar, 144
    relaxation, 157
    reverse, 136, 138, 145, 163
    ring, 77, 214
    San Andreas, 323
    scarp, 139
    shear, 155
    strike-slip, 136, 140, 148, 167
    surface evidence for, 233
    tear, 136, 141, 157
    thrust, 136, 155, 169
    wrench, 136, 140, 148

    en echelon folds and, 153
Feedlot effluent, 312
Felsic intrusive, 77
Ferromagnesian rocks, iron laterite formation and, 197
Field mapping, 169
Film characteristics, 29
Film pigments, 31, 33
Filters, 20, 23, 29
Fisheries, 264, 269
Flexures, 129
Flight lines, 28, 43
Flower structures, 136, 141
Fluted till plains, 131
Flying levees, 125
Flood control, 229, 264-265
Flood risk, 266
Flooding, mapping changes due to, 296
Floods, 324
Fluorescence LIDAR (FL), 316, 318
Fluorite, 85
Fluvial placer deposits, 196
Fluvial processes, 98
Focal length, 28, 95, 97
Folds, 98
    asymmetrical, 126
    buried, 104
    color anomalies indicative of, 106
    detachment, 156
    en echelon, 149, 150-151, 153
    fault bend, 156
    fault propagation, 156
    forced, 98, 103
    forms of, 104
    fracture patterns indicating, 112
    fractures and, 129
    halo, 112
    joints as indicators of, 130
    leading-edge, 162
    mineralization related to, 167
    reactivation and, 105
    surface, 103

symmetrical, 126
  topographic expression of, 105
  vegetation anomalies indicating, 110
Foliation, 79, 131
Forced folds, 98, 103
Foreshortening, 20, 25, 43
Fourier transform infrared spectroscopy (FTIR), 316
Fracture history, determination of, 206
Fracture intensity or density maps, 240, 241-242, 244, 249-250
Fracture maps, 235
  orientation data obtained from, 242
Fracture patterns, 112
Fracture zones, 129, 193
Fractured reservoir plays, productive fairway definition for, 209
Fractured reservoirs
  exploitation of, 233
  mapping location and extent of, 228
  oil exploration in, 206
Fractures, 128
  arcuate, 130
  coal, 244
  conjugate, 155
  contour maps of intensity of, 132
  density mapping of, 132
  evaluating, 234
  maps, 235
  mineralization related to, 167
  orientation, 135
  propagation mechanisms of, 112, 129
  silicified, 142
  surface, 244
  wrench-associated, 153
Fracturing, 25
  role of in secondary recovery program planning, 249
Frame grabber, 9
Fraunhofer Line Discriminator (FLD), 312
Frontier area analysis, 76
Frost heaves, 106

Galena, 195, 304
Gangue minerals, 195
Gas vents, 333
Gemstones, 196
Geographic information systems (GIS), 223, 275, 339
  categorizing habitats with, 296
  definition of, 273
  mine management using, 274
  monitoring landslide movement with, 322
  pipeline routing using, 274
Geologic mapping, 77
  thermal images and, 38
Geomorphic dip, 96
Geomorphology, use in power plant site selection, 287
GEOS, 67
Geostationary operational environmental satellite (GOES), 67
Geothermal resources
  locating, 229, 269
  mapping of, 258
Glacial terrain, buried folds in, 112, 135
Global positioning systems (GPS), 74, 223, 322, 339
GLORIA, 70
Glossy, 23
GMS-4, 67
GOES, *see* weather satellites
Goethite, 216
Gold, 195-196
  alteration, 210
  epithermal, 253-254
  fluvial placer deposits containing, 196
  native, 196
Goldfield, Nevada, alteration associated with gold at, 210
Gossans, 29, 74, 196
Graben topography, normal faults and, 141
Granodiorite porphyry, 214
Granzon, 197
Gravel deposits, 196, 278, 280-281

Ground control points, 21
Ground magnetic surveys, use in coal fire
　　detection, 301
Ground-penetrating radar
　　definition of and uses for, 296
　　detecting groundwater pollution with,
　　　312
　　limitations of, 298
Ground truth, 169
Groundwater, 38, 49, 142, 229, 258
　　locating, 278
　　pollution of, 308
　　ponding, 142
Groundwater contamination, law suits
　　concerning, 335
Growth faults, 136, 144
Gypsum, 84

Habitats, 4, 294
Halite, 84
Halo-type anomaly, 109
Hand-held orbital photography, 44
Hausmannite, 194
Hazard assessment, 281, 321
Hazardous waste site monitoring, 296
Hazards, marine, 331
Haze, atmospheric, 316
Haze suppression, 9
Hazy tonal anomalies, 106
Headlands, 268
Headless valleys, 149
Heat capacity mapping mission (HCMM), 52
Hematite, 194, 219
High resolution imaging spectrometer
　　(HIRIS), 68
Hingelines, 203
Historical photos, 294
Hogbacks, 95
　　relative height of, 101
Horizontal well drilling, estimating optimum
　　direction for, 238, 240
Horsetail faults, 155

Horst topology, normal faults and, 141
Horsts, 98, 102, 103
Huancavelica mining district, mapping of,
　　251
Hums, 83
Hydration, 85
Hydrocarbon exploration, 76
Hydrocarbon vapors, detection of, 316
Hydrocarbons
　　accumulations in groundwater, 308
　　generation of, 167
　　leaking, 108
　　locating onshore, 191
　　mapping location of, 233
　　oxidation of, 109
　　plants as indicators of, 111
　　seeping, 109, 111
　　vertical migration of, 155
Hydrologic applications of remote sensing,
　　229
Hydrology, 275
Hydrothermal alteration, 210, 252
Hydrothermal alteration mineral mapping, 176
Hydrothermal fluids, migration in faults, 85,
　　214
Hydrothermal sources, localization of, 144
HYMAP, 223
Hyperion, *see* EO-1 Hyperion
Hyperspectral data sets, 181
Hyperspectral digital imagery collection
　　experiment (HYDICE), 35
Hyperspectral imagery, 8, 179, 181, 213
　　application of, 184
　　diagnostic mineral recognition by, 197
　　image cube, 182-183
　　important minerals in, 85-86
　　mineral deposits and, 193-195
　　techniques of, 181
Hyperspectral mineral classification, effec-
　　tiveness of, 183
Hyperspectral remote sensing, 179
Hyperspectral satellites, 339

Hyperspectral scanners, 35, 35
    monitoring water pollution with, 307-308
Hyperspectral sensors, mapping of minerals associated with kimberlites, 221
Hypothermal fluids, 85

Ice floes, mapping movement of, 9, 284
Ice movement, patterns of, 284
Ice polygons, 283
Igneous rocks, 76, 78
    basalts, 78-80
    calderas, 78
    cinder cones, 78
    dikes, 76
    extrusive, 76, 80
    flows, 76
    intrusive, 76-78
    laccoliths, 78
    tuffs, 78
    volcanoes, 78
IKONOS-1, 7, 64
Illite, 216
Ilmenite, 196
Illumination, 26
Image interpretation, xiii, 20, 23
Image processing techniques, locating onshore hydrocarbons and, 191
Imagery
    airborne, 8
    cloud cover and, 9
    color vs. black/white, 9
    cost considerations of, 8
    coverage considerations of, 8
    digital satellite, advantages of, 47
        disadvantages of, 48
    frontier analysis and, 76
    geometrically rectified, 22
    haze suppression, 9
    heat capacity mapping mission (HCMM), 52
    hyperspectral, 179
        for endmember spectra, 86
    hyperspectral/multispectral, 195
    interpretation of, 20, 23
    Landsat, Paradox Basin study using, 204
    night surveys, 9
    radar, 9
        advantages of, 41
        disadvantages of, 43
    relief mapping, 10
    resolution of, 8
    satellite, 8
    seasonal considerations of, 10
    sonar, 69
    stereo coverage, 9
    thermal, 9
    timing considerations of, 8
    vegetation mapping, 10
    water-covered area mapping, 10
Imaging spectrometry, 179
Imbricate fans, 156
Imbricate thrusts, 163
Impact craters, 128
Incidence angle image, 187, 190
Index maps, 28
Indian remote sensing satellites (IRS), 6, 60
Indicator plants, 111
Infill well locations, choosing, 228, 230, 233
Infiltration, 82
Infrastructure mapping, 276
Initial production (IP), 133, 233
Instantaneous field of view (IFOV), 34
Instrument parameters, radar, 35
Intensity-hue-saturation (IHS) transformation, 50, 176
Interferogram, 43
Interferometric sonar, 70
Interpretation of images
    color influence, 24, 31
    scale influence, 23
    techniques of, 24
Intrusive rocks, 78
Inverted topography, 98
Iron, oolitic, 194

Iron formations, recognition of, 194
Isopach maps, oil field mapping and, 204
Isostatic rebound, 105, 114, 119
Isotherm maps, use in coal fire detection, 301

Jasperoid, 89, 195
JERS, 6, 57, 59
Joint orientations, 117
Joint zones, mapping for evaluation of
    fracturing, 234
Joints, 128, 130, 135-137
    stratigraphy of, 132

K-band radar, 40
Kakrapar atomic power plant, site selection
    of, 287
Kaolinite, 210, 216
    ground spectra of, 212
Karst, 83, 128, 282
    collapse breccias, 128 (*see also* circular karst sinkholes)
    collapse features, 328
    sinkholes, circular, 128
KATE-2000, 7-8
Kelsey Lake mine area, 219
Kettles, 112
KFA-1000, 7-8
Kimberlites, exploration for, 77, 216
Kinetic temperature, 38
Klippen, 155-156
Kobe, Japan, analysis of ground surfaces
    displacements near, 324
Kosmos spacecraft, 46
KVR-1000, 7

L-band radar, 40
Laccoliths, 80
Lag gravel, 80, 98
Lake Maricaibo, fractures around, 234
Lakes, pollution in, 306
Lamination, 23
Landsat, fault patterns mapped by, 148

Landsat digital satellite imagery, 49
    Paradox Basin study using, 204
Landsat enhanced thematic mapper (ETM),
    49
Landsat multispectral scanner (MSS), 4-5, 49
Fractured oil production fairways defined by,
    206
    interpretation of images from, 23
    spectral reflectance and, 175
Landsat return beam vidicon (RBV), 5, 49
Landsat thematic mapper (TM), 5, 49, 214
    color, 4, 8
Landsat TM imagery, 223
    detecting fresh water sources with, 262
    intensity-hue-saturation transformation in
        images from, 177
    mapping porphyry copper deposits using,
        214
Landslides, 83, 280, 282, 321
    mapping changes due to, 296
Laramide compression, 208
Laramide deformation and movements, 232
Large format camera (LFC), 46
Laser altimeters, 70
Laser fluorosensors, monitoring water
    pollution with, 306
Lateral ramp anticlines, 157
Laterites, 193, 196-197
Layover, 43, 54
Leading edge anticlines, 155-156
Lead-zinc deposits, Mississippi Valley type,
    88, 195, 253-254
Lead-zinc-silver mineralization, 304
Leaks,
    gas, 111
    monitoring, 299
Lease evaluation, 77
Least-cost analysis, 275
Ledges, 210
Levee preservation, 124
Light detection and ranging (LIDAR), 316
    advantages and limitations of, 318

atmospheric backscatter (ABL), 316
differential absorption (DIAL), 316, 318
fluorescence (FL), 316, 318
Light tables, 20
Limestones, 86, 195
fractured reservoirs containing, 206
gold mineralization in, 195
Limonite, oolitic, 194
Lineaments, 23, 155, 193, 205, 256, 259 (*see also* faults)
definition of, 129
northeast, oil producing well occurrence and, 209
oil field mapping and, 204, 207-208
surface, oil production and, 209
Linear escarpments, 323
Linear tonal anomalies, 201
Lineations, 79
LISS, *see* Indian remote sensing
Listric faults, 144, 169
Listric normal faults, 136, 157
Lithologic changes
mapping details of, 4
photography of, 29
Lithologic distribution maps, 182
Lithologic mapping, 116, 184
Lithologic offsets, 135
Lithologic units
erosional texture associated with, 174
lateral offset of, 148
mapping of, large regions, 186
Lithologies
changing across an area, 244
identification of, 78
Look direction, 40, 43
Low sun-angle photos (LSAP), 10, 28

Mafic intrusives, 77
Magnetite, 194, 196, 219
Magnifying glasses, 20
Manganese deposits, 70, 194
Manto deposits, 144, 195

Map projection, 22
Map symbols, photogeologic, 25
Mapping
change detection, 296
chronostratigraphic framework for, 184
contaminants, 180
detecting groundwater by, 259
erosion, 258
facies changes, 4
field, 169
fracture density, 132
geologic, 38, 77
geothermal resources, 258
habitat, 294
hydrothermal alteration minerals, 176
hyperspectral imagery, 195
lithologic, 116, 184
lithologic changes, 4
lithologic units, large regions, 186
mineral alteration, 254
mineral deposit alteration, 8
oil spills, 9
radar interferometry and, 43
reconnaissance, 272
seeps, 191
locating onshore hydrocarbons, 191
offshore thermal image detection of, 191
source rock, 191
stratigraphic, arctic example of, 186
surface, 135, 236
surface disturbance, 296
trails, 8
using reflectance curves, 174
vegetation, 10
vegetation patterns, 4
water-depth, 285
wildlife habitats, 4
Maps
bathymetric, 285
fracture intensity, 244
isotherm, use in coal fire detection,

301
  lithologic distribution, 182
  photogeologic, 174
Marine hazards, 331
Marine placer deposits, 196
Marine spills, 312
Marine waters, 312
Marls, 132
MASTER, 35, 37
Matte, 23
Mesas, 93
Mesothermal deposits, 87
Metal oxide-rich soils, 197 (*see also* laterites)
Metamorphic basement, characterization of, 130-131
Metamorphic rocks, 79, 81
Metasomatization, 86
Meteor impact craters, 128
Methodology, 205
Microcline, 86
Microseisms, 129
Microwaves, 9, 40
Microseepage, 108-109
Migration (fluid), 235
Mine reclamation, 301
Mine safety, 255
  Solving problems of, 228
Mine subsidence, 255
Mineral alteration, 109
Mineral alteration mapping, 254
Mineral City, Florida, titaniferous sands found at, 196
Mineral deposits
  extending known, 249
  mapping alteration of, 8
Mineral exploration, 76, 210, 221
  programs, 180
Mineral mapping, 174
Mineral mixing, 40
Mineral spectra, 175
Mineralization, 74, 85, 214
  lead-zinc-silver, 304
  mapping location of, 233
Mineralized zones, extending from known deposits, 228
Mineralizing solutions, precipitation of in faults, 214
Minerals, economic, 193
Mining,
  abandoned, 282
  slope stability monitoring, 275
Minus-blue filter, 29
MISR, 64
Mission Canyon, oil recovery from, 233
Mississippi valley-type deposits, 88, 195, 253-254
Mixing zones, 306
MK-4, 8
Moderate-resolution imaging spectroradiometer (MODIS), 35, 68
Modular optico-electronic multispectral scanner (MOMS-1), 63
Monazite, 85
Mongolia, mineral exploration in, 221
Monoclines, 157
Monticellite, 216
Montmorillonite soils, 283
MOS-1, 67
Mosaics, 28
Moses Rock diatreme, 216
MSU-SK, 59
Mule Ear diatreme, 216
Multispectral color radar, 43
Multispectral data, 33
  photogeologic interpretation of, 184
  spectral interpretation of, 184
Multispectral imagery
  forest canopy changes detected by, 197
  mineral deposits and, 195
Multispectral infrared and visible imaging spectrometer (MIVIS), 40
Multispectral scanners (*see also* Landsat)
  advantages of, 34, 40
  airborne, 33

thermal infrared, 40
Multitemporal images, 266, 282
Multivariate analysis, 294
Muscovite, degraded, 305

Namibia, diamond deposits of, 196
Natural gas, 244
Nested circular structures, 214
New Caledonia, nickel laterites found in, 197
Night surveys, 9
Night-time imagery, 262
Niobrara, oil recovery from, 233
Nitrous oxide emissions, detection of, 316
No data zones, 43, 69
Nome, Alaska, gold deposits at, 196
Nondispersive infrared spectrometry (NDIR), 316
Normal faults, 136
NS-001, 34

Oblique photos, 28
Offset
    indicators of, 235
    mineral deposits, 233
    reservoirs, 233
    streams, 142
Offshore drilling platforms, locating, 229
Offshore gas seeps, thermal image detection of, 191
Oil field exploitation, remote sensing and, 228
Oil production, fracture-controlled, 206
Oil recovery programs, enhanced, 228
Oil slicks, 191
    thermal scanner detection, 307
Oil spills, mapping of, 9
Oligocene Garrett Ranch volcanics, 200
Olivine, 216
Onlapping, Clinoform bedding as a clue to, 169
Oolitic iron deposits, 194
Open pit mines, slope stability in, 255-256
Opposing dips, 103

Optical sensor (OPS), 57
Orbview satellites, 64, 67, 68
Ordovician submarine volcanics, circular structure development in, 214
Ore minerals, 195
Organic matter, 177
Orthogonal joints or fractures, 77, 130, 148
Orthophotos, 22, 273
Oscillations, free, 129
Overhangs, 145-146, 148
Overlap, 9, 51
Overlays, 22-23
Overpressure, 130
Oxides, stable, 197

Pakistan, Salt Ranges in, 84
Paleobathymetric hingelines, stratigraphic traps and, 203
Paleochannels, 193, 259
    identification of, 77
Paleography, 234
Paleoshorelines, photointerpretation of, 203
Paleotopography, 230-231
Palm tree structure, 136
Panchromatic photography, 29, 263 (*see also* black and white photography)
Panochromatic imagery, SPOT high resolution, 8
Paradox Basin, Utah, stratigraphic traps at, 203
Parallax, 9
Parallax bars, 22
Parallel drainage deflection, 124-125
Particulates, suspended, 306
Patterned ground, 283
Pediment, 98, 102, 201
Pegmatites, 85-86
Pennsylvanian algal mounds, 203
Pennsylvanian Sierra Grande uplift, 232
Permafrost, 105, 228, 283
Permian Phosphoria Formation, 246
Permitting, 277

Perovskite, 216, 219
Perspective center, 28
Peru, mapping mines in, 251
Petroleum basin evaluations, 180
Petroleum exploration
    frontier, 200
    mature basin, 203, 206
Petroleum traps, 142
Phlogopite, 85, 216, 219
Phosphate processing effluents, 313
Phosphates, 193
    recognition of formations of, 194
Photobase, 98
Photochemical smog, 316
Photo finish, 23
Photogeologic interpretation of multispectral data, 184
Photogeologic map symbols, 25
Photogeologic units, 76, 174
Photogeology, 78
    oil field discovery by, 200
Photogrammetry, xiii
Photography
    airphotos, 29
    black and white, 29
    color, 29
    color-infrared, 31
    satellite, 44, advantages of 46
    Soyuzkarta, 8, 46
Photointerpretation, tools for, 20
Photostratigraphy, 184, 188
Phyllic alteration, 87
Piceance basin, coalbed methane resources in, 244
PIMA-II portable field spectrometer, 212-213
Pipeline leaks, detection of, 297
Pipelines, 275, 286
    effect of marine hazards on planning, 331
    ground-penetrating radar use for, 296
Placer deposits
    fluvial, 196
    marine, 196

Placers, 193, 196
Planar faults, 144
Planimeters, 22
Planimetric maps, 223
    interpretation of, 23
    scale changing devices and, 21
Platinum, 196
Poisoning, heavy mineral, 91
Polarization, radar, 40, 43
Pollution, air, 313
Ponds, pollution in, 306
Porphyry copper deposits, 87
    structural mapping as a guide to, 214
Porosity-feet, 249-250
Potassic alteration, 87
Powder River basin, coalbed gas resources in, 244
Power plants, 286
Precambrian belt supergroup sediments, 301
Pre-dawn imagery, 38, 259, 270
Principal components (PC) transformation, 176-178
    decorrelation stretch variation, 177
Propagation mechanisms, 129
Propylitization, 85
Propyllitic alteration, 87
Prudhoe Bay, ambient air monitoring at, 316
Public relations, 29, 336
Pyrite, 194, 195
    formation of, 109
Pyritization, 89, 91
Pyrolusite, 194
Pyrrhotite, ferrimagnetic, formation of, 109

Quartz-alunite, gold mining and, 210
Quartzites, 301
Quick Bird, 68

Radar
    advantages of, 41
    airborne, 40
    detecting groundwater with, 259

        detecting groundwater pollution with, 309
        disadvantages of, 43
        ground-penetrating, definition of and uses for, 296
        interferometry, 43
        monitoring water pollution with, 306
        multipolarization, 43
        multispectral color, 43
        oil slick detection with, 313
        polarization, 40
        Seasat, 53
        shadows, 10, 43
        shuttle imaging, 55
        side looking airborne, 40
        synthetic aperture, 57
        synthetic aperture radar, 40
        wavelengths, 40
Radar imagery, 9
Radar interferometry, 43, 284, 339
        use in monitoring landslides, 322-323
        use in monitoring volcanos, 325
        using in monitoring subsidence, 328
Radar satellites
        ALMAZ-1, 59
        ASTER, 64
        EO-1 Hyperion, 64
        EROS A1, 64
        ERS, 57
        IKONOS-1, 64
        Indian remote sensing satellites (IRS), 60
        MISR, 64
        modular optico-electronic multi-spectral scanner (MOMS-1), 63
        optical sensor, 57
        RADARSAT, 57
        Resurs-01, 59
RADARSAT, 6, 57
RADARSAT 2, 69
Radial distortion, 28, 273
Radial drainage, 119, 123, 232
Radial faults, 77, 214
Radiant flux, 35, 38

Radiant temperature, 38
Railroad Valley, *see* Trap Springs
Ratioing, 210
        definition of, 175
        mapping of hydrothermal alteration minerals and, 176
Reactivation, 105, 115, 119, 129
Reclamation, 275
Reconnaissance, 276
Reconnaissance mapping, 272
Rectified imagery, 22
Reef reservoirs, influence of facies changes in, 192
Reference curves, 35, 40, 175
Refinery cooling ponds, 312
Reflectance curves, lithologies and, 184
Reflectors, radar (corner, diffuse, specular), 40
Regional dip, 126
Regional extension, 129
Regulation compliance, 336
Reidell (R) shears, 153-154
Relaxation faults, 157
Relief, imagery considerations of, 10, 22, 28
Remote sensing data, some sources of, 11-15
Remote sensing web sites, 16-18
Replacement mantos, 193, 195
Resolution
        spatial, 8, 9, 34, 40
        spectral, 35
Restraining bends, 153
Resurs-01, 6, 59
Return beam vidicon (RBV), 49
Reverse faults, 136
Rhodochrosite, 194
Rhombic jointing, 131
Ring dikes, 216
Ring faults, 77, 214
River floodplains, identification of, 324
Riverbank erosion, sediment plumes due to, 306

Rivers, pollution in, 306
Roadcuts, 256
Rocks
    extrusive, 80
    igneous, 78, 216
    intrusive, 78
    Mesozoic, 93
    metamorphic, 81, 216
    sedimentary, 81
Roll, 28
Ronchi filter, 23, 25
Roof falls, 28, 255-256
Rose diagrams, use to display fracture set orientation, 235
    Frequency-weighted, 235
    Length-weighted, 235
Roughness, 40
Rule of V's, 95, 99
Runoff, 83, 264
Ruth Mountain #1, 246
Rutile, 196

Sag ponds, 152-153
Salinity, 261, 263, 266
Salt domes, 110-111
Salt glaciers, 84, 88
San Andreas fault, mapping creep along the, 323
San Juan basin, coalbed methane resources in, 244
Sand bars, 281
Sand,
    deposits, 196, 280
    drifting, 114
        folds occurring beneath, 114
Sand lenses, detection of, 297
Sandstone, 219
    Permian Tubb, 231
    stratigraphic mapping of, 186
    trending fluvial reservoirs, 192
    Triassic Chugwater, 109
    uranium-vanadium deposits in, 194

Sandstone reservoirs, northwest elongated, 192
Sandstones, 81-82, 132
    red bed, 109
Sandy soils, 282
Satellite digital imagery
    advantages of, 47
    disadvantages of, 48
Satellite imagery, frontier area, 76
Satellite imaging systems, comparison of, 5-7
Satellite multispectral scanners, improvements in, 339
Satellite photography, 44
    advantages of, 46
Satellites
    Landsat, 9
    weather, 8, 64
Scale
    definition of, 27
    influence of, 23, 34
Scale changing devices, 22 (*see also* transferscopes)
Scale rules, 22
Scarp, 144
Scour features, 131, 135
Sea ice, 333
    monitoring of, 284
Sea surface, 264
Seasat, 5, 53
Seasonal coverage, 10
SEBASS, 40
Secondary recovery operations, evaluation of, 246
Sedimentary rocks, 81
    carbonates, 83
    conglomerates, 82
    evaporites, 84
    sandstones, 81-82
    shales, 82
Seeps, 40, 69, 111
    effects of, 191

natural, 313
Seeps mapping, 191
    locating onshore hydrocarbons, 191
    offshore, thermal image detection of, 191
Seismic activity, 287
Seismic lines, 277, 280
Seismic zones, 280
Seismicity, 129
Semigloss, 23
Sericite, 87
Sericitization, 85
Serpentine, 216, 219
Settling, 129
Sewage effluents, 312
Shadows, 25
    radar, 10, 43
Shales, 82, 86, 132
    bituminous calcereous, 195
    Chainman, weathering of, 191
    copper ore containing, 195
    Cretaceous Niobrara, 112
    dewatering of, 105
    distinction from coal in airphotos, 194
    interbedded calcereous, 206
    mapping of organic sources of, 180
    Mississippian Chainman, 191
    stratigraphic mapping of, 186
Shear faults, 155
Shear, pure, 155
Shears, 153-154, 277
Shoals, 332
Shorelines, locating, 145, 203, 229
Short wave infrared (SWIR), 35, 49
Short wavelength infrared full spectrum imager (SFSI), 35, 213
Shuttle, 46
    hand-held photos, 8
    imaging radar (SIR), 5, 55
Side looking airborne radar (SLAR), 40
Sidelap, 21, 49
Siderite, 194, 305

Side-scan sonar, 69
Sierra Nevada, paleo-placers in, 196
Significance plot, 239
Silica, 305
Silicates, stable, 197
Silicification, 85, 183
Sillicate, 86
Siltstones, 301
Silurian marine sediments, circular structure development in, 214
Sinkholes, 128, 130, 282
Site selection, 279, 287
Siting, 272
Skarns, 77, 79, 86
Skylab, 7, 44
    S190-A, 44, 46
    S190-B, 44, 46
Slicks, oil or petroleum, 38
Slope stability, 228, 255-256, 275, 287
Sludge, 312
Slumps, 83
Smog, photochemical, 316
Smoke plumes, 316
Snowpack, monitoring runoff from, 264, 267
Soil erosion, mapping of, 268
Soils
    age dating of, 323
    engineering properties of, 282
    moisture, 43
    terra rosa, 197
Sonar, side-scan, 69
Source rock mapping, 191
Soyuzkarta KFA-1000 photography, 8
Space shuttle, 46
Spectra,
    mineral, 173
    rock, 174
Spectral classes, definition of, 294
Spectral curves, libraries of, 178
Spectral distortion, 28
Spectral end members, 182

Spectral interpretation of multispectral data, 184
Spectral libraries, 219, 252, 340
Spectral reflectance, 174
  measurements of minerals grouped by anions, 175
Spectral stratigraphy, definition of, 174
Spectral unmixing, 304
  definition of, 253-254
Spectrometers, multispectral infrared and visible imaging, 40
Specularite, 194
Sphalerite, 195, 304
Sphene, 85
Spills
  law suits concerning, 336
  monitoring, 299
Spinel, 216
SPOT, 5
  high resolution panchromatic imagery, 8, 51
  multispectral (XS) imagery, color, 4, 51, 53
  spectral reflectance ranges for, 175
Springs, 49, 135, 201, 258
  subsea, 262
  thermal, 269
Stacked vertical sequences, 193
Stefan-Boltzman constant, 38
Stepout well locations, choosing, 228, 230, 233
Stereo coverage, 9, 44
Stereo pair, 28, 97
Stereometers, 22
Stereoscopes, 21
Stratiform base metal deposits, 195
Stratiform deposits, 193
Stratigraphic changes, mineralization due to, 166
Stratigraphic mapping, arctic example of, 186
Stratigraphic remote sensing, methodology for, 184
Stratigraphic traps, 230, 246
  mapping of, 191
  mineralogic changes in, 109
Stratigraphic units, 193
Stratigraphic thickness, calculating, 22
Stream segments, braided, 123
Streams, fault recognition and, 135
Stress history, 208
Striations, 80
Strike-slip faults, 136, 148, 167
Structural form lines, 103-104, 107
Structural high, 98, 120, 201
Structural position, 230
Structural traps, mineralogic changes in, 109
Submarine slumps, 331
Subsidence, 228, 255, 328
  factors controlling, 330
  imaging for site selection, 282
Subsidence modeling, 76
Subthrusts, 163, 165
Sulfotaric alteration, 78
Supergene enrichment, 88
Supervised classification, 294
Surface analyses, trend, 126
Surface displacement, 126
Surface disturbance, 296
  determining reasons for change in, 298
Surface folds, 103
Surface mapping, 135, 236
Surface materials, imaging for site selection, 281
Surface slicks, monitoring of, 306
Surface slope, imaging for site selection, 282
Surface waters
  flood prediction monitoring, 264
  locating, 278
  monitoring, 229, 258, 263
Synclinal structures, drainage and, 119
Synoptic view, 8, 46
Synthetic aperture radar (SAR), 40, 43, 57

System Probatoire d'Observation de la
  Terre (SPOT 1), *see* SPOT

Talc, 219
Tantalite, 85
Tear faults, 136, 157, 161
Technique interpretation, 24-25
Temperature, 35
Terra rosa soils, 83, 197
Terraces, 93, 98, 102-103
Tetrahedrite, 195, 304
Thematic Mapper, *see* Landsat
Thermal anomalies, 328
Thermal contraction, 129-130
Thermal discharges, 306
Thermal expansion of minerals, coefficients
  of, 129
Thermal imagery, 9, 35
  detecting groundwater with, 259, 263
  monitoring surface temperatures with,
    313
Thermal inertia, 259
Thermal infrared multispectral scanner
  (TIMS), 34, 40, 178-179
Thermal infrared scanners, 269
Thermal maturity, evaluation of, 191
Thermal scanners, airborne, 35
Thermal sensors, detecting groundwater
  pollution with, 309
Thermokarst, 283
Thickness changes, 230
Three-point problem, 95, 99-100
Thrust belts, choosing imagery for, 4
Thrust displacement, 128
Thrust faults, 29, 136, 155, 169
Thrust plate thickness, 163
Thrusts, blind, 163
Till, 114
Timing, 8, 167
  of fold development, 74
Tin, fluvial placer deposits containing, 196
TK-350, 7, 46

Toluene, detection of, 316
Tonal anomalies, 106, 316
  diagenetic soil chemistry changes and,
    108
  hazy, 106
  soil, 108
Tonal bands, 93, 95, 101
Topaz, 85
Topographic anomalies, diagenetic soil
  chemistry changes and, 108
Topographic maps, 43, 95, 228, 296
  generation of, 273
  interpretation of, 23
  scale changing devices and, 21
Topographic highs, 110, 210
Topographic offsets, 135
Topographic profiles, 124, 126
  generations of, 22
Topography, digital, 339
Trails, mapping of, 8
Training sites, 294
Transfer zones, 142, 146
Transferscopes, 20-22
Transparencies, 20
Transport, 272
Transport direction, 160
Transportation, 284
Trap, hydrocarbon, 142
  stratigraphic, 203, 230
  structural, 200
Trap Springs, Nevada, structural traps at,
  200
Trellis drainage pattern, 95-96
Trend surface analyses, 126
Triangular facets, 144, 147
Truncation anticlines, 162
Tunnels, 286
Turbidity, 31, 263, 265, 269

U-shaped valleys, 82, 282
Ultramafic minerals, 216
Ultraviolet scanners, monitoring water

pollution with, 306
Undeformed terrain, 93
Unmixing algorithms, 340
Uplift, 129
Upper Cretaceous Laramie Formation, coals of, 328
Uranium deposits, 109
Uranium-vanadium, 193
    recognition of roll fronts of, 194

Variscan granodiorite, 214
Vegetation alignments, 130, 134
Vegetation anomalies, 110, 201
Vegetation mapping, 10
Vegetation patterns, 40, 43
    mapping details of, 4
    photography of, 29
Vein-type mineral deposits, 77, 144
Vergence, 156-157, 159, 163, 165
Vertical exaggeration, 21, 28, 95
Vertical migration, 155
Vertical photos, 28
Victoria field, Australia, paleo-placers in, 196
Vignetting, 28
Virginia Dale ring dike, 216
Visible-near infrared (VNIR), 35, 61
Vitrinite reflectance, 191
Volcanic craters, 128
Volcanos, 324

Water, 278
    clarity of, 306
    fresh, 262-263
    locating sources of, 229, 258
    marine, 312
    penetration, 10, 28
    shallow, radiometric model for evaluating, 263
    surface, monitoring, 263
    surface roughness, 263
    turbidity of, 263-264, 278

Water depth
    changes in, 306
    mapping, 285
    mapping for port facilities, 229
Water gap, 95
Water pollution, 306
Water table
    contamination, 308
    detection, 297
Waterflood operations, evaluation of, 246
Weather satellites, 8, 64
Weathering, 40, 79, 83
Web sites, remote sensing, 16-18
Well drilling, predicting direction of, 228
Wetlands, 312
    habitat mapping of, 294
WiFS, *see* Indian Remote Sensing
Wildlife habitats, mapping details of, 4
Williams Fork Formation, 244
Wratten 12 filter, 29
Wratten 89B filter, 29
Wrench faults, 136, 148

X-Band radar, 40
Xylene, detection of, 316

Yaw, 28

Zero edge of alluviation, 106
Zircon, 85, 196

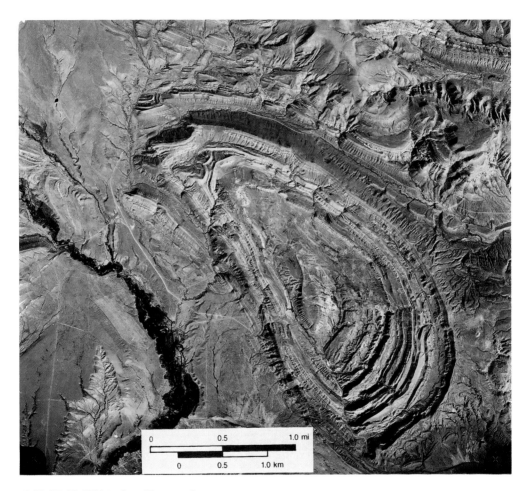

COLOR PLATE 1. See Chapter 3.
Figure 3.2, page 31.

COLOR PLATE 2. See Chapter 3.
Figure 3.3, page 32.

COLOR PLATE 3. See Chapter 3.
Figure 3.11, page 44.

COLOR PLATE 4. See Chapter 3.
Figure 3.12, page 45.

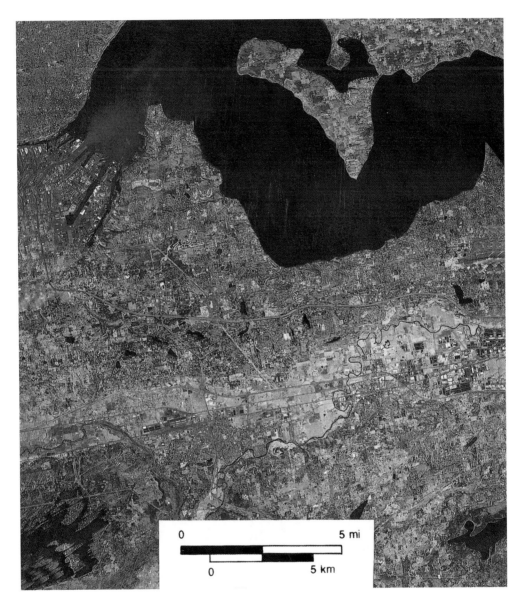

COLOR PLATE 5. See Chapter 3.
Figure 3.13, page 47.

COLOR PLATE 6. See Chapter 3.
Figure 3.15, page 50.

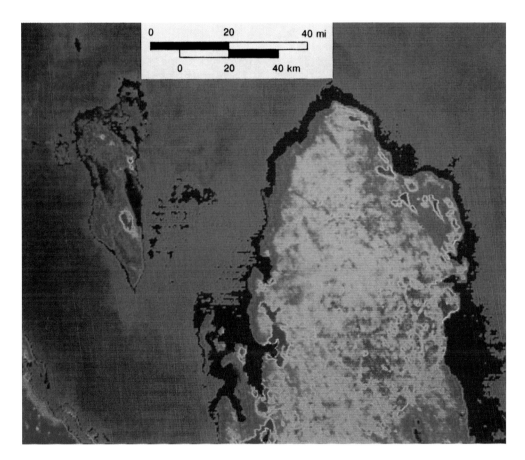

COLOR PLATE 7. See Chapter 3.
Figure 3.16, page 51.

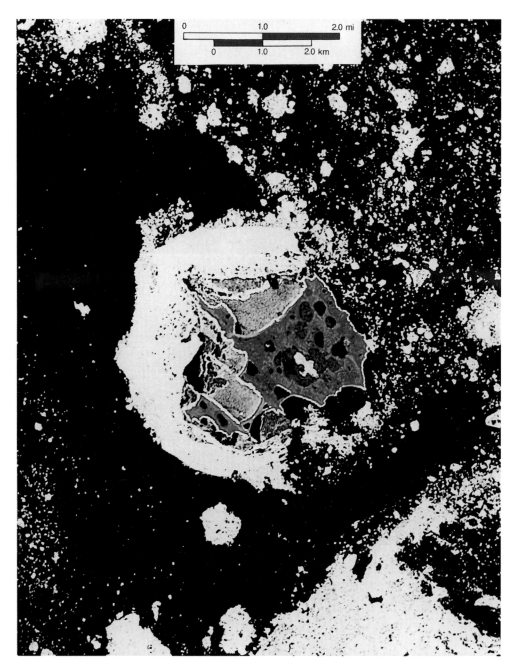

COLOR PLATE 8. See Chapter 3.
Figure 3.18, page 53.

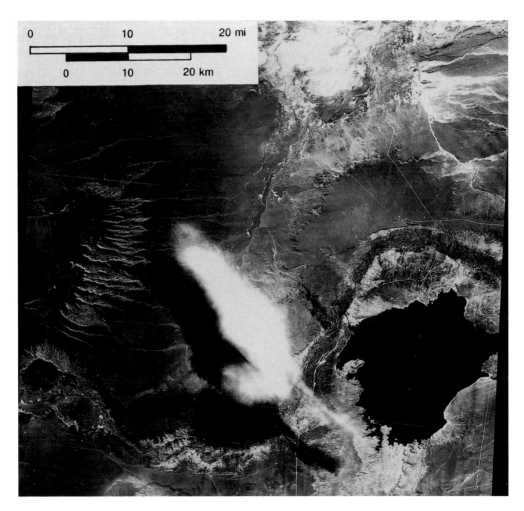

COLOR PLATE 9. See Chapter 3.
Figure 3.22, page 58.

COLOR PLATE 10. See Chapter 3.
Figure 3.26, page 62.

COLOR PLATE 11. See Chapter 3.
Figure 3.28, page 65.

COLOR PLATE 12. See Chapter 3.
Figure 3.29, page 66.

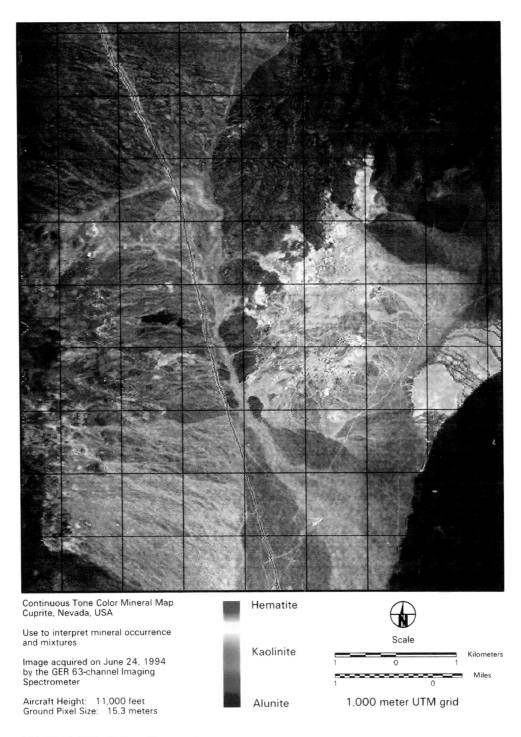

Continuous Tone Color Mineral Map
Cuprite, Nevada, USA

Use to interpret mineral occurrence and mixtures

Image acquired on June 24, 1994 by the GER 63-channel Imaging Spectrometer

Aircraft Height: 11,000 feet
Ground Pixel Size: 15.3 meters

Hematite

Kaolinite

Alunite

1,000 meter UTM grid

COLOR PLATE 13. See Chapter 4.
Figure 4.12, page 88.

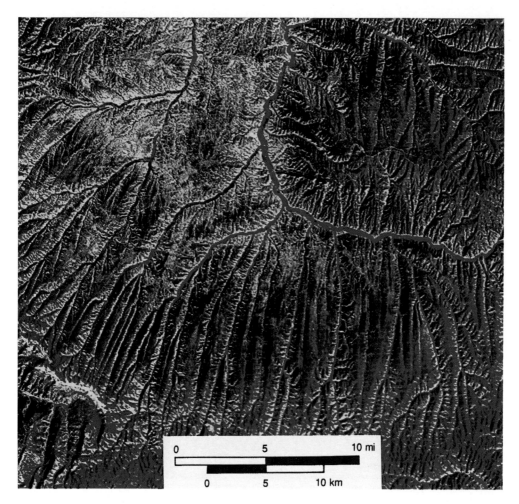

COLOR PLATE 14. See Chapter 5.
Figure 5.4, page 94.

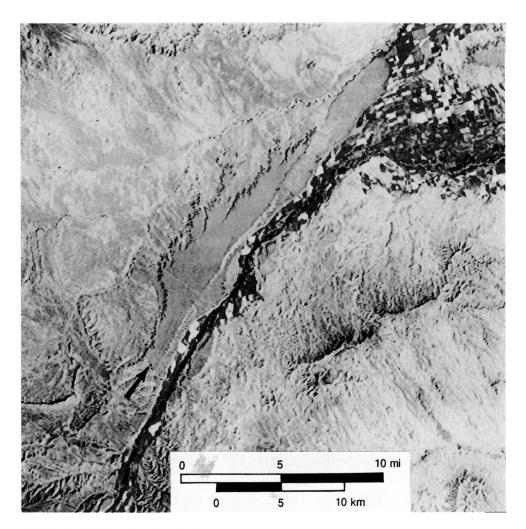

COLOR PLATE 15. See Chapter 5.
Figure 5.13, page 101.

COLOR PLATE 16. See Chapter 5.
Figure 5.20, page 108.

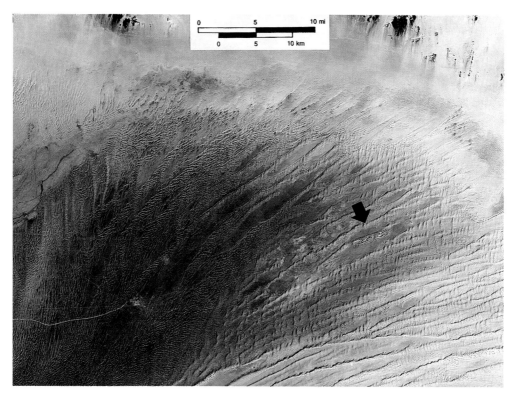

COLOR PLATE 17. See Chapter 5.
Figure 5.30, page 118.

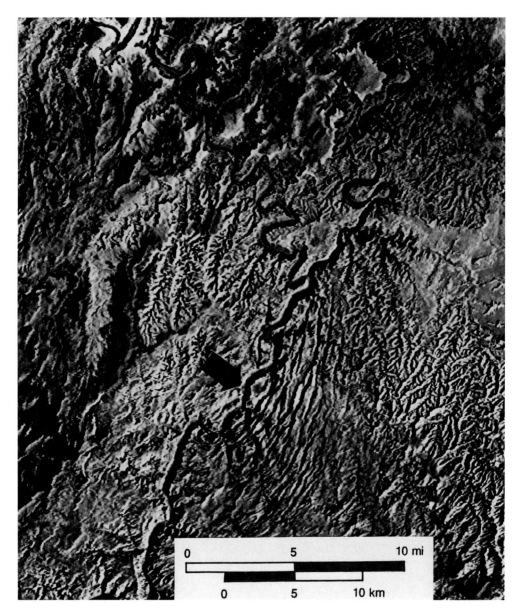

COLOR PLATE 18. See Chapter 5.
Figure 5.45, page 131.

COLOR PLATE 19. See Chapter 5.
Figure 5.47, page 132.

COLOR PLATE 20. See Chapter 5.
Figure 5.59, page 143.

COLOR PLATE 21. See Chapter 5.
Figure 5.65, page 147.

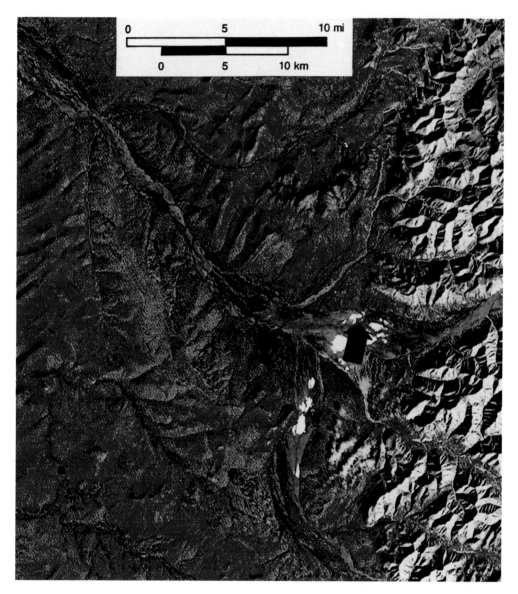

COLOR PLATE 22. See Chapter 5.
Figure 5.81, page 158.

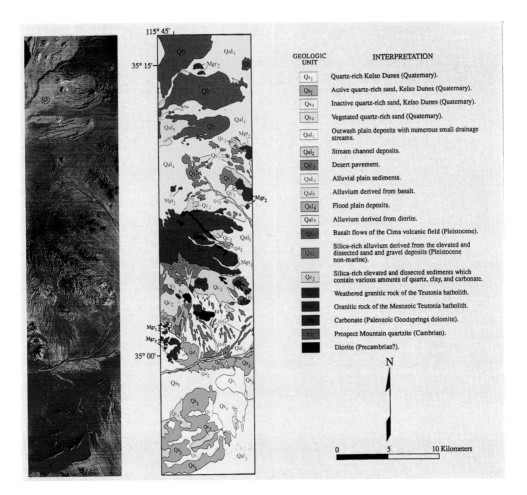

COLOR PLATE 23. See Chapter 6.
Figure 6.2, page 176.

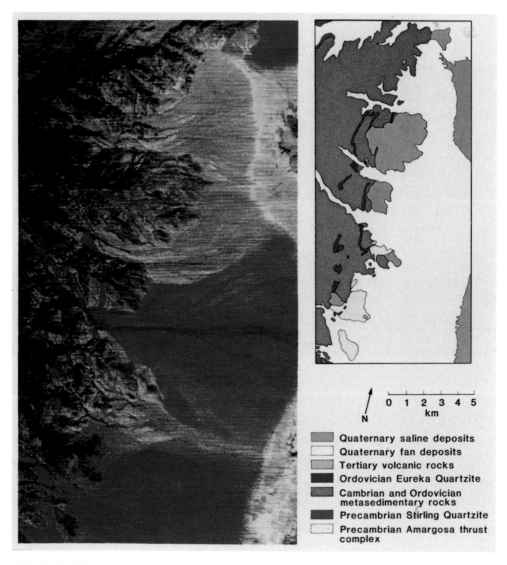

COLOR PLATE 24. See Chapter 6.
Figure 6.3, page 177.

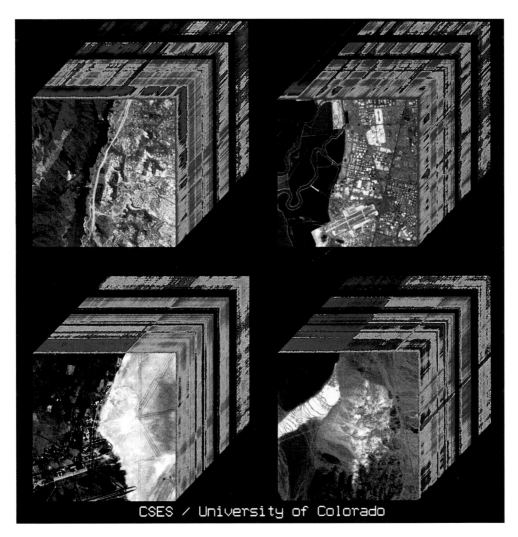

COLOR PLATE 25. See Chapter 6.
Figure 6.7, page 181.

COLOR PLATE 26. See Chapter 7.
Figure 7.14, page 211.

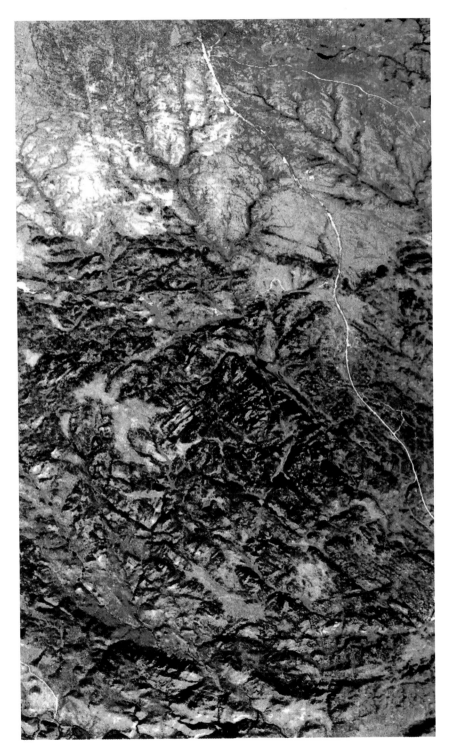

COLOR PLATE 27. See Chapter 7.
Figure 7.17, page 216.

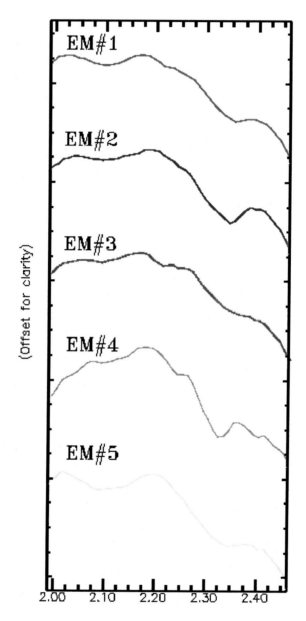

COLOR PLATE 28. See Chapter 7.
Figure 7.18, page 217.

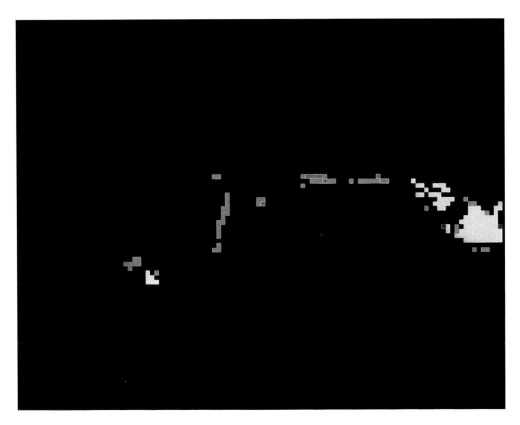

COLOR PLATE 29. See Chapter 7.
Figure 7.19, page 218.

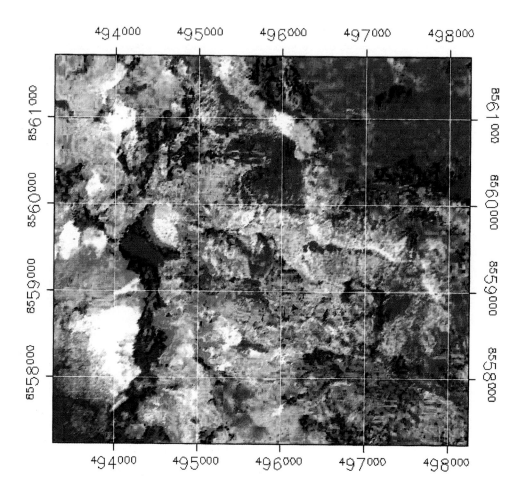

COLOR PLATE 30. See Chapter 8.
Figure 8.22, page 253.

COLOR PLATE 31. See Chapter 9.
Figure 9.3, page 259.

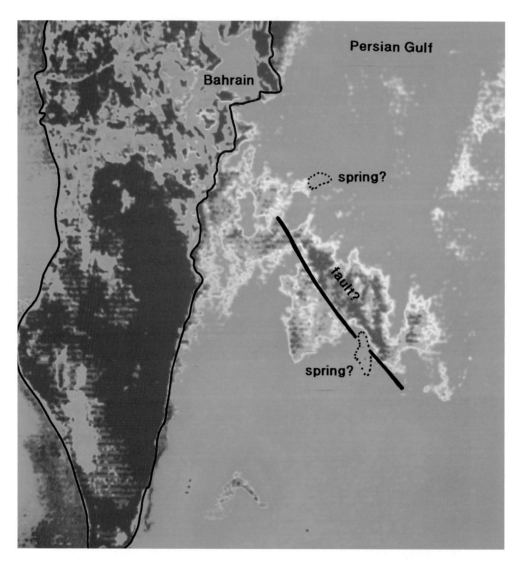

COLOR PLATE 32. See Chapter 9.
Figure 9.4, page 260.

COLOR PLATE 33. See Chapter 9.
Figure 9.6, page 263.

COLOR PLATE 34. See Chapter 9.
Figure 9.7, page 264.

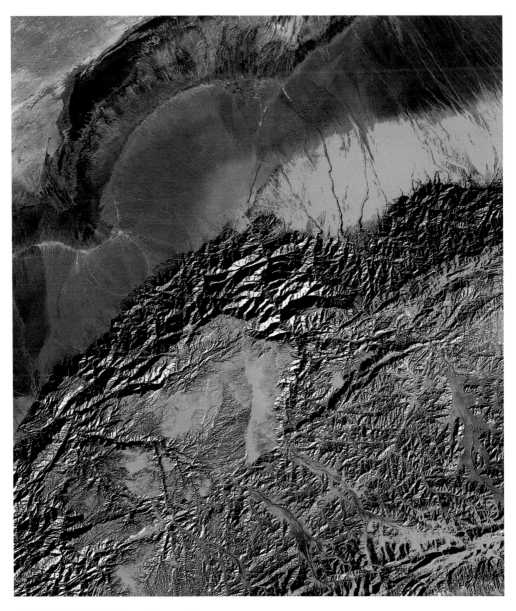

COLOR PLATE 35. See Chapter 9.
Figure 9.8, page 265.

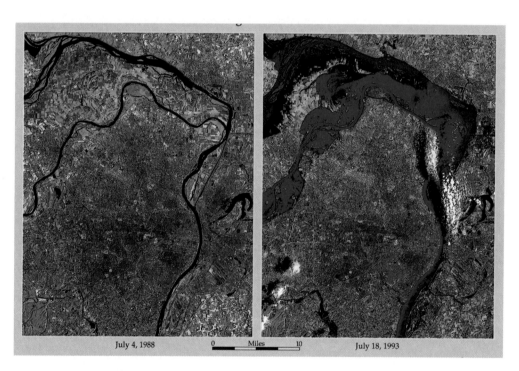

COLOR PLATE 36. See Chapter 9.
Figure 9.9, page 266.

COLOR PLATE 37. See Chapter 10.
Figure 10.1, page 272.

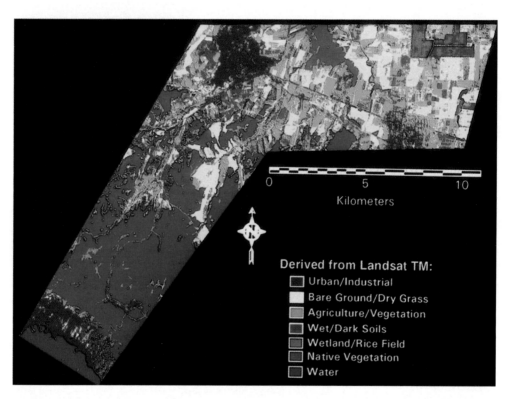

COLOR PLATE 38. See Chapter 10.
Figure 10.2, page 274.

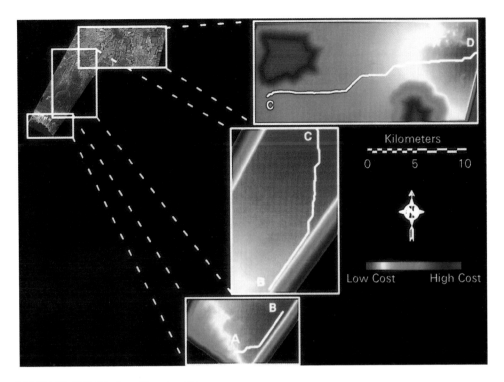

COLOR PLATE 39. See Chapter 10.
Figure 10.3, page 275.

North Cat Cay, Bahamas

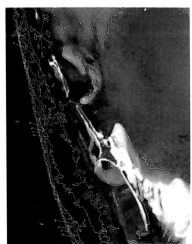

Contoured Depth Image
(Depths in Feet)

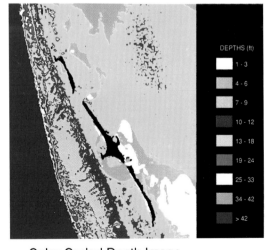

Color-Coded Depth Image

COLOR PLATE 40. See Chapter 10.
Figure 10.10, page 284.

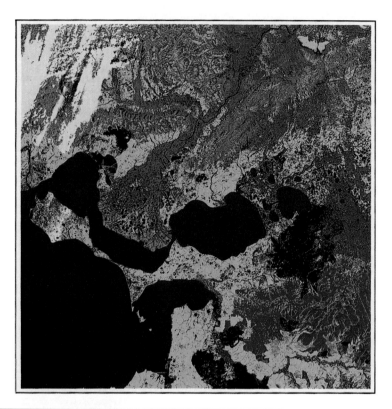
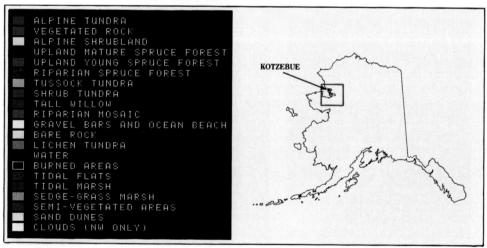

COLOR PLATE 41. See Chapter 11.
Figure 11.1, page 293.

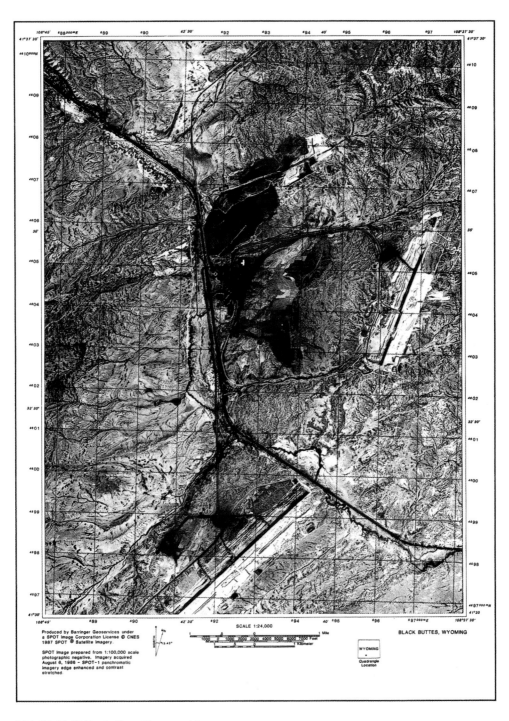

COLOR PLATE 42. See Chapter 11.
Figure 11.2, page 295.

COLOR PLATE 43. See Chapter 11.
Figure 11.4, page 297.

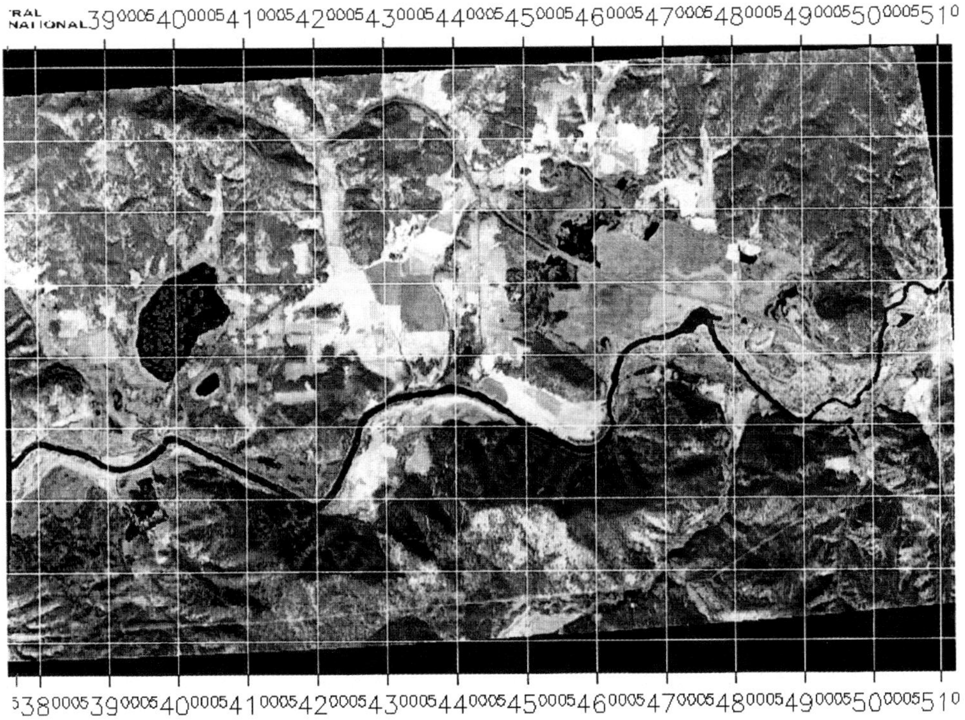

COLOR PLATE 44. See Chapter 11.
Figure 11.8, page 302.

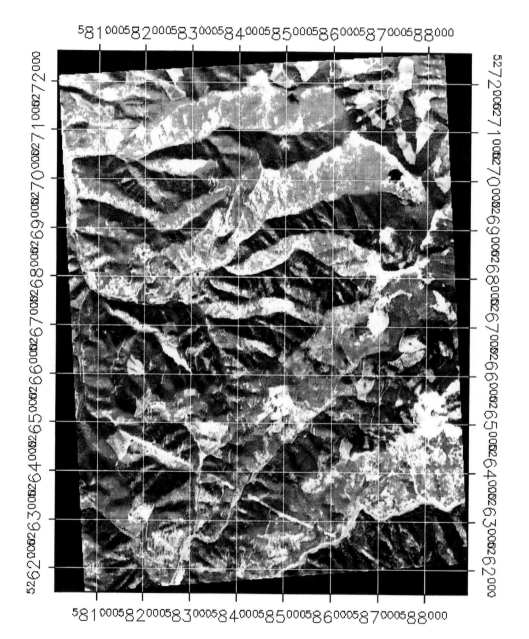

COLOR PLATE 45. See Chapter 11.
Figure 11.9, page 303.

COLOR PLATE 46. See Chapter 11.
Figure 11.10, page 304.

COLOR PLATE 47. See Chapter 11.
Figure 11.11, page 305.

COLOR PLATE 48. See Chapter 11.
Figure 11.12, page 306.

COLOR PLATE 49. See Chapter 11.
Figure 11.13, page 307.

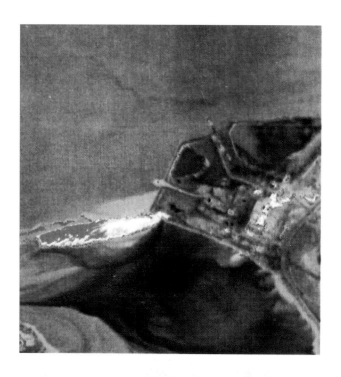

COLOR PLATE 50. See Chapter 11.
Figure 11.15, page 309.

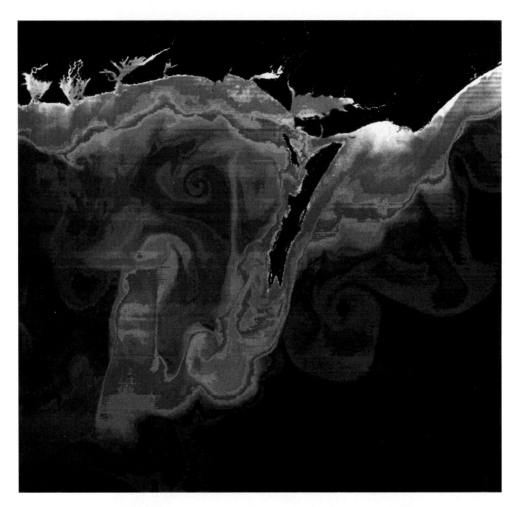

COLOR PLATE 51. See Chapter 11.
Figure 11.17, page 313.

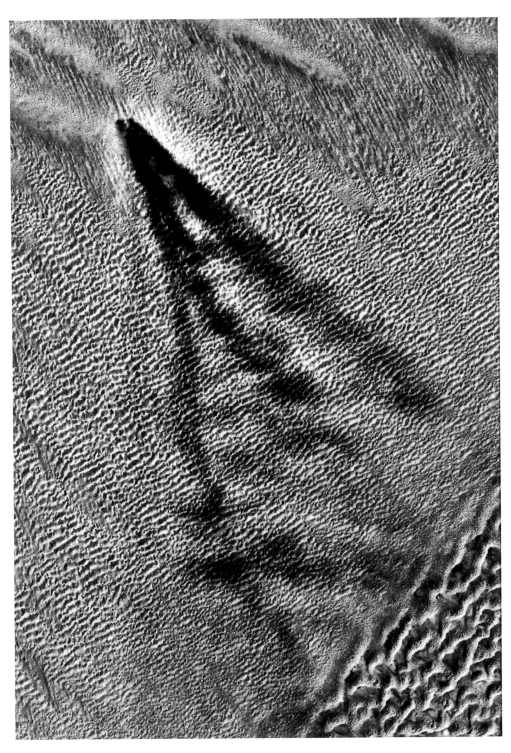

COLOR PLATE 52. See Chapter 11.
Figure 11.18, page 315.

COLOR PLATE 53. See Chapter 12.
Figure 12.1, page 320.

COLOR PLATE 54. See Chapter 12.
Figure 12.2, page 321.

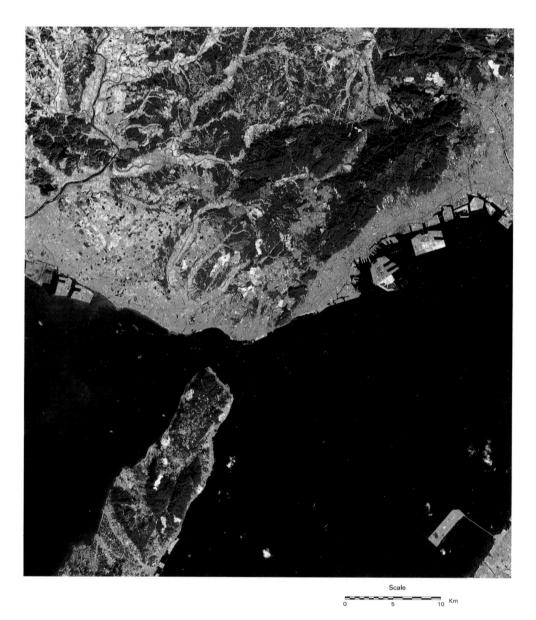

COLOR PLATE 55. See Chapter 12.
Figure 12.3, page 323.

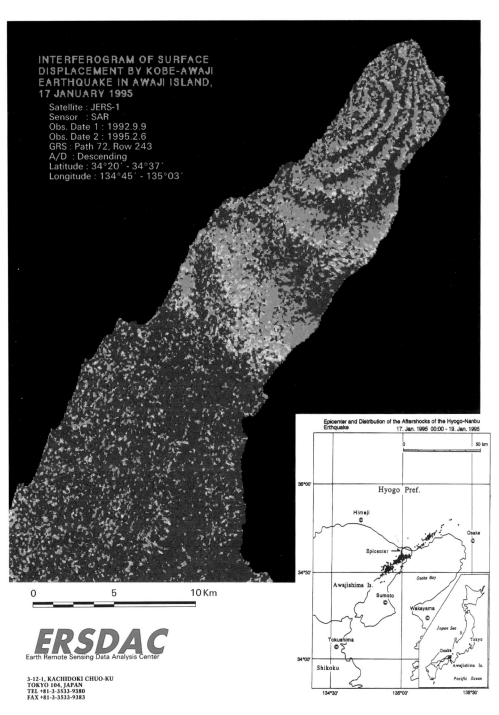

COLOR PLATE 56. See Chapter 12.
Figure 12.4, page 324.

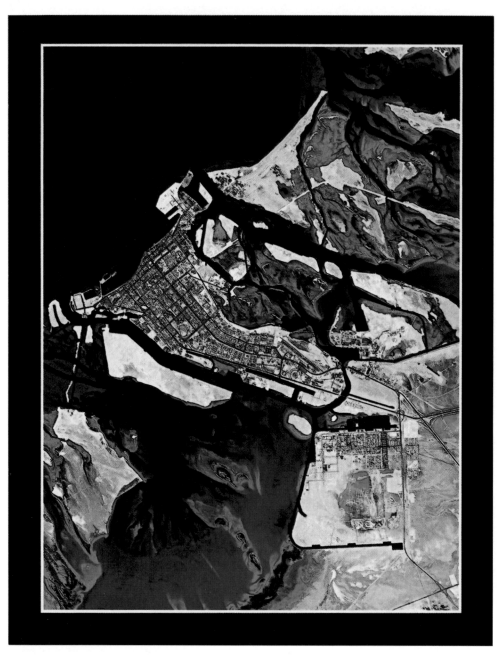

COLOR PLATE 57. See Chapter 12.
Figure 12.9, page 330.

COLOR PLATE 58. See Chapter 12.
Figure 12.11, page 335.